I0055344

Turbulent Origin of Chemical Particles

OTHER BOOKS BY NATHANAEL-ISRAEL ISRAEL

Get them at your local bookstore, or online (e.g. on Amazon, Science180.com/books)

Turbulent Origin of the Universe
There is Only One Scientific, Simple, Safe, Trustworthy, Unexpensive, Brave, Practical, Nonconformist, Universal, Verifiable Formula that Accurately Decodes the Universe Formation … But You Are Not Using It

Reconciling Science and Creation Accurately
What Science Accurately Teaches about Creation and God's Existence that Atheists, Freethinkers, and even Most Christians Don't Know … And How to Demonstrate it Without Falling into the Trap of Taking Sides Between Rationality and Faith

Origin of the Spiritual World
Top Secrets about the Origin of Everything in the Universe that Some Elites Have Hidden from You for Thousands of Years

From Science to Bible's Conclusions
How Decoding the Universe-Origin by Properly Revisiting Scientific Data—That Top Scientists Collected but Wrongly Analyzed—Bizarrely led to the 3500 Years Old Biblical Account of Creation

Turbulent Origin of Life
Why You Don't Have to Embrace Evolutionism or Check Your Brain at the Door in the Name of Faith or Science to Accurately Decrypt the Origin of Life Using the Historic Formula of the Universe Formation

How God Created Baby Universe
What Children Need to Scientifically Learn Early about the Universe Formation to Avoid Dangerously Abandoning God Later in Life Just Like Most College Students Who Embrace Evolutionism, Big Bang, and other Theories That Deny Biblical Creation

How Baby Universe Was Born
How to Scientifically Talk to Children about the Universe Formation and They will Know Forever How to Correctly Test the Intersection of Science and Faith

Science180 Accurate Scientific Proof of God
Can We Scientifically Explain the Formation of the Universe Through Natural Processes Without Evoking Evolution and Big Bang?

More books written by Nathanael-Israel Israel can be found at Israel120.com/books

NATHANAEL-ISRAEL ISRAEL, PhD

Founder of Science180: www.Science180.com
Father of Science180 Cosmology
www.Israel120.com

Turbulent Origin of Chemical Particles

Why You Don't Have to Embrace Evolution, Big Bang, or Deny God to Scientifically Prove the Formation of All Chemical Particles

Science180
Augusta, Georgia
United States of America
www.Science180Publishing.com

Copyright © 2025 by Nathanael-Israel Israel
Visit the author's website at Israel120.com

Turbulent Origin of Chemical Particles
Why You Don't Have to Embrace Evolution, Big Bang, or Deny God to Scientifically
Prove the Formation of All Chemical Particles

First edition: October 2025

Published by Science180
Augusta, Georgia (USA)
www.Science180Publishing.com

Book Cover and Illustrations by Nathanael-Israel Israel

ISBN: 979-8-9932150-3-7

Library of Congress Control Number: 2025920903

More books by the same author can be found at Israel120.com and Science180.com

For information about special discounts available for bulk purchases, please visit
Science180.com/discount for more details.

Science180 can bring authors including Dr. Nathanael-Israel Israel to your live or recorded
events. For more information or to book an event, please visit Science180.com/speaking

For any questions, please visit Science180.com/contact

To publish your book(s) with Science180 Publishing, go to Science180Publishing.com

To interview the author of this book, visit Israel120.com/interview
To donate, please visit Israel120.com/donate or Science180.com/donate.

Printed in the United States of America.

CONTENT

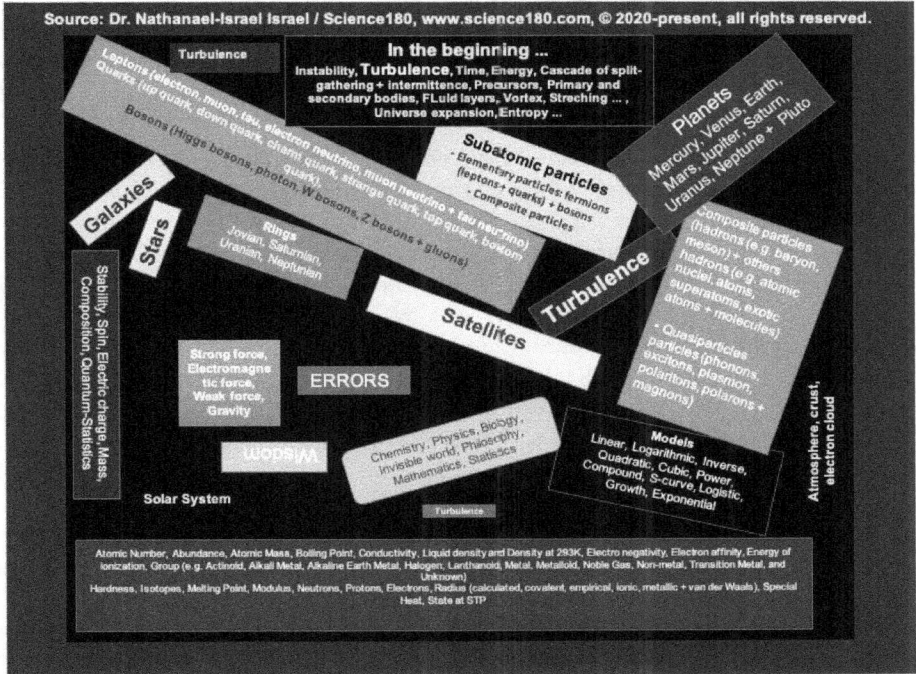

1. Can a life scientist turned physicist hold the key of the secrets that will save scientists' understanding of the origin of chemical particles? 1

2. Can chemists agree with physicists about the beginning of the universe and its chemicals without angering Big Bang proponents and their staunchest opponents? ... 8

3. If you don't know the origin of chemical particles, let the cascades of turbulent split-gathering and intermittence land you a very important key ... that the greatest names in chemistry have ignored 17

4. What do you know about the characteristics of the fluid layers in the precursors of particles that you need to change so your mind can become flexible enough to embrace the new chemical reality? 29

5. Unraveling the irrefutable system-additive variables and turbulent multipliers to understand turbulence at the chemical level—if you don't believe me, read this and you will see! 46

6. The incredible expert advice that is helping successful scientists to understand the breakup of the precursors of chemical particles even when the Big Bang nucleosynthesis can't–or can it? 52

7. What new insight can you get from the classification criteria of particles to pave your way to success even when people ignore the 24 most valuable variables necessary to crack the code of chemicals-origin? 58

8. What can elementary particles teach you that can quickly start pointing you toward the long-awaited physics beyond the Standard Model? 63

9. Why don't people who are giving the composite particles a bad interpretation listen to this expert who awakened the world to a new scientific reality? 82

10. Why the critics of the Standard Model of particle physics and the proton radius puzzle are no joke in an era when hypothetical particles can't help? 96

11. Why are scientific lies and confusion about the groups of chemical particles still tolerated? 103

12. Won't you refresh your mind with this summary about protons, electrons, neutrons, isotopes, mass, and the discovery year of chemical elements? 119

13. Can conductivity and special heat help you to debunk the origin of chemical particles? 129

14. Why don't people pay attention to the density of chemical elements? 131

15. Electronegativity and electron affinity of chemical elements 136

16. Hardness of chemical elements: A little variable you cannot ignore while demonstrating the origin of chemical particles 137

17. What is the modulus of chemical elements all about? 142

18. The unexplainable truth, joy, and accuracy in discovering the priceless secrets concealed in the ionization energy of chemical elements 148

19. Can the radius of chemical elements be a big player that bailed science out of the troubling and dangerous doubts about the decoding of the chemicals-origin? 170

20. Are the state and occurrence of chemical elements powerful enough to help us on our journey to crack the chemicals-origin? 182

21. Boiling point and melting point 187

22. Can a single variable make the others bow to its power by lighting in the heart of chemistry the lamp of understanding that opened the gate to unravel the origin of chemical particles? Discover what the abundance of chemicals is saying ... but scientists are not listening..188

23. If you are a scientist, pay attention to this crucial lesson from the comparison of the abundance of chemical elements–and avoid the dangerous mistake all chemists and physicists have made–unknowingly? 211

24. Understand the scientific power of the historic machinery of the formation of chemical particles so you can start asking the right questions that the future Nobel laureates in Chemistry could work on–I am not kidding! 243

25. The generic processes of the formation of elementary and composite particles that defied the Standard Model of particle physics and shook the foundation of chemistry 274

26. How physics and chemistry meet to explain the underlying causes of the diversity of celestial bodies in the universe 312

27. Discover the shocking story behind how the formation of the planetary ring systems including the Saturnian rings was finally decoded using chemistry and turbulence! 322

28. Remarkable scientific advancement that changed the rational explanation of the formation of galaxies, globular clusters, stars, and their chemical particles 331

29. Can the fundamental forces in nature and the mathematical errors applied to chemistry talk themselves out of doubting the truth about chemicals-origin (if we don't denounce them now)? 341

30. Why smart scientists are abandoning all existing theories on chemicals formation to embrace "Science180 Model of the Origin of Chemical Particles"..........................355

Next steps of the journey 362
References 369
Index...381
About the Author 396

CHAPTER 1

CAN A LIFE SCIENTIST TURNED PHYSICIST HOLD THE KEY OF THE SECRETS THAT WILL SAVE SCIENTISTS' UNDERSTANDING OF THE ORIGIN OF CHEMICAL PARTICLES?

Have you ever asked yourself how chemical particles came into existence? What was the process that molded them? Did one chemical particle birth another one? Or did the chemical particles just appear from nowhere, or did they just always exist and we do not need to seek for the underlying causes of their origin? Why is the world filled with various chemical particles? Where did all elementary particles and composite particles including atoms, molecules, minerals, and rocks come from? What are the fundamental factors, the machinery, and the generic processes that defined their formation and proprieties? What was the nature of their precursors at the beginning of the universe and what underlying processes shaped or molded them into the chemicals we know today? What was the primary cause of the abundance and diversity of chemicals in the celestial bodies in the universe? What is the accurate link between the formation of chemical particles and the formation of galaxies, stars, planets, asteroids, and satellites? What light can the origin of chemicals shed on the real cause and meaning of gravity and the other so-called fundamental forces in nature? How does the formation of the chemical particles fit into the big picture of the formation of the universe? Who or which process formed the universe in which the chemical particles are found? How does the formation of the chemical particles fit into the big picture of the formation of the universe? How can we answer these questions for sure using science, not philosophical speculations?

If you have ever asked any of those questions, and if because you were unsatisfied by the theories in the literature, you are looking for the answer from a fresh, original perceptive grounded in real science, then this book is for you.

Indeed, understanding the origin of the universe has been preoccupying human beings for a long time and it can affect your life. Using scientific data collected throughout the ages, but that were improperly analyzed by those who collected them, I found a historic, and well acclaimed solution to the question of the origin of the universe; therefore settling the disagreement between science and many philosophical origin views. Using a similar approach that led me to that breakthrough, I also unearthed and decoded deep secrets about the origin and formation of chemical particles; and I am very sure that mastering these secrets will help you to better understand how the chemical particles came into existence as part of the global process that formed the universe. This insight can also help you to better understand where the world is coming from and where it is going. I don't know much about you, but if I can make a guess, you picked up this book and want to read it because you want an answer for some of the above questions, and you want to know for sure how the matter we see all around us came into being. That is exactly what this book entails.

In this book, you will find all the reliable, convincing, scientific answers you need to successfully decode the origin of chemical particles safely. Why? Because, after studying these questions for more than 12 years, I, Dr. Nathanael-Israel Israel, discovered that the proper understanding of the origin of chemical particles is a very challenging but profitable task that requires original, scientific, mathematic, and philosophic efforts beyond the current state of modern science— until recently. The solution for all of these puzzling problems: *"Turbulent Origin of Chemical Particles"*, the straightforward and trustworthy book that will help you to quickly, cheaply, easily, and efficiently navigate everything you need to know to finally solve the hard problems about the origin, the formation, and the functioning of all chemical particles. Whether you are a chemist, a biochemist, any other scientist, an engineer, as long as you have a reasonable background in chemistry but ignore how to scientifically demonstrate the origin of all chemical particles, this marvelous book is for you!

Amazingly packed with eye-popping analysis, fantastic graphs, tables, and the historic formula that broke the universe-origin code, this book, *"Turbulent Origin of Chemical Particles"* will:

1. Scientifically explain to you the critical process that must be understood to correctly understand the turbulence that shaped the formation of chemical particles
2. Challenge the cosmological and chemical status quo and help you to embrace the real change that will disrupt all the cages that were holding you
3. Daringly contribute to setting on fire all false chemicals-origin theories or dogma that are enslaving humankind
4. Decode the DNA of the formation of chemical particles
5. Help you discover thrilling illustrations and unconventional explanations of the formation of all matter in the universe, written in a simple language that brings humankind much closer to the complete

deciphering of the mysteries at the very heart of chemistry, and open the way to a future of technology, innovation, discoveries, and breakthroughs

6. Equip you to bypass technical knowledge that restricts non-experts from accessing the origin-related secrets contained in the massive scientific data, and get to the bottom of origin-related mysteries regardless of your background so you can empower yourself to leave unforgettable marks in your field of expertise

7. Free yourself from boring explanations of the origin of chemicals and embrace the proven theory that opens doors to unparalleled opportunities

8. Give you a fresh chemicals-origin perspective that defies the existing norms and traditions that hold the true understanding captive

9. Give you inside secrets about how to locate flaws in origin-related theories so you can save time, money, and other resources to improve lives

10. Get practicable and undeniable proofs of the origin of chemicals so you can be fired up to become the best version of you, and to cause positive changes to your initiatives that will profit you today and forever

11. Get you ready to empower a new generation of thinkers

12. Give you relevant chemicals-origin stories that are specific to your field of expertise

13. Teach you outstanding value, insight, and lessons to assist you to accurately understand the true origin of chemicals so you can tap into that knowledge to improve lives perpetually

14. Make it easier than ever for you to properly understand, decrypt, and articulate the real origin of natural chemical particles in the universe, therefore freeing you from false and boring explanations of the origin of all matters, and embrace the proven theory that opens doors to unparallel opportunities

15. Equip you to participate in the global effort that is igniting an unquenchable fire under all nonsenses about the origin of the cosmos and its particles

16. Professionally teach you how to transform the true knowledge of the origin of chemical particles into insights that significantly add value to your life in less time, and successfully establish you as a symbol of freedom, power, creativity, and originality in your field of expertise

17. Satisfy your burning desire for freedom from beliefs and scientific theories about the universe-origin and chemicals-origin that suffocate you and bind your mind, faith, unbelief, heart, and education

18. Show you how to stay connected with practical tips about how to decode the origin of the universe and its chemicals and protect

yourself from wrong theories in the literature and the media. Learn more at Science180.com/chemical

With *"Turbulent Origin of Chemical Particles"*, the accurate decrypting and understanding of the formation of chemicals has never been profitable and easy. Hence this great book is THE ultimate how-to guide for great people wanting to correctly decode the origin of the chemicals and positively transform their lives.

But before going any further, let me tell you a little story about the origin of this book. Indeed, known as the nonconformist, rule-breaker, and accurate demonstrator of the universe-origin, and as the founder of Science180, the one-stop for answering the most crucial universe, chemicals, and life's origin questions, I, Nathanael-Israel Israel, have had the honor to be acknowledged as the fearless universe-origin decryption trailblazer (learn more at Israel120.com).

Truly, after years of investigation of the origin of the universe, I discovered several codes that were hiding mysteries related to the formation of the celestial bodies. I detailed those findings in my book titled *"Turbulent Origin of the Universe"*. In the first draft of that book, because the origin of the chemical elements is a piece of the huge puzzle concerning the origin of the universe, I spent hundreds of pages on chemical elements. By the time I finished the first draft of that book, it was about 2000 pages long. Because some readers trying to read such a large manuscript would be overwhelmed, I felt obligated to remove the chapters on chemical elements from that book and compile them into a separate book, which became this book. Doing a separate book on the formation of the chemical elements would also better appeal to those who are mostly interested in the chemical aspects of the formation of the universe, and also help them focus on their interest, rather than loading them with things more relevant to other disciplines.

After removing the chemical chapters from the first draft of the aforementioned book, I just did a few pages summary on chemistry in that book, hoping that those who are interested in knowing more about chemistry will follow up with this book devoted to chemistry. Consequently, many things I explained about the origin of the universe are not present in this particular book, but are in the other book I did on the origin of the universe. This also means that, to answer certain deep questions about the origin of the universe, you may have to refer to the book I devoted to that subject, just as those who have that book would also have to refer to this one to answer their advanced questions about the formation of chemicals.

In the book I wrote on the origin of the universe, I extensively presented my journey of writing that book, and the background and big picture of this book on chemistry. Nevertheless, in the first few chapters of this present book, I will review where and how the origin of the chemical elements fit into the full picture of the formation of the universe. Indeed, as I was working on the origin of the universe, I realized that the existing theories of the origin of chemical elements were wrong, for they failed to tackle the problem from the perspective of the mother of all turbulences, which took place at the beginning of the universe, and without which it is impossible to properly explain the origin of anything in the universe. As I was trying to learn about what was known on the genesis of chemical elements, I was left with so much hunger and thirst that I could not finish talking about the formation of the universe without addressing the formation of chemicals. Hence, like I did for the celestial bodies, I spent more than 7 years collecting data on the characteristics of chemical particles (e.g. subatomic particles, atoms, molecules, minerals, and rocks), then, by analyzing them with the perspective of the data available on the celestial bodies (e.g. galaxies, stars, planets, asteroids, satellites, and rings), I came up with the big picture of their origin, which, interestingly, has never been properly done before.

I was not shocked to realize that the same processes that explain the formation of the celestial bodies are also behind the formation of the chemical elements. I also understood that it could have been impossible for me to unearth the underlying processes of the formation of the chemical elements if I did not explore them by carefully paying attention to the information available on the celestial bodies where those chemical elements are found. In other words, the study of the variation of the characteristics of the chemical elements in conjunction with the variation of the characteristics of the celestial bodies that host them made it easy for me to pinpoint what it could have taken to form the whole world and its constituents. Therefore, as I address the origin of the chemical elements in this book, I will also be referring you to the book I did on the origin of the universe for more details on certain demonstrations.

In this book, I first reviewed some of the processes that took place from the moment the original matter in the universe was born until the precursors of the chemical elements were formed. Then, I reviewed the characteristics of the chemical particles and compounds before I detailed their formation. I also explored the similarities and differences according to the following types of particles:

- elementary particles:
    - fermions: leptons (electron, muon, tau, electron neutrino, muon neutrino, and tau neutrino) and the quarks (up quark, down quark, charm quark, strange quark, top quark, and bottom quark);
    - bosons: the scalar bosons (e.g. Higgs bosons) and the Gauge bosons (e.g. photon, W bosons, Z bosons, and the gluons).
- composite particles: hadrons (e.g. baryon and meson) and other hadrons (e.g. atomic nuclei, atoms, super atoms, exotic atoms, and molecules), and the
- quasiparticles (e.g. phonons, excitons, plasmon, polaritons, polarons, and magnons).

Some of the variables I investigated for the subatomic particles include: stability, spin, electric charge, mass, composition, quantum-statistics, etc. As for the characteristics of the chemical elements, I focused on:

- Atomic number

- Abundance (e.g. abundance in the universe or in our galaxy, abundance in the Sun, abundance in humans, abundance in oceans and sea waters, abundance in the Earth's crust, abundance in the atmosphere of the planets in the Solar System, abundance in the atmosphere of the Moon, and on the surface and in the crust of the Moon)
- Atomic mass
- Boiling point
- Conductivity (e.g. electric conductivity and thermal conductivity)
- Crystal structure
- Density (e.g. liquid density and density at 293 K)
- Discovery year
- Electro negativity (Pauling electro negativity)
- Electron affinity
- Electronic shell
- Energy of ionization (1st ionization energy to the 30th ionization energy)
- Group (e.g. actinoid, alkali metal, alkaline earth metal, halogen, lanthanoid, metal, metalloid, noble gas, non-metal, transition metal, and unknown)
- Hardness (Mohs hardness, Brinell hardness, and Vickers hardness)
- Isotopes
- Location in the human body
- Melting point
- Modulus (bulk modulus, shear modulus, and Young's modulus)
- Number of neutrons
- Number of protons/electrons
- Occurrence
- Period or number of energy levels
- Radius (e.g. calculated radius, covalent radius, empirical radius, ionic radius, metallic radius, and van der Waals radius)
- Special heat
- State at STP (standard temperature and pressure: 273 K)

Other variables that I studied but that I will not be reporting in this book include:

- Color
- Energy levels
- Standard potential
- State at standard temperature and pressure (STP): 273 K
- Valence

I devoted a chapter to most of the variables I just mentioned. In my attempt to address the origin of chemical particles, I also studied the Standard Model theory of particles, some hypothetical particles, the conversion of some particles into others, and the proton radius puzzle of the EMC debate. When addressing the interactions between particles and celestial bodies in the universe, people tend to use forces, some of which are termed fundamental forces in nature. As of 2020, four fundamental forces are believed to control all interactions in nature:

- strong force also called strong nuclear interaction
- electromagnetic force
- weak force also called weak nuclear interaction
- gravity

While the electromagnetic force and the weak force have been unified into the electroweak force, gravity still stands alone and is not much understood although the classical theory of Isaac Newton and the relativity theory of Albert Einstein claimed to address it. In the book, I reviewed what is known about those forces and the interpretation that I think better fits them in light of my understanding of the origin of the universe.

Throughout my writing, wherever you see "universe-origin", please know that I meant "origin of the universe" or "the origin of the universe". Likewise, wherever you see "life-origin", please understand that I meant "origin of life" or "the origin of life". In the same manner, wherever I mention "chemicals-origin", please know that I am referring to "origin of chemicals" or "the origin of chemicals".

In the first chapters of this book, I will present the big picture of the formation of the universe and where the chemical elements fit into it. Then, in later chapters, I will zoom into the details about the formation of the chemical particles. Because many people who are not specialized in chemistry are also interested in understanding the origin of chemical particles, I provided a lot of background information on chemistry in most chapters. Because this book is also intended for faculty and students who will use it as a reference and textbook, I also provided the background information because I know these users will appreciate them. Those who are proficient in chemistry may skip the background chapters or segments and read those which interest them.

CHAPTER 2

CAN CHEMISTS AGREE WITH PHYSICISTS ABOUT THE BEGINNING OF THE UNIVERSE AND ITS CHEMICALS WITHOUT ANGERING BIG BANG PROPONENTS AND THEIR STAUNCHEST OPPONENTS?

2.1. Appearance of the turbulent prima materia (the mother of all matter)

In *"Turbulent Origin of the Universe"*, I explained that in the beginning, an initial particle that I called the "turbulent prima materia", or the "turbulent original particle", or the "turbulent original substance", or the "foundational matter", which is different from any matter known today, meaning its characteristics are different from that of any current matter, mysteriously appeared in the universe out of nothing..., and through very complex, dynamic, and turbulent processes, that turbulent prima materia was progressively molded into all types of matter, bodies, systems or clusters of matter and of bodies known in the universe today. Because in *"Turbulent Origin of the Universe"* I extensively went over the potential characteristics of the initial matter that appeared at the beginning, I did not present those data here, but I did summarize a few aspects of them below.

After working more than 7 years on the data available on the bodies (small and big) in the universe, it appeared to me that the origin of all matter in the universe can be traced back to an initial matter that was transformed to yield the current world. I showed that, just as during the development of a living organism, an egg can go through steps of differentiation to become different types of cells, tissues, organs, apparatuses, or systems, which are all connected, so also the "turbulent prima materia" went through complex processes to birth all kinds of matters and systems of bodies in the world today. I demonstrated that the original matter in the universe was like an "undetermined" particle that had the potential of becoming anything just like the first cells at the beginning of the development of a living thing have the potential to become anything and are later specialized into specific cells because of some unique changes in the expression of their DNA, etc. Similarly, the initial matter in the universe had a kind of programming, which yielded different types of matter depending on the conditions it was subdued to. The main task I handled in this book is to explain the machinery of the processes involved into the shaping and distribution of that "turbulent prima materia" into the chemical particles (e.g. subatomic particles, atoms, molecules, minerals, and rocks) present in the universe today.

Unlike living organisms, which bodies usually start with one egg or cell that goes through many mitoses before starting its differentiation, in the case of the universe, things started with a huge bulk of a single kind of matter that filled the early universe. Just as undetermined cells can split into different parts with different functions, the bulk of the turbulent prima materia also went through fragmentations simultaneously associated with the modification of the characteristics of their daughter bodies. It appeared to me that the initial matter in the universe could have occupied a very huge portion of space, which size I did not bother to estimate. The "turbulent prima materia" was formless, leaving the process that shaped it after its appearance to form various matters seen in the world today. Although as of today, matter is usually found in 4 states: solid, liquid, gas, and plasma, it seemed to me that the "turbulent prima materia" was none of the 4 states of matter known nowadays. The original matter had to go through complex processes to form the current 4 states of matter. Because plasmas are less complex than gases, liquids, and solids, it is possible that the state of the "turbulent prima materia" could have been closer to that of a plasma, but not the plasmas known today. Because no particle in the early universe was left unchanged by the processes that shaped the world, nothing in the universe today is an exact photocopy of the "turbulent prima materia".

Immediately after its appearance, the turbulent prima materia could have been broken open by a violent event that can also be qualified as a burst asunder accompanied by a gigantic noise as that of an "explosion" due to factors such as its internal pressure and surface tension. Because I felt like what happened would not have been a mere explosion known as of today, I used the term "original mysterious scattering" instead of explosion as I do not want to confuse this with the explosion mentioned in other theories like the Big Bang theory. The original mysterious scattering split the turbulent prima materia into pieces or blocks of matter. The breaking apart of the turbulent prima materia could have exposed its content and "propelled" the broken pieces (which I also called the "major daughters of the turbulent prima materia") into various directions. Although I used the word pieces, some of them could have been thousands, millions, or billions of miles wide! The energy communicated to these "broken pieces of the turbulent prima materia" could have propelled them into motion and they started distancing themselves from one another and from the initial position of their mothers, therefore initiating the expansion of the early universe until today. In *"Turbulent Origin of the Universe"*, I established that some "major daughters of the turbulent prima materia" could have gone through additional "explosions" to yield smaller daughters, sometimes leading to a cascade of "explosions" until a point where the "multi-broken pieces of the turbulent prima materia" could no longer "explode". Meanwhile, changes were occurring in the bulk of matter that were breaking apart on different scales. The size of the broken bulk of matter varied tremendously. Some of them later became clusters of galaxies, while others became the precursors of stars. In the end, the organization of the universe into clusters of galaxies of various shapes and sizes was caused by the cascade of fragmentation and gathering together that the turbulent prima materia had to go through to birth the bodies in the universe (Fig. 1). Based on my work on the bodies in the Solar System, I established that the fragmentation of their precursors was not by means of a mere random explosion, but according to specific laws involving fluid flow, turbulence, shearing, diffusion, and other complex phenomena related to the mechanics and dynamics of fluids.

In *"Turbulent Origin of the Universe"*, I established that the broken pieces of the turbulent prima materia went through changes and processes of differentiation (e.g. breakup or fragmentation, squeezing or compression, and others), which constitute the basis of the story I voiced in that book about: fluid flow and instability, fluid breakup, formation of the precursors of bodies; birth, split, and transfer of energy; initiation of movement (e.g. revolution and rotation), formation of fluid layers, initiation and development of turbulence, fluid mixing, fluid gathering, squeezing of precursors, birth and strengthening of fundamental "forces", etc. Because I already detailed all of the aforementioned processes in *"Turbulent Origin of the Universe"*, I did not expound too much on them in the current book. However, I focused on aspects pertaining to chemical particles. I also found and proved that some of these processes I listed above affected the formation of the chemical particles just as they did for the celestial bodies. I also established that many of the processes involved in the formation of the universe are also found in some biological systems, reactions, and pathways seen today. For instance, during the development of most organisms, the first or initial cells are undetermined, but as they go through stages of development, biochemical modifications involving cellular division, gene expression, transcriptomic, translational, and epigenetics modifications intervene to transform the initially undetermined cells into determined cells, which are not always able to reverse back to other kinds of cells including the initial undetermined cells. Similarly, not only was the turbulent prima materia undetermined, but it was also transformed into specific daughter bodies, most of which are unable to reverse back to their precursors. Although some chemical elements can be transformed into others, the transformation of particles of one kind to another do not generally usually occur naturally on a large scale today. For the energy and processes involved in shaping things in the universe have been costly and it is not easy to reverse natural laws and things. Even on the level of human beings (who have a will), it is not easy to

Nathanael-Israel Israel: Member of the American Chemical Society

change human nature. Else, using their free will, human beings could have been more easily changing their mind and nature to adapt to new requirements or to become whatever or whosoever they want to be. In other words, natural laws were very strongly forged and cannot be easily broken using part of the energy that formed them.

Fig. 1: Breakup of the turbulent prima materia into pieces of major daughters, which started spreading across the universe and became the precursors of clusters of galaxies, stellar systems, stars, and other bodies and particles found in the universe

Source: Dr. Nathanael-Israel Israel / Science180, www.science180.com, © 2020-present, all rights reserved.

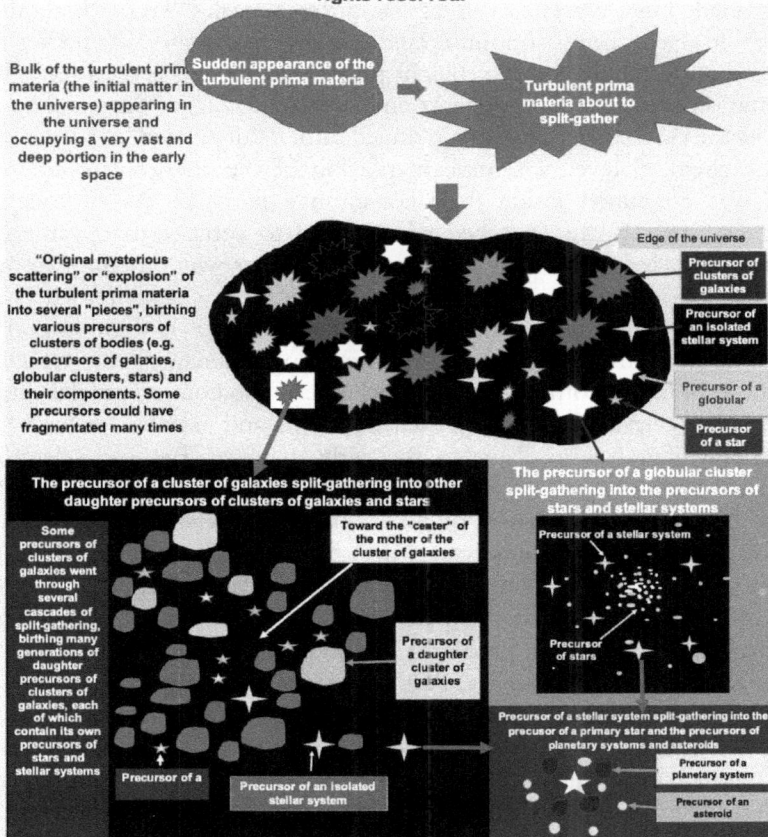

Each of the precursors mentioned above went through complex, dynamic, and turbulent stages of differentiation before becoming the corresponding body or system of bodies found in the universe today. Similar processes also occurred at the microscopic levels to yield subatomic particles, atoms, and their clusters.

To download a prettier and color version of this graph, visit www.Science180.com/TurbulentPrimaMaterialBreakup

TURBULENT ORIGIN OF CHEMICAL PARTICLES

Now, I would like to say a few words about the origin of the energy in the bodies in the universe. Indeed, from the smallest things to the biggest ones, from the living things to the nonliving things, everything in the universe is moving, though some movements may not be easily seen with the naked eye or using telescopes or microscopes. Each body (small or big) in the universe is a store of energy, which, under certain conditions can be released to do work or produce heat, or to manifest in other types of things or beings. I extensively explained how the bodies in the universe got their energy. I noticed that, by the time that the bulk of the turbulent prima materia "exploded", energy occurred in the universe and the processes that followed determined how the energy in the universe by this time was divided into clusters of bodies on different scales. To put it another way, energy could have been communicated to the precursors of bodies by the processes that destabilized the turbulent prima materia and ignited its movement and separation into small components embedded into bigger ones and so on and so forth as the cascade of breakup continued until it "dissipated".

At the chemical level, a significant portion of the energy used to form the particles was contained inside their constitutive particles. As of today, when chemical compounds are scratched or damaged to some extent, some of their energy can be leaked or released. This partially explains why rubbing two materials against each other can cause them to heat up and even catch fire. The heat that can be felt by rubbing two hands against each other for a while is part of the energy released from the elements that are being "damaged" or "destabilized" during the rubbing. In other words, particles (e.g. subatomic particles, atoms, or molecules) and celestial bodies (e.g. stellar systems and galaxies) are just "stores" of the energy that was available in the early universe. Because scientific laws suggest that energy cannot be created or destroyed, it appeared to me that the energy distributed in the clusters of bodies in the universe was born at an early stage of the formation of the universe. In other words, every type of matter was produced by the transformation that the turbulent prima materia went through during the turbulence that shaped the universe. The energy that was communicated to a mother precursor of bodies as it was being molded was trapped inside its daughter bodies. Part of the energy found in the bodies in the universe today was affected by their interactions with other bodies in their vicinity. At certain stages of their development or differentiation, the precursors of some bodies could have been like a flaming fire. This can also explain why all matters burn when a significant amount of fire is applied to them. For fire was present at some stages of differentiation that some, if not all of all the precursors of the chemical particles had to pass through. While some chemical elements like gold are not affected by fire but purified by it, others can be denaturized, but not damaged enough to yield other chemicals.

Nathanael-Israel Israel: Member of the American Chemical Society

2.2. Beginning of the instability and its development into turbulence in the universe

As the "broken pieces of the turbulent prima materia" started moving and stretching under the influence of the "original mysterious scattering", the whole universe was destabilized and a major turbulence broke out. I found that when fluids are heavily destabilized, their constituents rearrange differently into "pockets" or compartments, which composition and size can depend on the intensity of the instability. For instance, when creamer is poured into coffee, an instable mixture can be seen. Similarly, when water or any other liquid is put in a bowl and then thrown into the air, depending on the intensity used and the way it is thrown, that liquid not only can leave the bowl but, as it starts moving in the air, it can be scattered into clusters of water of different sizes, speeds, shapes, etc.

After looking at the types of matter known in chemistry, biology, and physics, from the smallest to the biggest, and after carefully reflecting on the thousands of pages I wrote about their characteristics, it appeared to me that the origin and movement of matter can be traced back to a major instability that occurred during the early stage of the formation of the universe. Under the initial instability, the state of the turbulent prima materia started changing. At one point, the state of the matter was plasma-like or fluid-like. As the initial matter was moving and mixing, vorticial structures were born. The fluid-like matter could have moved in a manner compatible with the direction of the force that started the instability of their mother.

As the instable fluids in the early universe were moving, they split and collected themselves into precursors of bodies of all sizes ranging from small to large scales. These precursors of bodies became the precursors of other bodies and so on and so forth until the smallest bodies or clusters of bodies were formed. The movement of some fluids was like a flow. The instability of these flows played a huge role in the organization, size, location, and structure of subatomic particles, atoms, molecules, planetary systems, stellar systems, clusters of galaxies and the universe as a whole. From the subatomic particles to the galactical systems, bodies in the universe were assembled into different groups based on how their precursors were moved around by the events that "disturbed" the turbulent prima materia.

As the precursors of matter started flowing, fluid layers were formed. Stacked on top of one another, these fluid layers could have started moving at different speeds. Due to the variation of their speed and according to their size, a turbulence started in them and soon after, patterns or re-arrangements of their matter emerged. The intensity of the turbulence is not always the same from one layer to the other and from one precursors of matter to the other. The pattern of the turbulence at the origin of the universe had structure within it at all imaginable scales, from the invisible or the tinniest imaginable scales to the biggest conceivable ones. Each of those scales of turbulence went through different developmental stages and led to the diversity of the bodies and systems of bodies

13

in the universe today. The fluids in the daughter bodies were subjected to different intensities of turbulence, leading to the formation of diverse bodies accordingly. When a small amount of energy was communicated to a daughter precursor, the level of its turbulence could have been smaller.

During the formation of the universe, turbulence occurred on different scales, with different intensities, at different positions, and led to the formation of different celestial bodies and chemical particles in the universe. The scales of turbulence that I will focus on in this book are about:

- Subatomic scale: scale of current elementary particles
- Composite particle scale: scale of composite particles smaller than atoms
- Atomic scale
- Molecular scale
- Mineral scale
- Rock scale

In other words, in this book, I will not dwell on the turbulence of the celestial bodies, but on that of chemical particles.

'Science180 Academy' Success Strategy:
SCIENCE180 ACADEMY OVERVIEW

Science180 Academy is a training, speaking, consulting, and mentoring program designed to groom and empower people of all backgrounds in the truth about the origin of the universe, life, and chemicals. According to their background and interest, trainees are taught different levels of scientific facts to grasp a deeper understanding of the origin of the universe, how to properly think to unearth mysteries hidden in the massive scientific data collected across the globe, but which is unfortunately less analyzed. If you want to be enlightened and equipped so you can cause positive changes in your respective field of expertise, then Science180 Academy program is for you.

Science180 Academy does not confer college credit, grant degrees, or grade its attendants, participants, or students. It is not an accredited university or college, but is the one-stop-destination for universe-origin, life-origin, and chemicals-origin experts. It is where scientists and laypeople get all their origin-related questions properly answered. It is the only place where the accurate interpretation of the universe-origin, life-origin, and chemicals-origin data matters a lot.

Science180 Academy brings together Dr. Nathanael-Israel Israel (the Founder of Science180) and other experts to deliver outstanding value, insight, and lessons to assist you to accurately understand the true origin of the universe, chemicals, and life, so you can tap into that knowledge to improve lives perpetually. Nathanael-Israel's goal is to give you practicable and undeniable proofs of the formation of the universe so you can be fired up to become the best version of you, and to cause positive changes to your initiatives that will profit you today and forever.

For Nathanael-Israel, decoding the origin of the universe and everything in it is not a job, but his life mission, and helping others to fully understand that is his mission. Visit Science180Academy.com today to start.

If you are still wondering if Science180 Academy is for you, let me also inform you that some of Science180's clients and prospects have a profound technical knowledge and background in science, while others don't. Some are creationists (e.g. Science180 creationism, Young Earth creationists, Old Earth creationists, Intelligent design proponents), others are anti-creationists. Some are believers, others are freethinkers (including atheists, humanists, rationalists, agnostics, nontheists, nonreligious, skeptics, nonbelievers, religiously unaffiliated, spiritual-not-religious, ex-believers, and doubters). Regardless of their background, belief, or disbelief, Science180 works with each of these people to figure out their needs, priorities, and the products and services that best fit them. Science180 improves their knowledge, experience, performance, and answer their questions (related to the universe-origin, life-origin, and chemicals-origin) by crafting a personalized program that perfectly matches their interests, needs, and things that are dear and meaningful to them whether it is to:

- Protect yourself and loved ones by keeping all of you secured and empowered with the true knowledge of the origin of chemicals, universe-origin, and life-origin
- Have a reliable access to the world's authority on origin-related matters and get your chemicals-origin questions professionally answered with the truth step-by-step
- Connect with practical tips about how to decode the origin of chemicals, universe, and life, and protect yourself from wrong theories in the literature and the media
- Get inside secrets about how to locate flaws in chemicals-origin, universe-origin, and life-origin related theories so you can save time, money, and other resources to improve lives
- Bypass technical knowledge that restricts non-experts from accessing the chemicals-origin, universe-origin, and life-origin related truth contained in the massive scientific data, and get to the bottom of scientifically-locked origin-related secrets regardless of your background.
- Free yourself from boring explanations of the origin of chemicals and embrace the proven theory that opens doors to unparallel opportunities
- To register or to learn more, visit Science180Academy.com today.

CHAPTER 3

IF YOU DON'T KNOW THE ORIGIN OF CHEMICAL PARTICLES, LET THE CASCADES OF TURBULENT SPLIT-GATHERING AND INTERMITTENCE LAND YOU A VERY IMPORTANT KEY ... THAT THE GREATEST NAMES IN CHEMISTRY HAVE IGNORED

3.1. Generations of turbulent split-gathering of the precursors of chemical particles

I established in *"Turbulent Origin of the Universe"* that, the precursors of chemical elements were just one of the 3 major kinds of precursors of the bodies in the universe:

- Precursors of the celestial bodies
- Precursors of the chemical particles
- Precursors of spiritual or nonphysical things and beings

I also proved that the precursors of the atoms and subatomic particles were not a product of the same generation of breakup. I showed that as some bigger bodies were forming, smaller ones inside of them were also going through their own process of fragmentation and gathering together. Under other circumstances, some mother precursors must have gone through some fragmentations before their daughter bodies could have gone through their own. For instance, unlike human beings whose children have a developmental growth younger than their mother's, meaning that mothers birth children, who must grow up to become adults before birthing their own children (therefore giving grandchildren to their mothers), the fragmentation and gathering together of matter and clusters of bodies was different. All daughter bodies in the universe could not have waited for their mothers to finish their development before going through their own. When the mothers of some bodies (including chemical elements) were going through their developmental stages, some of their daughters could have already birthed their own children, which could even be growing and taking good care of their own children before their grandmas finished their formation or development. Part of this dynamic and complex process is due to the fact that the same turbulent prima materia was used to mold both the children, their mother, and their grandma.

Furthermore, the diversity of the precursors of the bodies belonging to each generation of breakups can explain the diversity of the bodies in the universe according to their types. For instance, the diversity of the chemical elements in the universe is a consequence of various precursors being born at different locations in the universe following similar stages of breakup of their mother precursors. In other words, the plurality of chemical elements in the Solar System for instance is due to the fact that many precursors of chemical elements were split-gathered during the similar stages of breakup of such precursors.

Considering the organization, characteristics, and the current classification of the clusters of bodies and clusters of matter in the universe, I established that about 9 generations of breakup and generations of gathering together of precursors could have occurred during the formation of the universe. Because I already handled those generations of breakup in *"Turbulent Origin of the Universe"*, I will just list them here:

- 1st generation of turbulent split-gathering, leading to the cascades of precursors of clusters of galaxies and clusters of globular clusters
- 2nd generation of turbulent split-gathering, leading to the precursors of stellar systems
- 3rd generation of turbulent split-gathering, causing the precursors of stellar systems to form the precursors of primary stars and the precursors of the bodies that would orbit them (e.g. planetary systems and asteroid systems)
- 4th generation of turbulent split-gathering, causing the precursors of planetary systems to birth the precursors of their primary planets and their satellites, while the precursors of asteroid systems birthed the precursors of primary asteroids and their satellites
- 5th generation of turbulent split-gathering, causing the precursors of satellite systems to form the precursors of individual satellites and rings
- 6th generation of split-gathering leading to the precursors of minerals, mineraloids, and rocks
- 7th generation of turbulent split-gathering, leading to the precursors of atoms and their clustering into molecules and chemical compounds
- 8th generation of turbulent split-gathering, leading to the formation of the precursors of subatomic particles
- 9th generation of turbulent split-gathering, leading to the precursors of the smallest particles that may never be scientifically discovered

18

CHAPTER 3: CASCADES OF TURBULENT SPLIT-GATHERING AND INTERMITTENCE

At this point, I would like to elaborate a little bit more on the last 4 generations of breakup of precursors (from the 6th to the 9th generations of breakup). Details about the first 4 generations of breakup can be found in *"Turbulent Origin of the Universe"*. Let's start with the 6th generation of split-gathering leading to the precursors of minerals, mineraloids, and rocks. Indeed, as the cascade of breakup of precursors was progressing from the upper levels of clusters of galaxies to the lower levels such as that of the satellites and asteroids which can be considered as the lowest level of organization of celestial bodies, internal clustering occurred inside the precursors of celestial bodies (e.g. asteroids, planets, satellites, and stars) which were unable to break into any other celestial body. If I can use the term "unsplitable" to describe the ability of the precursors of asteroids, planets, satellites, and stars to hold together without breaking into any other celestial bodies, I would define the 6th generation of split-gathering as the internal clustering of precursors of "unsplitable" celestial bodies into precursors of minerals, mineraloids, and rocks. For instance, in the precursors of the crust of the terrestrial planets (e.g. Earth), precursors of rocks were assembled and each of them birthed specific daughter bodies like rocks, minerals, and mineraloids. Although some rocks and minerals could have been born recently, meaning long after the formation of the universe, the organization of minerals and rocks in the crust of the Earth for instance implies that their precursors went through dynamic and turbulent rearrangement, clustering, and mixing that affected their purity and other characteristics. Some chapters in this book are devoted to rocks and minerals.

Talking about the 7th generation of turbulent split-gathering (leading to the precursors of atoms and their clustering into molecules and chemical compounds), I would like to say that, as the precursors of minerals, mineraloids, and rocks were being split-gathered, precursors of composite particles such as atoms, molecules, and chemical compounds were formed. The precursors of some atoms and molecules were split-gathered differently to form specific atoms, some of which are not ordinary or conventional atoms, but exotic ones, having characteristics not seen in the ordinary atoms on Earth. In other stellar systems beyond the Solar System, some atoms can have unfamiliar and strange characteristics. Because the inventory of atoms in the Solar System is not even exhaustive yet, we are very far from knowing the diversity of the precursors of atoms that could have existed in the universe. The plethora of isotopes on Earth alone illustrates the many combinations or arrangements of the precursors of atoms that could have existed in the universe.

As for the 8th generation of turbulent split-gathering (leading to the formation of the precursors of subatomic particles) is concerned, I would like to report that, as the precursors I mentioned above were splitting and collecting themselves together into clusters, the microscopic matter (not limited to atoms only) inside of them were also split into precursors of subatomic particles and precursors of other kinds of particles smaller than the currently known subatomic particles. If the vast number of subatomic particles in nature could be revealed, we would be shocked at how very little science has scratched the surface of the gigantic chemical iceberg. Therefore, I expect the details of the precursors of subatomic particles to be very limited. This also calls for tolerance toward those who mistakenly invented hypothetical particles not seen yet and which will probably never be discovered. As more particles are found and additional details understood, some particles like electrons (which are still considered as one of the smallest elementary particles) would be proven to consist of other particles.

The 8th generation of turbulent split-gathering is about the breakup and gathering together of all discovered and discoverable subatomic particles. Because of their small size, I think that particles which precursors belong to the 8th generation of turbulent split-gathering will cause many problems to the scientific community. Although more of these kinds of particles will be discovered, scientific equipment, existing and probably uncorrectable mistakes in some existing theories and dogma locked into some scientific strongholds will probably be limited in their quick and real understanding. In other words, these particles will cause so many problems and confusion to the scientific world and they will be a field full of contradictions and controversies due to the huge variation of their characteristics and the inability of scientists to have a consensus on their nature, functions, origin, and changes. The confusion in quantum physics is an illustration of this!

Finally, the 9th generation of turbulent split-gathering is about the precursors of all particles smaller than the currently known elementary particles which exist in nature and which, due to their size and the limits of scientific enquiries, will never be detected by physical equipment. The characteristics of the precursors of the particles under the influence of the 9th generation of turbulent split-gathering could have been close to those of the turbulent prima materia, the initial matter in the universe. These particles may belong to the spiritual or invisible spectra beyond that accessible by the most advanced and sophisticated science yet to come.

The precursors of some types of matter found in the spiritual world can also belong to the 9th generation of turbulent split-gathering. I had to address these "invisible particles" of the 9th generation of split-gathering to agree with certain things I believe in, but which I cannot demonstrate. I wish I can address here the kinds of matter involved in the formation of some spiritual entities (e.g. the prophetic, magic, miracles, and witchcraft) that some people have experienced and testified to have seen, touched, but which a mere human being cannot explain by modern physics. For although scientists will keep searching for the meaning of things in nature, I know that science will never discover certain details of matter, for they are hidden and can be unlocked only by codes that the modern scientific methodology has been rejecting for centuries and will likely continue to discard or disregard.

Nathanael-Israel Israel: Founder of Science180, the Accurate Universe-Origin, Life-Origin, and Chemicals-Origin Decoder

3.2. Generations of turbulent intermittence

Radius intermittency is the presence of smaller bodies between bigger or larger ones. For instance, on the microscopic scale, larger chemical elements are mixed with smaller ones just as large celestial bodies are separated by small ones. On the chemical level, size intermittence could be due to how the turbulent prima materia and its daughters were broken up and gathered together into clusters of particles. Intermittence could have been caused by other factors including how the cleavage lines, cleavage points, interfaces of separation, or the interfaces of breakup of the precursors of the particles and their clusters were not neat or sharp, therefore leading to the formation of smaller particles or systems of particles between larger ones. I also classified the turbulent intermittence into 9 generations accordingly to the generation of split-gathering:

- 1st generation of turbulent intermittence (leading to the presence of smaller clusters of galaxies between larger ones and smaller clusters of globular clusters between larger ones)
- 2nd generation of turbulent intermittence: intermittence of stellar systems
- 3rd generation of turbulent intermittence: intermittence of stars
- 4th generation of turbulent intermittence: intermittence of planetary systems and asteroids system
- 5th generation of turbulent intermittence: intermittence of satellites
- 6th generation of intermittence: intermittence of minerals, mineraloids, and rocks
- 7th generation of turbulent intermittence: atoms, molecules, and chemical compounds
- 8th generation of turbulent intermittence: intermittence of subatomic particles
- 9th generation of turbulent intermittence, leading to the precursors of the small invisible bodies between and within larger ones, which will never be scientifically discovered.

Below, I will expound more on the 6th to 9th generation of intermittence. Indeed, the 6th generation of intermittence can explain why smaller amounts or deposits of minerals, rocks, and mineraloids exist between larger ones. Rocks have different sizes and between bigger rocks, smaller ones exist, pointing to the distribution of their precursors during the formation of bodies in the universe. Some rocks composed of complex compounds are separated by simpler rocks because of the 6th generation of intermittence. Likewise, minerals are not equally distributed. Although some minerals are pure and composed of the same chemical elements, data from ore and well digging suggested to me that smaller cluster of minerals exist between bigger ones.

As for the 7th generation of turbulent intermittence (which concern atoms, molecules, and chemical compounds), I would like to pinpoint that, atoms and molecules come in different sizes. Some atoms and molecules are very big, while

21

Science180: Accurately Understand Chemicals-Origin. Be Happy Forever!

others are small. Likewise, bigger atoms and molecules are mixed with smaller ones. For instance, soil analysis suggests that in most soil samples, many kinds of atoms and molecules can be found and mixed. Even in most pure samples, impurities are found, meaning that the precursors of some atoms and minerals were separated by the precursors of other kinds of atoms and molecules, not always of the same size. The 7th generation of intermittence explains the presence of small atoms, molecules, and chemical compounds between larger ones.

The 8th generation of turbulent intermittence (intermittence of subatomic particles) is also responsible for the presence of smaller subatomic particles between larger ones. In fact, within a given atom or composite particle, the subatomic particles are not usually of the same size. Even in the nucleus of atoms, the particles do not have the same size. Protons, neutrons, and gluons are different. Although all electrons are believed to have the same size, based on the trends in the data I analyzed, I think that some electrons are bigger than others. In other words, there would be an intermittence in the size distribution of electrons as well as most of the other subatomic particles. All of this would have been caused by how their precursors were split-gathered together during the formation of these subatomic particles.

Finally, although I could not see them with the naked eye, I think that the size of some invisible particles that science will never discover could be intermittent. Nevertheless, the closer the invisible particles may be to the turbulent prima materia, the more similar their size would be. The detail of the 9th generation of intermittence can be appreciated only by those who can see in the spirit, meaning beyond what the most sophisticated scientific tools can ever perceive.

3.3. Inclusion levels of primary and secondary particles

In this segment, I will talk about the split and organization of the precursors of the systems of particles into precursors of primary bodies (e.g. nucleus) and precursors of secondary bodies (e.g. electrons). In *"Turbulent Origin of the Universe"*, I showed that the clusters of matter in the universe mostly consist of a kind of "central" body, which can be called the primary body "around" which secondary bodies are usually positioned. For instance, just as electrons were believed to orbit the nucleus and/or move in shells or orbitals around the nucleus, so also do satellites orbit their primary planets. I also noticed that most primary bodies are not "elementary" bodies for they generally consist of other bodies, which, when carefully analyzed, seem like entities of matter squeezed, wrapped, and organized into different manners appealing to a phenomenon similar to what could have "ejected" away or split the precursors of secondary bodies from the precursor of the primary bodies. In other words, the precursors of secondary chemical particles (e.g. electrons) could have escaped the precursors of their primary bodies (e.g. proton and neutrons forming the atomic nucleus).

By studying the variation of the characteristics of the secondary bodies across various systems of bodies, I realized that the precursors of bodies and of systems of bodies were early split-gathered together into the precursors of up to 2 types of daughter bodies:

- the precursor of the primary body or the precursor of the system of primary bodies and sometimes
- the precursor of the secondary body or the precursor of the system of secondary bodies (if any)

To put it another way, when a mother precursor was broken up or fragmentated, it yielded the precursor of a primary daughter and the precursor of a secondary daughter (Fig. 2). The proportion of the mother precursors that landed into the primary bodies and into the secondary bodies was not by chance, but followed specific laws that I will explain very soon. At the atomic level, the precursors of some atoms were split-gathered into the precursor of the nucleus and the precursor of the electrons, which would later "orbit" the nucleus. Another way of saying this is that the precursor of the system of secondary bodies of each atom is what was split-gathered into the electrons around ordinary atoms.

Fig. 2: Fragmentation of the mother precursor of a system of particles into the precursor of the primary particles and the precursor of the secondary particles

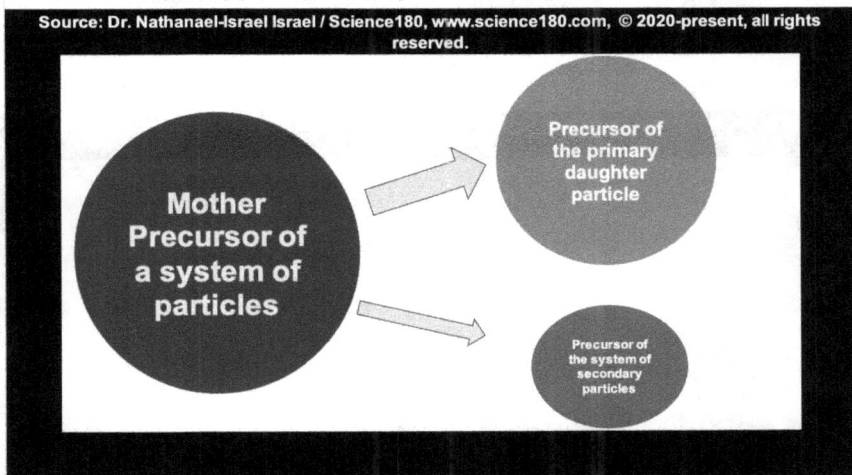

Applied to the formation of atoms, the Fig. 2 looks like the following (Fig. 3).

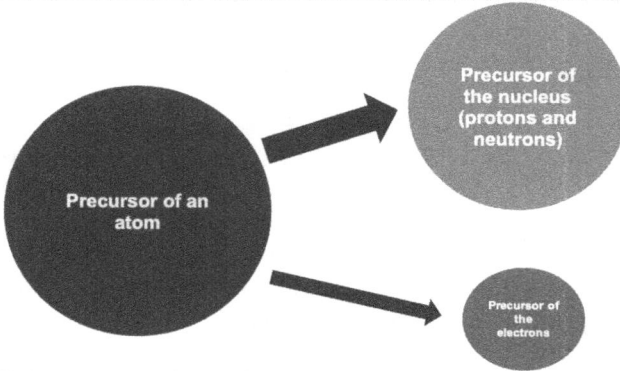

Fig. 3: Fragmentation of the precursor of an atom into the precursor of the nucleus (the primary body) and the precursor of electrons (the secondary bodies)

After splitting from the precursor of the nucleus, the precursor of the electrons was split gathered into individual electrons as sketched in Fig. 4.

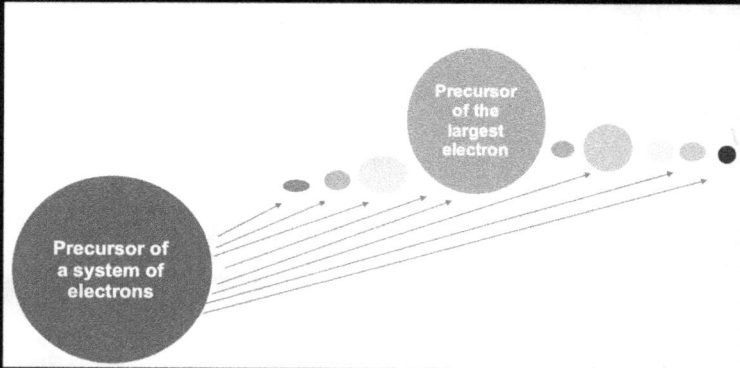

Fig. 4: Split-gathering of the precursor of a system of electrons

The precursor of the system of secondary bodies in each atomic system was split-gathered into electrons placed around the nucleus (Fig. 5-6).

Fig. 5: Precursor of an atom split-gathered into a nucleus surrounded by electrons

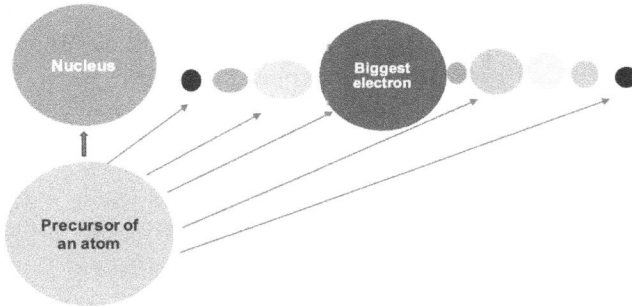

Fig. 6: Layout of the nucleus and electrons born from the precursor of an atom

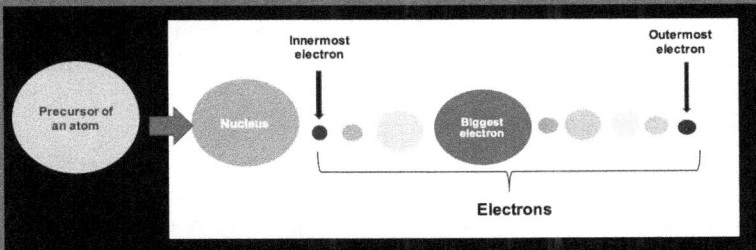

Just as some asteroids and planets have no satellite, so also some chemical particles have no electron "orbiting" their nucleus. In other words, while atoms are known to have a nucleus surrounded by at least an electron, many other types of particles found in nature are not orbited by others. For instance, electrons and other types of leptons (e.g. muon) can also be found in nature not around a nucleus, but in free and isolated forms.

As of today, most secondary bodies (e.g. satellites) orbit their primary body (e.g. planet), while other secondary bodies (like electrons) are said to be in a shell around their primary body (e.g. atomic nucleus). However, when the bodies were being formed, it took some processes before their movement (e.g. rotational and orbital) were appointed by turbulent events that their precursors had gone through and that shaped and defined many of their characteristics. Putting it another way, secondary bodies did not just start "orbiting" their primary bodies overnight or by chance, but complex mechanisms were deployed so that, as the mother precursors were being molded into their daughter bodies, the changes that the daughters went through caused them to acquire specific movements in the end. In *"Turbulent Origin of the Universe"*, I also detailed how specific movements by the bodies in the universe were acquired and maintained until today.

Considering the systems of bodies that I studied, I classified the bodies in the universe into 9 groups that I called inclusion levels of primary bodies and of secondary bodies. Inclusion level 1 is about the organization of the invisible world. Inclusion level 2 is about the microscopic scales such as that of atoms (e.g. electrons and other leptons surrounding the atomic nucleus). Indeed, inside any type of celestial body (e.g. planets, satellites, asteroids, and stars), atoms are the main ordinary systems of particles and they are organized as primary bodies orbited or surrounded by secondary bodies. Although some molecules and other chemical compounds can be organized like clusters of atoms at the center around which other atoms are arranged, these compounds do not seem to be like primary bodies orbited by secondary bodies. Therefore, I used the configuration of electrons (leptons) around the nucleus of atoms to define the 2nd level of primary bodies and secondary bodies.

As I illustrated later in this book, the nucleus of most atoms contains many kinds of particles such as the nucleons (e.g. protons and neutrons), which are the most known by the general public, and other particles like bosons (e.g. Higgs bosons, W bosons, and gluons) which are less known by the public. The number of these particles varies according to the type of atoms. Although the nucleus of the protium (the ordinary hydrogen) is said to have only one proton, the nucleus of most atoms is not just a single or simple primary body, but a system of primary bodies. For instance, each cluster of neutrons, protons, and bosons are systems of particles. Another way of explaining this is that, unlike celestial bodies (e.g. stars, planets, asteroids, and satellites) for which the primary body is a single body, for most atoms, the primary bodies (which can be perceived as the nucleus) is a system of bodies. Because no one has ever observed any subatomic particles with the naked eye or with a microscope yet, it is difficult to say whether the types of primary bodies inside the nucleus of atoms are separated or mixed all together. In other words, the cluster of protons may not be necessarily separated from that of the neutrons and from that of the bosons. Instead, all protons, neutrons, and bosons may just be mixed all together. No matter how the particles in the nucleus are organized, they are complex and diversified.

While many subatomic particles are found in atoms and other composite particles, others are not found, but are isolated. Some of those isolated subatomic particles can be sometimes found in association with others in atoms. For instance, protons (a particle usually found in the nucleus) can also be found in isolation and not orbited by any electron or any other lepton. Similarly, electrons and other types of leptons usually found around the atomic nucleus can be found

Nathanael-Israel Israel: Founder of Science180, the Accurate Universe-Origin, Life-Origin, and Chemicals-Origin Decoder in isolation and not in orbit or in orbitals around a nucleus. For instance, some free electrons exist in nature and in artificial equipment as electronic current, which is the basis of certain forms of electricity. In other words, some subatomic particles are primary bodies not orbited by secondary bodies, while others are "secondary" bodies not orbiting a primary body. The other levels of inclusions (which I approached in *"Turbulent Origin of the Universe"*) are about:

- satellite system scales (e.g. satellites orbiting planets and asteroids)
- planetary system scales (e.g. planets and asteroids orbiting a star)
- stellar system scales (e.g. stars orbiting a galaxy)
- galactic system scales (e.g. galaxies orbiting a cluster of galaxies)
- scales of clusters of galaxies (e.g. clusters of galaxies orbiting a supercluster)
- all other higher order scales including sets of superclusters
- scale of the whole universe (all largest clusters of galaxies embedded in the global system formed by the universe).

- Why in spite of the massive amount of scientific data collected on living things, scientists have misunderstood the formation of life until now, and then uncover in a simple language the one thing that was needed to accurately crack the code of life but that scientists have missed and that has been causing them headaches, overwhelm, and burnout

- Step-by-step pathway to decode the origin of life and get the power, freedom, and boldness to take advantage of the opportunities that accurate understanding of the origin of life creates (Science180.com/life)

- The high connection between the code of the universe formation and the process by which life on Earth was formed so you can become a fulfilled thought leader in your field of expertise

- Tools to stand as a lighting bolt that electrifies those who are still struggling to understand the formation of all forms of life in the universe

- Strategies to push the boundaries of human abilities to properly understand what is perceived as un-understandable, mysterious, supernatural, unimaginable, impossible, and unthinkable that hold people back

- Scientific approach to holistically detect, correct, and remove all misinformation, ambiguity, and misleading claims and theories surrounding the origin of life

Whether you are a scientist or a layperson, a believer, or a skeptic, you cannot afford to ignore the greater, better, faster, simpler, cheaper, easier, and accurate formula unlocked in this important book that successfully decoded the origin of life. Get *"Turbulence Origin of Life"* today and change lives! Don't wait!

Dr. Nathanael-Israel Israel is the Father of Science180 Cosmology and the Founder of Science180 Academy. He is fortunate to be known as the source of unconventional wisdom and knowledge that help people accurately crack the code of the formation of the universe, of life, and of chemicals. Get some resources by visiting his personal website at Israel120.com.

CHAPTER 4

WHAT DO YOU KNOW ABOUT THE CHARACTERISTICS OF THE FLUID LAYERS IN THE PRECURSORS OF PARTICLES THAT YOU NEED TO CHANGE SO YOUR MIND CAN BECOME FLEXIBLE ENOUGH TO EMBRACE THE NEW CHEMICAL REALITY?

In *"Turbulent Origin of the Universe"*, I devoted more than 100 pages to the fluid layers of the precursors of the celestial bodies and particles. In this chapter, because I could not replicate everything here, I will introduce how the characteristics of the fluid layers of the precursors of bodies shaped the features of the subatomic particles, atoms, molecules and the space or voids between them. In short, this chapter lays the foundation of many variables I will address later in this book.

4.1. Fluids layers of the precursors of chemical particles

As the precursors of the chemical particles were being split-gathered, their fluids flowed because they were not a solid that could move altogether in a block. During their flow, fluid layers were formed and moved according to their location. As I showed for the celestial bodies, the fluids in the precursors of chemical particles consisted of layers sliding over one another with each layer moving faster than the one beneath it. The precursors of some particles could not have been initially organized into layers but were just formed as the tube or filament or ligament.

The fluid layers of the precursors of the chemical particles did not have the same size, thickness, or depth. Some could have been tiny and others deep. Between some larger layers, smaller (or tiny) ones could have been found. Some characteristics of the fluid including their viscosity could have affected the intermittence of the size of the layers. My findings with celestial bodies suggested to me that, in the precursors of the bodies orbiting the nucleus (considered as primary bodies), the fluids in the top layers of the bulk of the precursor of electrons were molded into the innermost electrons while the lowest or bottom layers were shaped into the outermost electrons. The presence of the layers of fluids in the precursors of the chemical particles could also explain the layers of particles or clouds of particles in atoms.

Part of the speed of the fluid layers in the precursors of the chemical particles was translated into orbital and rotational speed of the particles. Indeed, the fluid layers of the precursors of the particles did not move at the same speed. Some layers moved fast, while others were slow. The top layers could have had a speed higher than that of the bottom layers. The speed of the layers of the precursors of some chemical particles could have been converted into an "orbital" speed, rotational speed, and other types of speed or movement. The direction of the flow of the fluids of the layers could explain the direction of the movement of particles. It is possible that the motion of particles would be prograde while that of others is retrograde just as in the case for celestial bodies.

The topological and structural changes of the fluids of the precursors of chemical particles could explain some of the differences in their characteristics. These topological changes of the fluid of the precursors of particles contributed to the formation of diverse types of matter in the universe. The interface of the fluid layers of the precursors of particles could have birthed the formation of very unstable structures, which could have become bigger structures as they were mixing and interacting with others. The duration of the topological changes that the precursors of particles was limited according to the characteristics of the layers. As the clusters of bodies of different scales were interacting with one another, their degree of freedom could have decreased until these clusters could no longer rearrange much, therefore locking themselves into a sort of "equilibrium" stage, which also defined their "final" characteristics (e.g. position, movement, composition, and speed). From the invisible scales (smaller than the microscopic scales) to the astronomical ones, topological changes occurred in the early universe. The characteristics (e.g. size, speed, and outcome) of the topological changes are responsible for some of the foundations of the formation and diversity of the particles in the universe. The precursors of the particles in the universe could have birthed different kinds of daughter particles if the turbulence they went through was different. Putting it another way, the phenomena that "destabilized" the early universe and transferred different intensities of turbulence in the fluid layers of the precursors of particles was responsible for the characteristics of most particles in the universe today.

As I carefully studied the bodies in the universe, I realized that their particles were molded according to the "impartation" of proprieties they received from their mother precursors. In other words, the particles were formed after some laws inherited from or "dictated" by their mothers. Unlike nonliving things, children of human beings do not always act like their mothers or parents. In addition to what they receive from their parents, children and even adult human beings make their own choices, which can change the course of their lives. In contrast, the course of the genesis or destiny of matter was mostly defined by their mother precursors, the environments they were formed in, and the changes their precursors went through. For instance, if the intensity of the perturbation that later broke and separated the fluids in the precursors of bodies was different than what it was, the universe would have been shaped differently. In other words, the intensity of the perturbation of the mother precursors influenced the size of their daughter precursors. The instability that the daughter precursors went through also affected their shape, composition, and positioning, etc. To some extent, if the turbulence that occurred during the formation of the universe was not measured or "calibrated", a different world could have been formed, meaning that the precursors of the universe would have been fragmented into other types of bodies, which may not match what they are today. This also implies that in different stellar systems, it is possible to find other types of bodies or matters different from what is present in the Solar System. Therefore, it can be very dangerous for scientists to model the universe or its constituents just by basing their simulations on experimental data collected only on Earth or in the Solar System.

As I proved in *"Turbulent Origin of the Universe"*, the fluid layers of the precursors of bodies in the most developed turbulences can be divided into regions that I called turbulence zones. In the case of the satellites for instance, I distinguished and focused on 5 turbulences:

- Turbulence Zone 1 (where the innermost secondary particles are found)
- Turbulence Zone 2, a transition zone
- Turbulence Zone 3 (where the largest secondary particles are found)
- Turbulence Zone 4, a transition zone
- Turbulence Zone 5 (where the outermost secondary particles are usually found)

I think that similar turbulence zones could exist for chemical particles (e.g. atoms) as well. Considering what I did for the celestial bodies, it is possible that the fastest electrons may be found in Zone 1 of atoms. Likewise, the largest electrons may be found in Zone 3. In other words, I do not think that the electrons size or cloud will be the same for all atoms. The difference in the turbulence of the precursors of the electrons could explain why all electrons are not and do not behave the same. In the same manner, the slowest electrons could be in Zone 5, which are the outermost electrons. For the atoms which precursors have had a fully developed turbulence. In the next graph (Fig. 7), I sketched the organization of the precursors of some electrons of an atom.

Fig. 7: Layers of fluids in the precursor of the electrons of an atom having 5 turbulence zones

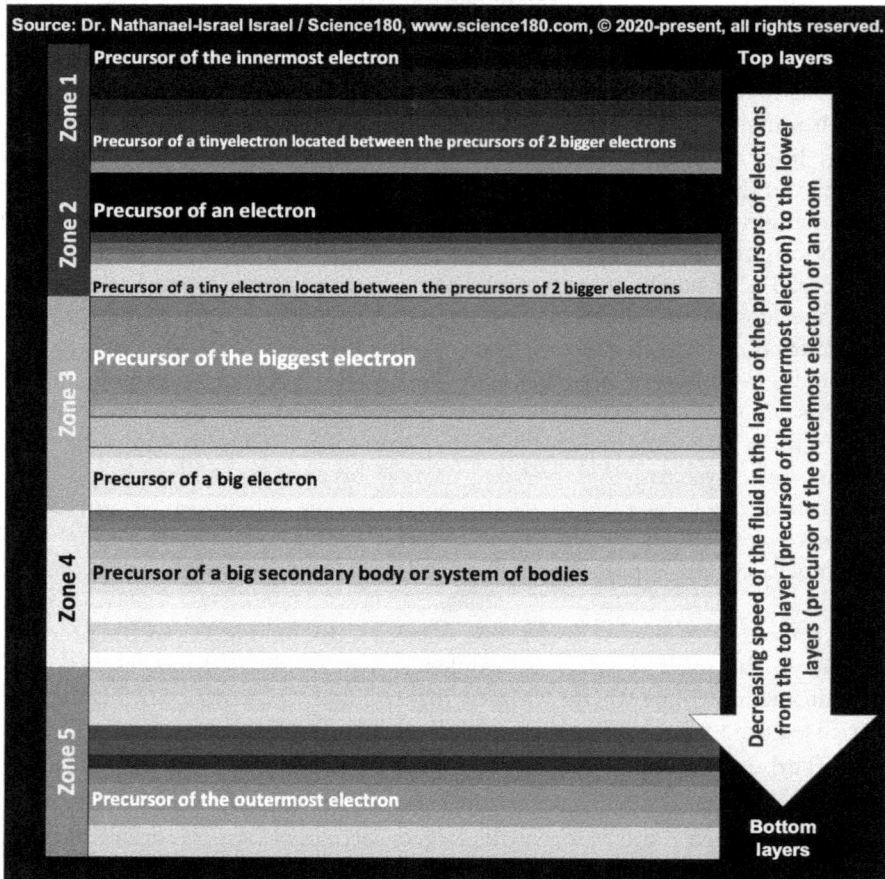

Considering the differences in the presence and abundance of chemical elements in the atmosphere of the celestial bodies in the Solar System (I detailed that later in this book), I think that all of the chemical elements in the precursors of the bodies could not have been formed and completely differentiated before their separation. The turbulence that each precursor of bodies went through after splitting from their mother precursors defined the "final" chemical composition of their daughter bodies. Therefore, the similarities and differences between the chemical composition of the bodies in the universe is due to the processes that molded their precursors.

One of the last steps of the formation of chemical particles was the gathering together of the fluids in the layers of their precursors into a unique body or system of bodies after their separation. The organization of the fluids of the precursors of the particles as they were moving under the influence of the turbulence caused

32

them to collect together. The chemical particles were not collected together the same. Some were wrapped up tightly, while others were gathered together loosely (Fig. 8).

Fig. 8: Gathering together of fluid layers of the precursors of particles into elementary particles and composite particles including chemical elements

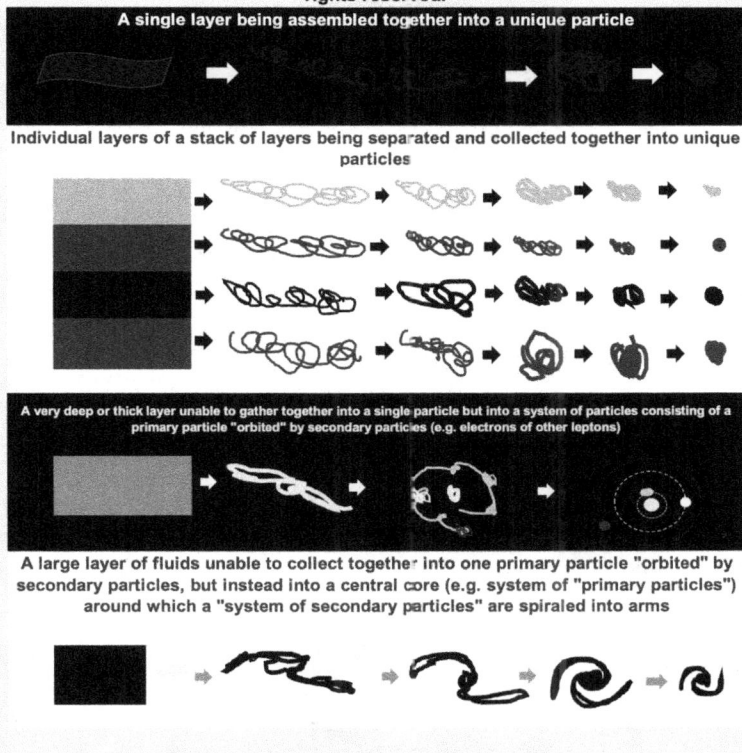

A single layer being assembled together into a unique particle

Individual layers of a stack of layers being separated and collected together into unique particles

A very deep or thick layer unable to gather together into a single particle but into a system of particles consisting of a primary particle "orbited" by secondary particles (e.g. electrons of other leptons)

A large layer of fluids unable to collect together into one primary particle "orbited" by secondary particles, but instead into a central core (e.g. system of "primary particles") around which a "system of secondary particles" are spiraled into arms

As some vortices or eddies were being "rolled" to form circular or spherical bodies, they were squeezed. The energy in the precursors may have also affected the squeezing intensity or the applied pressure. As the layers of fluid were being wrapped, they could have applied a squeezing force to their daughter particles. The squeezing of the vortices could have also affected the density and other characteristics of the particles. The processes that gathered together the particles before and after the layers of their precursors were separated played a crucial role in the establishment of some fundamental "forces", shape, rotational, and "orbital" movement of the particles.

Considering the organization of the subatomic particles, atoms, molecules, chemical compounds, minerals, rocks, crust of terrestrial planets, the atmosphere of the celestial bodies, the plasma on the Sun, and the variables I studied in this book, I felt like the gathering together of the chemical particles in the universe could have involved processes associated with the following phenomena: amalgamation, bending, braiding, clumping, kinking, knobbing, lumping, netting, pulling, pushing, shearing, spiraling, twisting, winding, and many other processes I already discussed about the fate of the broken piece of the turbulent prima materia.

Depending on its size, spatial localization, speed, the characteristics of its constituents, the same precursor of particles can lead to the formation of different final daughter particles. For instance, a liquid can be converted into ice by lowering its temperature below a certain critical value. However, depending on the duration of the temperature lowering, a same water can lead to the formation of various types of ice with different structures or different states of matter. For instance, amorphous water is formed by quickly lowering the temperature of water. Although it is a kind of ice, amorphous water does not have regular repeating crystal structure like normal ice. Similarly, the outcome of some precursors of particles was highly affected by the duration of the process they went through.

4.2. Precursors of subatomic particles, atoms, and molecules

The precursors of atoms were split-gathered into the precursors of the nucleus and that of the secondary bodies "orbiting" them or found in shells or orbitals around them (e.g. electrons) (Fig. 9).

Fig. 9: Split-gathering of the precursor of an atom into the precusor of the nucleus and the precursor of the electrons

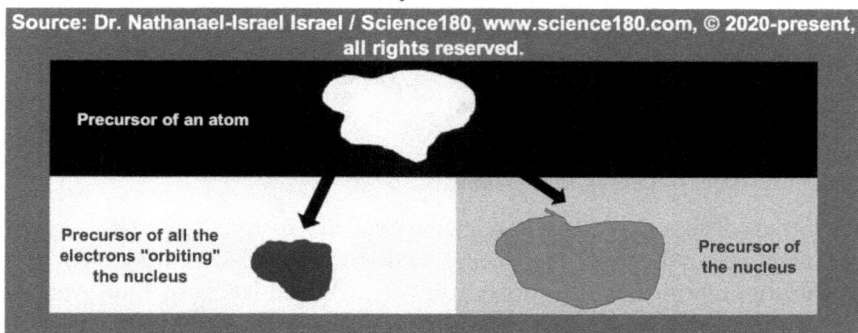

Likewise, looking at some subatomic particles, similar processes of split-gathering could have occurred. Due to the lack of reliable experimental data on the size of all subatomic particles and atoms, and the fact that no one has ever seen any of these particles, I just relied on my understanding of the data I analyzed to address the precursors of microscopic particles. The complex and sometimes

CHAPTER 4: IMPORTANT CHARACTERISTICS OF THE FLUID LAYERS IN THE PRECURSORS OF PARTICLES

controversial and contradictory data in the quantum physics of subatomic particles also points at how scientists are still struggling to characterize particles and much less, how they do not understand them.

The precursors of electrons around the nucleus could have also been organized into tiny layers according to the scale of the turbulence that occurred at their level. Current knowledge in classical and quantum physics does not provide sufficient information to properly model the size of all electrons according to their position around the nucleus.

Because some atoms do not have many electrons, the turbulence of the fluids in the precursors of the electrons around nuclei may not have had fluids in all 5 turbulence zones. However, for bigger atoms, the fluid layers of the precursors of their electrons could have existed in all turbulence zones. Unfortunately, very little is known about the characteristics of the innermost electrons of the heaviest atoms. For instance, when I was studying the ionization energy of atoms (more details coming up later), I realized that most literature did not go beyond the 30th ionization energy, suggesting that ionization energy data could not have been collected on electrons closer to the nucleus than that responsible for the 30th ionization energy. Or among all of the variables I studied on electrons, ionization energy was the one that gave me a better insight into potential behaviors of layers of electrons around the nuclei according to their potential position. In other words, data does not exist to model the distribution of electrons and much more the way their precursors could have been organized into layers or turbulence zones.

Unlike celestial bodies where layers of precursors of secondary bodies were mostly converted into secondary bodies orbiting their primary bodies, it is the case for electrons, the fluid layers of their precursors could not have always been fully wrapped around or gathered together to form unified electrons. Another way of expressing this is that the fluids layers of the precursors of some, if not of all, electrons could have failed to wind up, spiral, or cluster into one body which can orbit the nucleus, therefore they could have been scattered into a shell around the nucleus. On the macroscopic scale, this trend is also found with the precursors of some celestial bodies such as the main belt asteroids, which, because it failed to aggregate into one body, was split into many asteroids which cover the space of the main belt located between Mars and Jupiter. The inability of some layers of the precursors of electrons to collect together into a single body could explain why electrons are postulated not as orbiting the nucleus but spread over an electronic shell. However, as I explained later in the chapter on radius and rings, when some precursors of secondary bodies were gathered into their daughter bodies, some of their fluids were left out, and consequently they became rings surrounding some bodies such as satellites. Similarly, it is possible that even in the electronic shell, some electrons may be surrounded by some smaller particles which once belonged to their mother precursors and which were not compressed or collected together with the electron itself. In other words, even in an electronic shell, the size and

Science180: Chemicals-Origin Formula Accurately Made Easy

particles of the constitutive bodies may differ according to their position with respect to the nucleus of their atoms.

The precursors of the atoms could have gone through processes similar to those of the precursors of the systems of celestial bodies, but this time on a microscopic scale. Because as of 2020, about 118 chemical elements, accounting for more than one thousand isotopes (see the chapter on chemical elements) where discovered, the turbulence which shaped these atoms could have been different, hence different atoms were obtained.

4.3. Precursors of the space between particles

At this point, I would like to address the precursors of the medium or space between particles in an atom or other systems of particles. Indeed, just as in the case of the celestial bodies, a space or void is found around and between chemical particles. I labeled those spaces as:

- Inter mineral space (space between minerals)
- Inter rock space (space between rocks)
- Inter molecular space (space between molecules and chemical compounds)
- Inter atomic space (space between atoms)
- Inter subatomic particles space (space between subatomic particles)
- Invisible world space (space between spiritual things and beings invisible to the naked eye and even to the most advanced scientific equipment).

Some of those spaces are filled with tiny particles, dusts, or clouds that I defined as:

- Inter molecular dust
- Inter atomic dusts
- Sub-subatomic dusts
- Atomic dusts
- Subatomic particle dusts
- Other types of cosmic dusts

I used the term "subatomic dusts" because I believe that some particles tinier than the known subatomic particles may be filling some space between atoms, molecules, and even subatomic particles. Those sub-subatomic particles may be tinier than dust. Moreover, some particles found in space today could not have been there since the beginning of the universe, but were formed after some particles reacted with others. In *"Turbulent Origin of the Universe"*, I talked more about the space and particles found around and between celestial bodies.

As some fluid layers originally stacked one on top of the other were separating, a space appeared between them. To put it another way, after their separation, 2 originally consecutive fluid layers became surrounded by a space. Above the space between them could have been the layers that were on top and below the space between them was the fluid layers that were below. Some fluid bodies of the precursors could have leaked into these layers, therefore causing some spaces to be filled with particles, while others can be "empty". The number of particles between bodies in space could have been defined by the amount of fluid which was leaked into them or which landed into them. By the time the fluid bodies were gathered together, the space that appeared between them became the space where the secondary bodies of the layers of fluids were born. As the fluid layers of the precursors of the secondary particles were going through their own split-gathering into its daughter secondary particles, additional space was created between their layers. These new spaces became the precursors of the space around each secondary particle.

Unlike some celestial bodies that have a gigantic atmosphere around them, atoms and subatomic particles do not have much space around them. Yet, when these bodies were being shaped, "empty voids" were formed between their constitutive particles. In the case of atoms surrounded by electrons, the space around the nucleus where the electrons are located is called "electron cloud". If there is nothing in that space, it is empty space. The radius of atoms and subatomic particles is usually very small, billions or trillions fold smaller than a millimeter; consequently, the space around the "nucleus" of their primary bodies is also very small. Likewise, the space cloud of matter between electrons are extremely small. Although these particles cannot be seen, a space exist between them. For instance, the space between chemical elements affects the length of their bonds. Just as I explained for the celestial bodies, when the tiny layers of the constitutive particles of atoms and subatomic particles were being split-gathered, a tiny space was also formed between the daughter bodies. For each particle, the size of that space defined the size of what could be called "atmosphere", which in the case of microscopic particles is the space between the electrons and the nucleons. Just as the atmosphere around a celestial body is under the influence of the later yet all the space between 2 celestial bodies is not always under the influence of neither of them, so also between 2 atoms or subatomic particles, a space near them can be under their respective influence, but another space existing between them is controlled by none of them. The length of the space between atoms and subatomic particles can affect their "bonding" to one another and their reactivity with others.

Regardless of the size of the particles, the space between them was initiated since the split-gathering of the fluid layers of the precursors of these bodies, and then "amplified" by the expansion of the universe. In the chapter on the abundance of chemical elements, I dedicated dozens of pages to the chemical composition and origin of the atmosphere of the celestial bodies. In those chapters, I also delved into details regarding the space between microscopic particles. Because of the differences in the environment where the precursors of the atmosphere or of the clouds around particles were formed, different particles were formed. The amount of fluids that landed into the precursors of those atmospheres or space defined the abundance of particles found in them. With time, some particles could have leaked or volatilized from the celestial bodies after their formation to station in the atmosphere. Likewise, other particles in the atmosphere were formed after some reactions including solar radiations. Adapted from my work on the formation of the atmosphere of celestial bodies (see *"Turbulent Origin of the Universe"*), Fig. 10 outlines how the space in a system of particles (e.g. atoms) could have been formed after the fluid layers of its precursors separated.

Fig. 10: Precursor of the space between particles

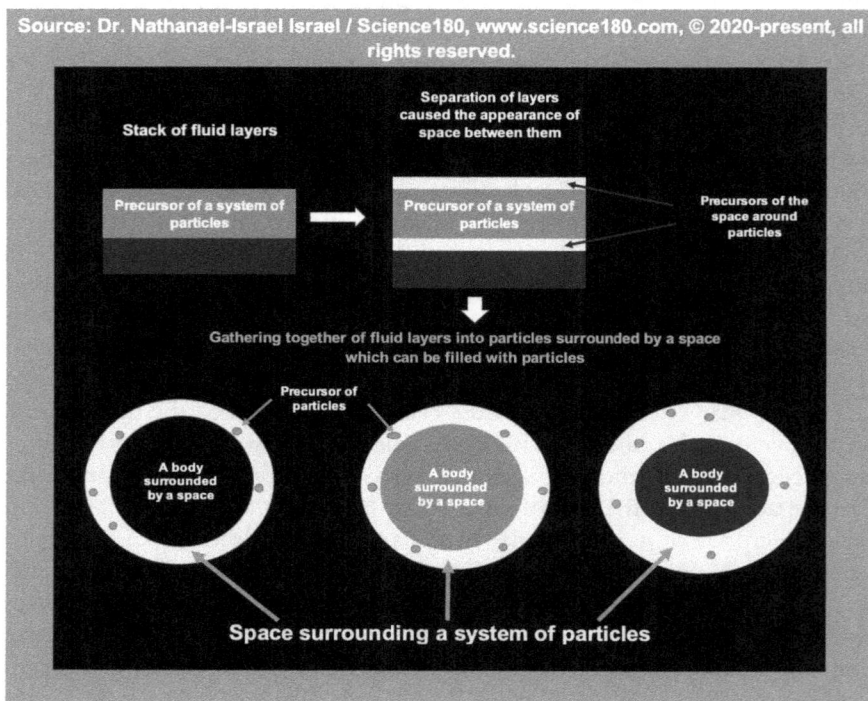

At this point, I would like to mention that, although I am aware of the existence of other types of particles unknown to the chemical community, I did not handle them in this book. For instance, considering the stories and revelations in many religious books, other beings that are not made with the same chemical elements constituting Earth's biosphere (e.g. plants, wild animals, and human beings) also exist in nature. For example, angels and other forms of spiritual beings or spiritual things (believed in all cultures across the world) are not common to mere human beings, yet some of them are real and made of special materials. Even those who believe in their existence, cannot properly articulate their characteristics beyond what is in their religious book. Therefore, to know my viewpoint on the kind of particles constituting those supernatural beings, you may want to consult the book I devoted to them.

4.4. Positioning of chemical particles at specific distances and speed

As the precursors of particles were being assembled and their movement set, they were dynamically led into orbit or orbital, where they would spend the rest of their "lifespan". This implies that, when chemical particles were being formed, the position of their precursors was changing in space and at the end of their formation, the position of their "orbital" movement inside their system was "locked down". The time it took to position the particles at their position in orbit or orbitals could have been affected by the longitudinal stretch and detachment of the ligaments of their precursors.

Subatomic particles, atoms, and molecules are not static, but are moving although the precision of the measurement of their motions can be hard to reach. However, the rotation and revolution of chemical particles and the speed associated with them were established as the bodies were being separated by the split-gathering of their precursors and then put into orbit or into the location of their orbitals. Before each particle was placed into its orbital, there was a time when nothing could have been called rotation and revolution yet. Between the time the precursor of a particle was born and the time it reached its "final" orbit, the speed it had was different from the speed it got once in orbit or in its orbital or in its cloud of particles. As the precursors of the particles started moving under the influence of the push they got from their mother, the movement of their fluids progressively birthed the precursor of their movement that later matured into the "orbital" speed and "rotational" speed. The way the vortices of the precursors of chemical particles could have been stretched, twisted, and tilted could have affected the inclination of their particles. In *"Turbulent Origin of the Universe"*, I detailed many facts connected to the stretching, inclination, and tilting of precursors of bodies.

My research on the origin of the universe also taught me that the speed of clusters of particles bound together would be smaller than that of individual particles of the same kind. For instance, it was recently demonstrated that the speed of photons bound together was smaller than the speed of individual photons (Liang et al., 2018). In addition to other evidences, this behavior of photons made me think that the clustering of particles can slow them down. Indeed, for a long time, scientists have been forming molecules by mixing atoms. However, efforts to stick photons together to make bigger and heavier photons have been vain until a few years ago. For scientists, individual photons that make up light do not interact. Part of the problem can be due to the high speed of photons that constitute light. In 2018, a group of researchers at the Massachusetts Institute of Technology (MIT) and Harvard University, said that they managed to cause photons to stick together and form heavier and larger triplets of photons. Once they managed to put the photons together, they noticed that "the heavier photon molecules" were considerably less nimble, and rather than moving at about

300,000 kilometers per second (the speed of light), they were moving 100,000 times slower than normal noninteracting photons" (Liang et al., 2018). Some of the members of this same group of scientists mentioned in the previous reference were able to put together 2 photons about 10 years ago, suggesting that, under certain conditions, it is possible to form bigger bodies from smaller ones, even from very tiny particles like photons that scientists have assumed for so long to be massless and incapable of interacting with others particles. In the same manner that the clustering of the photons into heavier compounds slowed down their speed, so also the speed the precursors of particle clusters could have been reduced as individual precursors were being gathered together to form clusters of bigger ones, or as smaller particles were being formed or shaped inside larger ones. In other words, the speed of the precursors of chemical particles could have been much higher than the speed of the particles that can be observed in nature today.

4.5. Orbital and rotational speed of chemical particles

Just like celestial bodies, chemical particles are also moving, and understanding the cause of their motion can help to better understand their origin. Here, using what I learned from the celestial bodies, I discussed two types of speed for chemical particles: orbital speed and rotational angular speed. In the chapter where I will discuss the ionization energy, I will also provide more information on the relation between the speed and energy of chemical elements.

In *"Turbulent Origin of the Universe"*, I extensively talked about orbital speed. Here, I will just pinpoint a few data relating to chemicals. From the celestial bodies to the microscopic particles (e.g. electrons) passing by the planets, satellites, asteroids, stars, planetary systems, and galaxies, everything in the universe is moving. The orbital movement is the translational movement of an object around a primary object. The orbital speed of the celestial bodies is not constant but varies with their position in orbit. In general, celestial bodies reach their maximum orbital speed at their perihelion (the closest point of approach of a celestial body orbiting a primary body), whereas they reach their minimum orbital speed at their aphelion (the most remote or the farthest point of approach of a celestial body orbiting a primary body). Between the aphelion and the perihelion, the orbital speed changes.

Data available on chemical particles did not allow me to deeply study the variation of their orbital speed. Additionally, some of the modern models of configuring atoms for instance seem to reject the notion of electrons orbiting the nucleus, but as particles in a cloud or orbital around the nucleus. In other words, the term "orbital speed" may not even be accepted by some people. Here, I defined the orbital speed of the particles as their speed on their trajectory no matter if the trajectory is perceived as an orbit, orbital, or a cloud. In the following lines, based on what I know about celestial bodies, I deduced a few things about the speed of chemical particles (e.g. electrons and nucleons). Indeed, I do not think that all electrons have the same speed. The data I studied on turbulence

CHAPTER 4: IMPORTANT CHARACTERISTICS OF THE FLUID LAYERS IN THE PRECURSORS OF PARTICLES

showed that the uppermost layers had the fastest velocity, while the bottom layers had the smallest speed. The energy with which the precursors of chemical particles "pushed" away their daughter particles could have been one of the main forces that communicated a shearing force or ability of the fluids of the daughters. The force that acted on the precursors of the daughter particles changed as they were being shaped. Consequently, the precursors of particles could have been "propelled" or launched into motion and could have also taken the form of a shear stress as they were being pushed, "wound up", or wrapped together by the processes that molded them. At the same time, the fluid layers that were on top were also acting on those that were beneath them. Just as I explained for the celestial bodies, it is possible that some fluid layers of the precursors of particles were also subdued to the weight of the fluids laid on top of them. Because of the stress due to the weight above them, the fluid layers could have pressed some bottom layers more than the top layers. This could later translate into the variation of some characteristics of the chemical particles. For instance, if the stress is strong, strain (which is a change in form or shape of a body after a stress is applied) can occur. Strain could have deformed, crushed, broken, or squeezed some precursors. The stacking of the fluid layers could have also affected the velocity gradient of the precursors of chemical particles.

The orbital speed of chemical particles (e.g. electrons) around atomic nucleus is a footprint of the speed with which their precursors were flowing or scattered into the layers of fluids from which they were born. As the precursors of the primary particles started flowing, the turbulence and the reorganization or rearrangement of their constitutive particles affected their orbital speed. For instance, because of their size, the layers of the precursors of the primary particles did not flow at the same speed as they were being gathered together to form the primary speed. Consequently, the speed of the primary particles could have been slowed down a little bit as these particles were being gathered together as a whole. The fluid layers in the precursors of the secondary particles (which are the particles "orbiting" the primary ones) moved at different speeds as they were being split from one another. More information on the speed of chemical particles are in *"Turbulent Origin of the Universe"*. In that book, I explained that, as they were moving, the fluid layers of the precursors of bodies started collecting themselves and progressively, they acquired a rotational movement under the influence of the turbulence that defined them. I also showed that rotation was imparted onto the bodies as a consequence of the spiraling, winding up, and other aspects of the turbulence of the fluids in the precursors. I also showed that a very strong and negative relationship exists between rotational angular speed and semi major of the satellites in Zones 1, 2, and 3. However, in Zones 4 and 5, no significant relationship exists between the variables.

Likewise, I felt like the variation of the angular speed of the chemical particles can be linked to the position of their precursors in the space where their turbulence took place. The chemical particles that were formed in higher

Science180: Chemicals-Origin Formula Accurately Made Easy

turbulence zones may have a smaller rotational angular speed. It appeared to me that, as the precursors of the fluids were spiraling to yield each body, forces were induced or applied to them, and could have contributed to creating and maintaining some fundamental forces. Subsequently, some fundamental forces are the result of how the bodies were formed and the field that was created around them. The scaling of the regression between the rotational angular speed of electrons and their distance from the nucleus can be controlled by their kinetic energy (total, translational, and rotational) and the rotational angular momentum.

4.6. Sequestration of energy in chemical particles

The processes involved in the split-gathering of the precursors of particles required energy. By the time the formation of the universe was "completed", energy was gathered into entities of matter. As I established in *"Turbulent Origin of the Universe"*, part of the energy in the turbulent prima materia was used to form matter, others were stored in different components, including chemical bonds. Energy was also used to tilt, incline, or flip the orbital plane and rotational axis of some particles. Likewise, chemical energy was stored in atomic bonds and released or absorbed during chemical reactions when existing bonds are "broken", or altered, and new ones are formed. Energy is also stored in the form of nuclear energy, meaning stored in the nucleus of atoms. Nuclear energy can be released during nuclear reactions, which usually alter the fundamental identity of chemical elements. During chemical reactions, atoms are believed to be reorganized differently to form new compounds, whereas during nuclear reactions, atoms are believed to change into other types of atoms leading to different compounds (Wikipedia, 2020a). Fission reactions, fusion reactions, and radioactive reactions involve energy. Indeed, a fission is a nuclear reaction through which atoms split apart. Radioactive particles and photons can be produced by some fission reactions. In contrast, fusion is a nuclear reaction during which atoms "join" together to form new atoms. According to the previous reference, the amount of free energy contained in some nuclear fuels is millions of times that contained in a similar mass of a chemical fuel such as gasoline. In other words, a lot of energy is condensed in the atomic nuclei. In addition to the energy they release, nuclear fissions produce electromagnetic radiation for instance in the form of gamma rays (Wikipedia, 2020a), implying that electromagnetic fields could have been one of the early daughter bodies of the initial matter that was molded into the various chemicals in the universe. It is possible that under certain conditions, some electromagnetic fields could have been condensed into some dense blocks, which could have been used as a component of atoms or other types of matter. However, by-products of nuclear fission can be polluted, radioactive, and destructive, therefore causing a nuclear waste problem. This is because waste of nuclear fission are compounds which are not normally meant to be in the environment where they are produced. Therefore, once they are formed, they are unstable and by trying to go through some reactions to stabilize themselves, they can damage other chemical elements and living organism in their neighborhood. Similarly, if human beings or other living organisms are placed on certain celestial bodies, they can die and become sick because of the presence of chemical elements they are not used to on Earth. For the characteristics of elements were made to match the environments in which they were formed.

Scientific efforts have been made to try to explain the formation of chemical elements with fusion reactions. For instance, although science has shown that hydrogen atoms can be fused to produce helium, a reaction during which a lot of energy is released, I strongly believe that most chemical elements in nature were not produced by just fusing atoms together. Later in this book, I explained the real processes that were involved in the formation of chemical elements. Scientists have shown that elastic energy can be stored mechanically in compressed fluids, coiled objects, stretched elastic objects, etc. Similarly, during the formation of the universe, energy could have been stored in matter as their precursors were wrapped and shaped differently according to the turbulence they went through. The mechanical form of energy can explain some of the energy stored in chemical and nuclear bonds. How matter was wrapped could have also determined the amount of energy it can contain. This can explain why some of the biggest chemical elements like Uranium and Plutonium are some of the most energetic ones.

As atoms were being linked together during their formation, some bonds could have absorbed energy as they were "stressed", and that energy can be released when these atoms are disconnected from one another, or when their bonds are relaxed. A careful analysis of the energy released by chemical and nuclear reactions can hint at the amount of energy involved in the formation of chemical elements. Because energy is needed to form chemical particles, it is possible that the speed of their precursors could have been higher than the speed of the elements today. The formation of the chemical elements that constitute the celestial bodies required energy, which could have been withdrawn or taken from the precursors as they were being shaped or molded into daughter bodies.

To some extent, the ability of energy to be stored in chemical bonds can challenge the famous formula of Albert Einstein: $E=mc^2$. For the same mass of two chemical elements can contain different amounts of energy depending on how their precursors were molded or configured during their formation. For instance, the same mass of a plastic that is relaxed cannot contain the same amount of energy as a plastic that is expanded or that is highly coiled or tied. To put it in other words, the energy in a sling which is extended is different from that in the sling which is relaxed. Yet their masses are the same.

Although electrons are said to be very light (not heavy), they are responsible for some forms of electricity, which are a very powerful source of energy. Similarly, most electromagnetic fields are not heavy, yet, they can carry a lot of energy. The nature of the carriers of energy can vary from one environment to the other, but in the end, the nature of the work that energy can do can be the same.

This suggested that most energetic things can be the manifestation of the same thing but just in different forms according to how these energetic things were formed.

In the Solar System, the Sun is by far the largest energy reservoir. When it releases its energy in the form of photons for instance, plants can capture and transform it into chemical energy, which they use to build organic matter and other types of biological compounds. Additionally, plants can extract water and certain chemical elements from the soil and the atmosphere to build other compounds. As they break down these chemical elements to construct new ones, plants also use the released energy to heat themselves, build their matter, grow, and maintain themselves. At their turn, other livings organisms depend on the energy in the primary producers (e.g. plants) and on others compounds they eat or drink. As they digest their food and drink, animals get the necessary energy they need to live. All living organisms know how to adjust to the changing environmental conditions so their metabolism and ecology can match the requirements of their existence. Otherwise, they also feel pain or suffering. Finally, nonliving things or matter, have been formed in environments that match their states, and when the environmental conditions change, the characteristics of the matter can also change at many levels, including chemical and nuclear levels. The ability of living organisms and nonliving things to change their shape or constitution in order to adapt or satisfy their "needs" or respond to environmental constraints also proves that everything in nature was formed in specific conditions and are required to respect and maintain certain behaviors or proprieties in order to keep existing. Because non-living things cannot (or we think they cannot) make certain decisions like living organisms, they have been placed on a path (e.g. orbit or location) that helps "maintain" their structure and energy levels to some extent. In other words, without a clear path or trajectory, and process that "preserve" their energy, nonliving things could have also lost most of their identity or been transformed into other things already. The entropy and all other types of decay in nature are forms of the degradation that nature or the environment is feeling from its inhabitants and its internal constituents.

Considering some findings in electromagnetism, I perceived that, if sufficient energy could be deployed, anything in the universe could be displaced. For instance, under certain circumstances, photons (i.e. particles of light) are able to knock out electrons from some materials. For instance, when light is shone onto a material, electrons or other free carriers can be emitted by a process called photoelectric effect for which Albert Einstein won the 1921 Nobel Prize in Physics. The classical electromagnetism theorized that the photoelectric effect can be caused by the transfer of energy from light to electrons. Scientists have shown that, if an electron absorbs the energy of a photon, that electron can be ejected from its initial position around a nucleus (Wikipedia, 2021a). Researchers have also shown that, if the photon energy is too low, the electron can be unable to escape from the atom it was "orbiting". The energy of the released electrons was proven to depend on the energy of the incident photons. Considering the photoelectric effect, it is possible that when sufficient energy is applied to some bodies (small or big), they can be removed from the initial orbit and sent to a different position. This also implies that the position or location of the particles and even celestial bodies in the universe depend also on their energy and size. Of course the history of the formation of the bodies in the universe also influenced their characteristics. However, considering the energy in the celestial bodies, their displacement may be very difficult, but possible if the right energy can be applied to them. Furthermore, it is important not to rule out that the systems of bodies in the world can be destroyed or displaced instantly if the right energy is applied to them from the right system. Just as electrons can be displaced by applying energy to them, so also celestial bodies or systems of celestial bodies can be displaced if the right amount of energy can be applied to them the right way and in the right direction. Furthermore, just as a powder can be scattered and its elements lost in the air, so also celestial bodies and even the whole universe can be scattered if the right amount of energy can be applied to them from the right perspective. Although these possibilities may sound like a fiction, they should not be rendered impossible. This also suggests that some religions that talk about the end of world should also be taken seriously for, even the scientific data suggest that the world can also end one day. I deferred this philosophical digression in one of my books of the origin of the universe. Furthermore, one of the challenging questions in science and particularly in cosmology is how energy came into existence. Like I told the participants of Science180 Academy (www.Science180.com), this question cannot be fully answered without referring to some supernatural facts, some of which I addressed in my book called *"Reconciling Science and Creation Accurately"*, the Biblical version of my book titled *"Turbulent Origin of the Universe"*, in which, I extensively elaborated on the following energies: rotational kinetic energy, translational kinetic energy, total kinetic energy, square of orbital speed, and turbulent kinetic energy.

CHAPTER 5

UNRAVELING THE IRREFUTABLE SYSTEM-ADDITIVE VARIABLES AND TURBULENT MULTIPLIERS TO UNDERSTAND TURBULENCE AT THE CHEMICAL LEVEL–IF YOU DON'T BELIEVE ME, READ THIS AND YOU WILL SEE!

5.1. The 9 system-additive variables and the 99% vs. 1% rule

When I was studying the origin of the celestial bodies, I invented the term 9 system-additive variables to designate the following variables:

- mass
- orbital angular momentum
- orbital moment of inertia
- rotational angular momentum
- rotational kinetic energy
- rotational moment of inertia
- total kinetic energy
- translational kinetic energy
- volume

In *"Turbulent Origin of the Universe"*, I established that, during the split-gathering of the precursor of the Solar System, more than 99% of the value of its 9 system-additive variables went into the precursor of the Sun and less than 1% went into the precursors of the bodies orbiting the Sun. After going through some turbulent changes, the precursor of the Sun "finalized" its formation, becoming the Sun, which kept more than 99% of the value of each of the 9 system-additive variables of the bodies in the Solar System.

During the split-gathering of the precursor of the bodies orbiting the Sun, more than 99% of the 9 system-additive variables went into the precursors of the planetary systems and less than 1% went into the precursors of the asteroids. More than 99% of the 9 system-additive variables of the planetary systems is accumulated in the 4 giant planetary systems. During the split-gathering of the precursors of the planetary systems, more than 99% of each of the 9 system-additive variables went into the precursors of the primary planets and less than 1% went into the precursors of their satellites. Therefore, I felt like the 9 system-additive variables for the atoms could not be the same if data can be collected on them to test them. When I studied the characteristics of the subatomic particles, I also felt like the size, energy, mass, and angular momentum of even the electron around the same particles may not be the same as modern science suggests. In the incoming chapters, I expounded more.

In *"Turbulent Origin of the Universe"*, I also proved that, during the split-gathering of the precursors of the satellites, more than 99% of the 9 system-additive variables were accumulated in their most turbulent zone, usually Zone 3. For instance, more than 99% of the total kinetic energy of the satellites is accumulated in the most turbulent zone, usually Zone 3. Likewise, more than 99% of the rotational kinetic energy of the satellites is accumulated in the most turbulent zone, usually Zone 3. Similarly, more than 99% of the translational kinetic energy of the satellites is accumulated in the most turbulent zone, usually Zone 3. I also demonstrated that more than 99% of the rotational angular momentum of the satellites in each planetary system is accumulated in the most turbulent zone, usually Zone 3. Most of the 9 system-additive variables of the precursors of the satellites was concentrated in a few satellites. Considering the trends that I established for the celestial bodies, I think that, eventually, scientists will come to realize more variations among the characteristics of the subatomic particles from one chemical element to the other, more so than they have suggested thus far.

For celestial bodies, I showed that the general rule of the split and gathering together of the precursors of the systems of bodies is that, more than 99% of the 9 system-additive variables was pushed into the precursor of the primary body and less than 1% was pushed into the precursors of the systems of secondary bodies. As the less than 1% in the precursors of the systems of the secondary bodies started flowing and mixing together, different daughter precursors of secondary bodies were born and the 9 system-additive variables in them depended on various factors. The precursor of some secondary daughter bodies went through their own breakup and gathering together of their own bodies by still generally following the 99% and less than 1% rule of the amount of the 9 system-additive variables that they pushed into their primary bodies and secondary bodies.

Considering the current estimation about the mass and energy of subatomic particles, it sounds like most of the mass and energy is concentrated in the nucleus, which, according to the terminologies I used in this book, is a system of primary bodies. I wished to also study the split-gathering of the 9 system-additive variables of the atomic systems just like I did for the celestial bodies, but I could not because little work has been done on these 9 variables of the electrons. Efforts

47

Science180: Bringing People Together Through the Power of the Accurate Decoding and Understanding of the Origin of Chemicals

have been made to estimate the mass, volume, and speed of an electron, of course. But such works have not been done for each of the electrons around each of the chemical elements. Such efforts could have allowed to better know whether the characteristics of the electrons are the same from one orbital or electronic shell to the other or from one atom or chemical element to the other. The small size of subatomic particles is one of the things that have been preventing scientists from looking into some of their details. Therefore, because such data do not exist, I was limited in how deep I could go with my analysis of the 9 system-additive variables of atoms.

Since the mass of atoms is assumed to be that of their nucleus, the total kinetic energy of atoms can be assumed to be mostly that in the nucleus. Yet, electrons are also assumed to be moving at a higher speed in their orbital or in the shell more than nucleons. Nevertheless, due to the small mass of electrons, the kinetic energy of electrons around the atomic nucleus is assumed insignificant. Moreover, although many electrons are postulated to orbit atoms, their speed has not been properly measured. Considering what I learned from the movement of the primary bodies and secondary bodies in the various systems of celestial bodies that I studied, I think that the speed of electrons in their orbital or electronic shell would not be the same, not even among the electrons within the same atoms. The innermost electrons could be moving faster than the outermost ones. Similarly, considering the variation of mass of the secondary bodies around their primary bodies, and knowing that the bulk of the mass is usually concentrated in the most turbulent zone, usually Zone 3, I think that the mass of electrons would not be the same. Some electrons located in what could be called Zone 3 may be heavier and bigger. Because of the lack of detailed information on electrons, it was impossible for me to elaborate much on the distribution of energy of electrons around atoms using scientific terms. In the segments related to ionization energy in the chapter on chemical elements, I talked more about energy in atoms.

For the celestial bodies, I studied how the 9 system-additive variables of the primary bodies relate to those of all of their secondary bodies. Indeed, one of the major assumptions in physics and in other scientific disciplines is that physical laws or laws of nature and the distribution of bodies in the universe are the same everywhere. Chemists also seem to assume the same things regarding chemical laws. However, from my understanding of laws surrounding for formation and sustenance of celestial bodies, I do not think that laws of nature and of chemical particles are the same everywhere. When I investigated the relation between key variables collected on primary bodies and the bodies orbiting them, the trends did not suggest a similarity or constancy of natural laws across all systems. In fact, I established that the ratios of the primary bodies to that of the bodies orbiting them is not a constant but depends not only on the type of variables, but also on the type of bodies. The variation of these ratios suggests that the interaction between the primary bodies and the bodies which orbit them depends on the system. For instance, the interactions between a planet and its satellites depend on the planetary system just as the interactions between a star and all bodies orbiting it may depend on the type of star. This data can also help explain the diversity of chemical elements and their diverse characteristics in nature. For the diversity of the splitting or fragmentation of the precursors of matters into different clusters of matters during the formation of the universe conferred to each cluster or body unique characteristics, which are differently scaled from one cluster of bodies to the others and from one environment to the others. The trends explored in this chapter point to the fact that laws of nature are not the same nor scaled the same across the universe.

5.2. Turbulent multipliers of the 9 system-additive variables

Before addressing the multipliers, I would like to first quickly explain why I invented the term "turbulent multipliers". Indeed, a random multiplier is a variable used to explain fragmentation of bodies. I studied it to explore how the precursors of the celestial bodies may have been fragmentated to yield the bodies in the universe. As my understanding of the origin of the universe was increasing, I realized that the multipliers or fractions of the variables in the celestial bodies were NOT random, but were precise and governed by meticulous laws beyond chance, accident, or randomness. I therefore became extremely uncomfortable with qualifying these multipliers as random, for saying so seemed to me like a major mark of ignorance or unconsciousness. Hence, I chose to call them "turbulent multipliers". I used the adjective "turbulent" because I discovered them by studying turbulence. Subsequently, I inverted 9 turbulent multipliers of the 9 system-additive variables:

1. mass
2. orbital angular momentum
3. orbital moment of inertia
4. rotational angular momentum
5. rotational kinetic energy
6. rotational moment of inertia
7. total kinetic energy
8. translational kinetic energy
9. volume

In *"Turbulent Origin of the Universe"*, I established that satellites did not descend from their primary planets, but from the common precursor of the primary planet and all its satellites. Therefore, it may not be correct to model satellites after the characteristics of their primary planet as if the planets birthed the satellites. Likewise, because the electrons did not descend from the nucleus of the atoms where they are found, it is unwise to model the electrons after the nucleons. I explained this by using the analogy of the relationship between siblings. For instance, imagine a family consisting of siblings of various sizes and ages. It can be misleading to try to model the siblings in that family by modeling the younger or smaller siblings as dependents of the larger or older ones. Although they can influence one another, the younger or smaller siblings do not depend on the older

49

Science180: Bringing People Together Through the Power of the Accurate Decoding and Understanding of the Origin of Chemicals

ones, and vice versa. However, both the older or larger siblings and the younger or smaller ones depend on their parents. Therefore, the best way to model the origin of those siblings is to refer to their parents and not to one another. Although some parents may be dead, by carefully explaining the characteristics of the siblings, it may be possible to explain what the parents may have looked like before their death. Similarly, although the precursors of the planets (considered here as the bigger or older siblings) and of the satellites (considered here as the smaller or younger siblings) are dead, for they no longer exist, it is possible to properly estimate the characteristics of these precursors by carefully examining the properties of the planets and satellites and by considering them as descendants of the same precursor, which is the precursor of their planetary system. The difficult part of this modeling is how to explain the process by which that precursor was split into its daughter bodies having different outcomes. Likewise, electrons should NOT be considered as dependent on the nucleons, but on the precursors of the atoms (which include the nucleons and the electrons). As a general rule, it is misleading to model secondary bodies orbiting a nucleus of an atom after the characteristics of the nucleons. For in a specific system, both the primary body and its secondary bodies descended from the same precursor which was the precursor of their system.

In each system, the turbulent multiplier with respect to a variable of a body was calculated by dividing the value of that variable for each body by the sum (total) of the same variables for all of the bodies in the system. By tracing the sequential cascade of breakup of the precursors, I was able to estimate how the precursors were fragmentated from a mother precursor into its daughters and so on so forth. For instance, the turbulent multiplier with respect to the energy of a body in a sequential cascade of breakups can be calculated by dividing its energy by the energy of its precursors which is approximately the energy of the bodies in its system. Knowing that some matters were compressed or contracted, while others were expanded during the formation of the universe, volume may not have been conserved from the mother precursors to their daughter particles. In other words, it can be very misleading to talk about volume conservation while addressing the formation of the particles. Therefore, the dynamics of incompressible fluids can be very limiting to address the formation of the chemical particles in which the fluids in the precursors were changing (e.g. compression and expansion) as they were undergoing their genesis.

I realized that the distribution of the turbulent multipliers is not uniform or identical from one generation of the cascade of breakup to the other. For instance, for the same variable, the turbulent multipliers for the planets with respect to the Sun (the primary bodies) is not the same for the satellites with respect to their primary planets (which are the primary bodies of the satellites). Even when I considered the satellites only, the turbulent multipliers of the satellites depended on their types. For the 9 turbulent multipliers that I studied, the highest variance in each planetary system were observed in Zone 3, while the smallest values were usually in Zones 1 and 5 followed by Zones 2 and 4. In general, the biggest variation was found with the biggest systems in the cascade. The variance of the turbulent multipliers seems to decrease as the cascade goes down, suggesting that the series of breakup of the mother precursors into daughter bodies, which at their turn broke into smaller daughter bodies and so on and so forth could have been a way to minimize the variation of the 9 system-additive variables among the daughter bodies. By this time of my analysis, I felt like scientists run a big risk by trying to explain the universe based on the limited experiments done on Earth which are not even enough to address most things in the Solar System.

At the early stage of my investigation of the data in perspective of turbulence, the turbulent multipliers were some of the first variables that quantitatively inspired me about how the characteristics of the precursors of the planetary systems could have affected those of their satellites and primary planet. They allowed me to suspect that the precursors of different stellar systems could have differently impacted their daughter stellar systems (stars and the planets and asteroids orbiting them). The variation of the turbulent multipliers across the planetary systems was one of the facts that supported my idea that the laws found in the Solar System may not be exactly the same as that in other stellar systems. Likewise, the precursors of different chemical particles (e.g. subatomic particles and atoms) could have affected their daughter particles differently. Hence from one celestial body to another one, even in the same stellar system, the characteristics of the chemical particles is different. You can learn more about the turbulent multipliers in my other books.

CHAPTER 6

THE INCREDIBLE EXPERT ADVICE THAT IS HELPING SUCCESSFUL SCIENTISTS TO UNDERSTAND THE BREAKUP OF THE PRECURSORS OF CHEMICAL PARTICLES EVEN WHEN THE BIG BANG NUCLEOSYNTHESIS CAN'T–OR CAN IT?

The information in this chapter is crucial to understanding most of the processes involved in the shaping of the precursors of the particles into various particles having different characteristics. In the first section, I will explain how the stretching, tilting, inclination, elongation, rotation, size, and density of the vortices in the precursors affected the characteristics of the chemical particles. Then, I will approach how the fluid ligaments and necks connecting the precursors were pinched off, broken without forgetting the impact that the viscosity of the fluids had on the formation of the chemical particles.

6.1. How the stretching, tilting, orbital inclination, axial tilt, eccentricity, rotation, radius, and density of the vortices in the precursors affected the characteristics of chemical particles

In my study of the celestial bodies, I spent years to investigate the relationship between the turbulence, rotational period, size, orbital inclination, axial tilt, and eccentricity of their precursors. I understood that a lot of secrets are hidden behind the physics of vortex and jets. A vortex is a region in a fluid in which the flow is rotating around an axis line, which can be straight or curved. Present on the scale of the universe as well as on the subatomic length scales, jets are streams of matter (usually fluids) having approximatively a columnar shape. Since the beginning of this book, most of the demonstrations I did on spit-gathering or fluid breakup are highly enrooted in the physics of jets and vortices.

CHAPTER 6: BREAKUP OF THE PRECURSORS OF CHEMICAL PARTICLES

I discovered that the movement and interactions between the precursors of consecutive bodies (e.g. vortex) in a stratified fluid layers played a crucial role in the characteristics of their daughter bodies. I established that the way the fluid layers of the precursors of the bodies passed one another affected the deformation rate of their daughter bodies. The changes in the orbital inclination, axil tilt, and eccentricity of the fluids can be explained by the deformation of the fluids in their precursors as the fluid layers flowed passed one another, or as they were being collected together to form their daughter bodies. I established that the squashing or stacking of the fluid layers in Zones 1, 2, and 3 contributed to explaining why their rotational angular speed decreased as the distance from the primary body increased. In contrast, once the weight or pressure of the bodies in Zones 1, 2, and 3 was removed from the precursors of the bodies in Zones 4 and 5, their column or fluid layers could have stretched more. In other words, as soon as the fluid layers in Zones 1, 2, and 3 were removed or split from the layers of Zones 4 and 5, the vortices in the layers in Zones 4 and 5 could have been stretched more and consequently, their rotation rate increased. These same trends could exist with chemical particles and help explain some of their characteristics due to the changes in the configuration of their constitutive particles. For instance, the way that the precursor of the electrons of an atom were organized around a nucleus can significantly affect the properties of that atom.

I showed that the tilting of a vortex can change its morphology in certain directions and the location of the tilt can depend on the strain in the fluid. The location of some tilted particles may be explained by the strain in the fluid of their precursors and that of the precursors downstream or upstream of them in the environment where they were formed. The stretching of some vortices could have increased the strain and shearing of their fluids. The forces (e.g. shearing) that acted on some fluids could have been strong enough to overthrow or tilt them. Hence the orbital of some particles (e.g. electrons) can be tilted. Considering that the speed of the vortices of a precursor of bodies can be divided into 3 components (i.e. X, Y, and Z components), I established that the change (e.g. increase or decrease) of the speed in any of these components could have affected that of the other components. As the precursors of the particles were being slowed down in their X and Y axis (e.g. as their orbital speed and shearing rate was decreasing), the Z component of the speed (or vorticity) could have increased and could have been translated into the tilting of the orbital plane and/or rotational axis of some bodies impacted by the pulling and pushing.

Considering my discoveries on the origin of the universe, the turbulence, shearing, deformation, stretching, separation, tilting, and elongation of vortices of fluid of the precursors of chemical particles could have diversified the axial tilt, eccentricity, inclination of orbital, and the variation of the rotational angular speed of the chemical particles. The stress applied to the vortices of the precursors of particles and the resulting strain could have shaped some daughter particles. The variation of the he characteristics of the particles could have been caused by the deformation of the vortices of their precursors and how they collected themselves together after breaking from other precursors. The rate of change of the angles or the sides of the vortices could have defined the scale of deformation and the change of shape of their resulting particles. Some daughter particles could have been flipped over, while others could have been rotated by the forces acting on them during their formation. Microscopic vortices could have formed in the fluids of the precursors of particles according to the proprieties of these fluids and the intensity of their turbulence. Based on what I established in *"Turbulent Origin of the Universe"*, well-defined chemical particles could have been formed in the location where the turbulence was higher, while in less turbulent locations, less defined particles or clouds of particles could have formed. This can explain why some particles can be very dense, while others are loose and organized as clouds of particles.

Based on the demonstrations in *"Turbulent Origin of the Universe"*, I think that the stretching of the vortex of the precursors of some chemical particles could have also contributed to decreasing their radius and increasing their rotational angular speed, which in the end could have affected their spin, electron affinity, electro negativity, and many other properties. The stretching or lengthening of vortices in a fluid flow of the precursor of some chemical particles could have increased some components of their vorticity for instance in the direction of the stretching. The slowing down of the rotational angular speed of the bodies in some layers could have increased the stretching or inclination of the axial component of the shear vorticity of the structures formed in their layers and/or those downstream of them. The axial vorticity of the vortex of the precursors of some chemical particles could have been strong in such a way that it could have been flipped over just as I showed for the celestial bodies in Zones 4 and 5.

Like I established in *"Turbulent Origin of the Universe"*, the split-gathering of the fluids in the precursors of the particles could have also involved the following:

- formation, thickening, thinning, and amalgamation of braids, and
- co-rotation, stretching, merging, interaction of vortices and braid cores with others as the flow carrying them was being pushed or pulled by the stress or forces communicated to it by its mother precursor.

As some vortex filaments were forming, they could have weaved or interweaved together with those near them as they were moving, therefore forming braids. Some of those braids could have been amalgamated, wrapped, or collected together into bigger ones while others could have been shattered or scattered, leading to the formation of smaller chemical particles. Some braids could have also been formed by the interaction of vortices formed by fluids in tinny adjacent layers, which interacted as their precursors were being moved by the turbulence around them. The movement of the fluid layers as a whole could have wrapped, curled, amalgamated, or collected together most of their constituents with a speed in the same order of magnitude of the speed of their chemical bodies.

After the precursors of the particles in each fluid layer separated, they could have continued moving independently until their final shape and position of orbit were "completed". After breaking from their mother and the ligament that connected them to other neighboring precursors, the precursors of the bodies could have continued their movement (including stretching) before breaking up later. The rotating eddies in the precursors could have gathered together and reached their "final" shape, while some of their constituents were mixed with one another. The precursors of some particles consisted of the amalgamation, wrap, or collection of many vortices, hence some particles ended up being composite particles, while others are elementary. Some larger vortices could have contained smaller ones, which, at their turn, contained smaller vortices, and so on and so forth until the smallest vortices, usually found with the smallest elementary particles. This hierarchy, inclusion, or fragmentation of the precursors can also explain the presence of smaller particles or systems of particles inside bigger and more complex ones. That is why for instance, a rock contains minerals, which at their turn contain molecules, which are made of atoms, which are made of subatomic particles, which are made of other particles, which are a modified version of the turbulent prima materia, the original particle or mother of all particles. In *"Turbulent Origin of the Universe"*, I devoted dozens of pages to explaining how similar processes caused the organization of the universe into systems of celestial bodies (e.g. clusters of galaxies containing galaxies organized into stellar systems, which can be organized into planetary systems and asteroid systems, which can be organized into primary planets or primary asteroids orbited by satellites).

In other words, just as a vortex is a compact region of vorticities, the precursors of chemical particles were compact regions of vortices. The magnitude of the squeezing, compaction, or compression of the vortices of the precursors affected the density and other parameters of the chemical particles born from them. Chemical particles which were strongly squeezed are denser than those which were loosely squeezed. The precursors of some chemical particles were very turbulent, birthing large particles, but, because their vortices were not highly squeezed, some of them are less dense than others. However, because the correlation between density and radius of the bodies is not perfect, other factors besides radius can explain the density of the particles. While some vortices could have merged, yielding larger particles, others did not. While the matter in some vortices solidified, that in others were like liquid, gas, and plasma. In *"Turbulent Origin of the Universe"*, I also explained how the conditions in the precursors of the celestial bodies affected the destiny of their constitutive particles.

6.2. Ligament, neck, breakup, and viscosity of fluids during the formation of chemical particles

In *"Turbulent Origin of the Universe"*, I devoted dozens of pages to the formation of fluid ligaments, neck, pinching off, and breakup of the precursors of celestial bodies. But here, I will just address a few points related to chemical particles. I will explain how fluid breakup affected the distribution and the distance separating chemical particles.

A neck of fluid is a thread like area connecting two fluid bodies that are about to break up. When the precursors of the chemical particles were splitting from one another, fluid ligaments were formed between some of them. As the fluids of the precursor of the secondary particles were leaving their mother precursor to form the precursors of the secondary particles, a neck was formed to connect the ligament to the secondary bodies to the ligament of the primary body. That neck could have stayed in place for a time to allow the fluid of the mother precursor to flow into the ligament of the secondary bodies until a moment when the ligament of the secondary body pinched off from the precursor of the primary body and the neck broke. After breaking or pinching off from the larger fluid mass of its primary particle broke, the fluid in the precursor of the secondary particle(s) could have stayed as a single mass or was broken up into many particles (as was the case of atomic nuclei "orbited" by electrons) according to many factors, including the recoil disturbance imposed on it by the separation of the main droplet or other environmental factors.

During the split-gathering of the precursor of an atom, a neck was formed to connect the precursor of the nucleons (primary bodies) and the precursor of the electrons (secondary bodies). After the ligament of the precursor of the electrons split from that of the nucleons (e.g. protons and neutrons), it went through other stages until all of the electrons were formed. In other words, by the time the precursor of the electrons pinched off from the precursor of the nucleons, all that was left for the mother precursor of the corresponding atom was the precursor of the nucleus. Put another way, after the precursor of the electrons pinched off to be molded, the precursor of the nucleons was released and had to go through some shaping to yield its constituents (e.g. protons and neutrons). By the time that all of the precursors of the particles were separated, the flow which birthed them stopped.

From the time they were separated from their mother precursor and later from other daughter bodies adjacent to them (e.g. located upstream and downstream of them), the precursors of the chemical particles went through processes that "escorted" their daughter bodies into orbit. After their separation from the fluid ligament connecting them to others, the blobs and vortices of the precursors could have continued their movement until their configuration or shape was "settled" when the degree of freedom of their deformations was reduced enough for them to relatively "stop" changing as they have done since their beginning. By that time, some chemical particles (e.g. electrons) were positioned in the path of their orbital and a sort of equilibrium was found between their characteristics and environmental conditions. Although chemical particles have been changing their position since the early moment of their genesis, there was a time or position at which their current identity or characteristics were almost reached. The time it could have taken to complete the pinching off or the breakup of the neck connecting the ligament could have depended on the intensity of their turbulence,

CHAPTER 6: BREAKUP OF THE PRECURSORS OF CHEMICAL PARTICLES

the speed, size, volume, viscosity of their fluids, and other factors involved in their stretching and breakup including the forces involved.

After breaking up, the way the end of a neck recedes, recoils, or retracts, can affect whether or not and how it will break to lead to smaller drops between bigger ones. This means that the destiny of the fluids near the breaking point of the precursors of the chemical particles was not always the same and depended on many factors. For a ligament to break up, its fluid jet must first thin to a point whereby it has to break.

These trends suggest that some smaller particles found between larger ones could have been formed around breaking points of the precursors of particles.

As I proved in *"Turbulent Origin of the Universe"*, the nature of the turbulence in the fluid flow of the precursors of bodies could have impacted:

- the smoothness or roughness of some fluid ligaments,
- the stability and destabilization of the neck connecting the bodies,
- the detachment of the vortices according to the scales, and
- consequently, the number and size distribution of the daughter bodies.

I have established that the intensity of turbulence in the precursors of the bodies affected the smoothness or roughness of their ligaments and defined the number, diversity, and distribution of their daughter bodies after breakup. Consequently, the size, number, and distribution of the electrons in each atom could have been affected or defined by the roughness and smoothness of the fluid ligament that birthed them. Because the turbulence of their precursors was different, the fluids in the ligaments of the precursors of the chemical particles could have had different smoothness and roughness. Smooth ligaments of fluids could have produced more mono-disperse chemical particle, while rough ligaments could have produced particles with a broader collection of size. This means that the size and diversity of the chemical particles in an environment were defined by the turbulence that their precursors went through.

Moreover, the viscosity of the precursors of some particles could have affected the size and the distance between their daughter bodies. The impact of temperature and pressure on viscosity could have caused the precursor of some particles to be less viscous than that of others. The density and the number of electrons of atoms could be connected to what could be called the "viscosity" of their precursor. Finally, I also established that the theories of turbulence that assume that turbulence dissipated at the molecular level where/when viscosity is said to control the dissipation of energy, are partial and neglect to consider the turbulence that occurred and is still occurring on the scale of atoms and subatomic particles and scales smaller than them.

CHAPTER 7

WHAT NEW INSIGHT CAN YOU GET FROM THE CLASSIFICATION CRITERIA OF PARTICLES TO PAVE YOUR WAY TO SUCCESS EVEN WHEN PEOPLE IGNORE THE 24 MOST VALUABLE VARIABLES NECESSARY TO CRACK THE CODE OF CHEMICALS-ORIGIN?

To properly explain the origin of the chemical particles, I needed to also explore the root causes of their diversity. For that purpose, the classification of the chemical particles can help to recapitulate some key traits. In this chapter, I will present some of the key factors used to classify chemical particles. Later, I will use this information to address the formation and the variation of the characteristics of the chemicals. Indeed, human beings are accustomed to seeing big objects on the Earth and in space. On Earth, some of the things that can be easily seen are living organisms (human beings, wild animals, plants, etc.), and nonliving things such as rocks, minerals, water (in rivers, oceans, and lakes), etc. When they look up to space during the day, they can see the Sun and during the night, they can see the Moon, the stars, some planets, asteroids, and even galaxies. They may not see the air, but they can feel it particularly when the wind is blowing. The discovery of the telescope and microscope have enhanced the number and details of things human beings can see, therefore allowing the visualization of previously invisible things. Everything that can be seen consists of matters, which are made of various types of particles. Also called corpuscle, a particle is a small localized object or entity that can be characterized by physical or chemical properties such as density, mass, or volume (AMS, 2015; Oxford Dictionary, 2005).

Centuries before the common era, ancient philosophers considered matter to be made of particles, but it is more recently that a better insight into the constitution of matter is gained, yet, advances are still being made and the

Nathanael-Israel Israel: Member of the American Society of Biochemistry and Molecular Biology composition and characteristics of matter are continually being reviewed and updated. According to their size, some particles are called:

- subatomic particles (particles smaller than atoms e.g. electrons, protons, neutrons, and others),
- microscopic particles (e.g. atoms and molecules),
- macroscopic particles (particles larger than atoms and molecules e.g. granular materials such as powders, dust, and sand)
- celestial bodies (despite their gigantic size, galaxies, stars, and their clusters are sometimes considered particles).

In general, subatomic particles are divided into 3 groups i.e. elementary particles, composite particles, quasiparticles, and other types of composite particles. However, when I use the term "particle" in my writings, I am more than likely not talking about celestial bodies. For I felt like celestial particles are too big to be called particles, which people usually tend to consider as tiny things. Moreover, the characteristics of particles is an intense field of study of nuclear physics and particle physics (also called high-energy physics). Particles can be classified according to their stability, spin, electric charge, mass, composition, quantum-statistics, etc. Below, I defined these concepts. Those who are experts in chemistry may skip this chapter if they do not want to read the background information that I provided to help non chemists understand some chemical terminologies I will use later in this book.

7.1. Stability

Stability relates to the ability of particles to decay or change their composition because of the gain or loss of some of their constituents, particles are classified as:

- Stable particles (e.g. electrons) and
- Unstable particles (e.g. Uranium)

7.2. Spin

The spin is an intrinsic form of rotational angular momentum, which together with the orbital angular momentum form the two types of angular momentum in quantum mechanics. Although scientists have not properly understood the spin, nor its cause, it is agreed that spin is for particles what an angular momentum is for large bodies, including celestial bodies. In other words, spin is equivalent to the angular momentum of a body spinning around its center of mass. While spin is traditionally expressed in (N·m·s) or (kg·m^2·s^{-1}), in particle physics, spin is expressed as a dimensionless quantum number because divided by what is called a reduced Planck constant ħ, which bears the same unit as angular momentum. The nuclear spin is a fundamental property used for some applications related to nuclear magnetic resonance (NMR).

It seems to me that the orbital angular momentum is usually reported when angular momentum of particles is concerned and sometimes it is used as a synonym of spin, which, normally, is much more assimilated to a rotational angular momentum. Recalling my findings on the angular momentum of the celestial bodies that I studied in the Solar System, the orbital angular momentum is 2615.5 times to 2.24E+16 times the rotational angular momentum and the ratio "Orbital angular momentum / Rotational angular momentum" depended on the types of bodies. These data obtained on the celestial bodies suggest that, with subatomic particles, the rotational angular momentum may also be very small as compared to the orbital angular momentum. Considering the small size of subatomic particles, it may be very difficult to distinguish between their rotational angular momentum and their orbital angular momentum. Hence, the accent is usually put on the orbital angular momentum, which of course, could be the most dominant. Considering how the rotational movement and the parameters associated with it (e.g. rotational angular speed and rotational energy) helped me to delimitate and understand the turbulence zones, particularly Zones 1, 2, and 3 where the rotational angular speed was significantly affected by the semi major axis of the bodies, I think that the lack of consideration or the inability to consider the rotational movement in some particle theories could have been inducing significant mistakes in the interpretation of scientific chemical data.

7.3. Electric charge

The electric charge is a propriety that allows particles to experience the effect of a force when they are placed in an electromagnetic field. Symbolized by the lowercase symbol q, the electric charge can be positive or negative or null. The movement of charged particles can cause an electric field, which can generate a magnetic field and both fields combined create an electromagnetic field responsible for what is called the electromagnetic force. In other words, electric current is a flow of electric charge through an object, suggesting that the movement of charged leptons can create an electromagnetic force, whereas the movement of the protons and other charged particles in the nucleus can be partially responsible for the "force" inside the nucleus. Above all, the turbulence that led to the formation of these particles hides the secret of the force binding them.

According to the field of study, the electric charge can be expressed in coulomb (C), ampere-hour (Ah), Faraday constant (equivalent of the charge of one mole of elementary charge), and in elementary charge (e) unit. The coulomb is the charge transported by a field of 1 Ampere for 1 second, while Ampere (A) is equivalent to the flow of electric charge across a surface at the rate of one coulomb per second. The unit of the electric charge mostly used in particle physics is the elementary charge ($e = 1.602 \times 10^{-19}$ coulombs), meaning that 1 coulomb corresponds to the amount of charge of about 6.24×10^{18} electrons.

7.4. Mass

Conventionally defined as the physical property which causes matter to resist acceleration, mass is usually expressed in kilogram (kg), Daltons (Da), or in

MeV/c² (millions of electron-volts relative to the square of light speed). Indeed, also referred to as the unified atomic mass unit (u), the Dalton is $1/12^{th}$ of the mass of carbon-12 and is estimated at about 1.66×10^{-27} kg. Furthermore, 1 u = 931.494 MeV/c². Mass is usually used to differentiate between particles belonging to a same family and which have the same electric charge and share other characteristics in common.

7.5. Composition
The composition of particles is related to the types of matter they are made of. According to their composition, particles can be categorized into two types:
- elementary particles or fundamental particles, thought to be indivisible, meaning not composed of other particles or any known component or substructure;
- composite particles consisting of other particles, and
- other types of particles including quasiparticles.

It would not be surprising that, in the near future, many particles known today as elementary would be found to be composite. As a reminder, a few centuries ago, atoms were thought to be indivisible. Yet, today, many types of smaller particles are found inside atoms. Only time and advances in science will reveal which particles deemed elementary today are truly elementary.

7.6. Quantum-Statistics
Quantum-statistics is a mathematical study of the probabilistic distribution of particles according to their states. For instance, quantum-statistics is used to divide elementary particles into what is called bosons and fermions. I will provide details later in this book. In the next chapters, I will deal with subatomic particles, which are usually divided into 3 groups: elementary particles, composite particles, and quasiparticles as well as other types of composite particles. Then, I will delve into atoms, chemical elements, minerals, and rocks in the universe.

7.7. Variables I used to characterize chemical elements
As my goal is to provide a global and comprehensive explanation of the origin of the universe and everything it contains, I needed not only to focus on the bigger bodies such as galaxies, stellar systems, planetary systems, satellites systems, and asteroids but also to explain the diversity and specification of the subatomic and atomic particles. To properly explain how the precursors of chemical elements were differentiated into different chemical elements, I needed to also consider some characteristics of the celestial bodies; therefore, I decided to introduce the celestial bodies before addressing the chemical elements. Else, I would have put the cart before the horse and therefore complicated the understanding of a complex story that is already hard to tell. By the time I better understood the origin of the chemical elements, I felt like instead of saying how they were formed, it would be better to lead the readers through the strategic thought process that

allowed me to discover it, therefore giving you the background that led me to this groundbreaking discovery.

Unlike the bodies that are bigger than satellites and asteroids, atoms and subatomic particles are small and less information is available on them and most of these chemical particles cannot be seen with the naked eye, telescopes, or microscopes. Therefore, I had to rely not on the physical information about them, but on the chemical data which is a small leap into the invisible world. I studied whatever I could on the chemical properties of these particles and then, I drew some conclusions based on my perspective of the global trends of the data. To avoid making mistakes on the formation of the celestial bodies (galaxies, stars, planets, asteroids, and satellites), I reviewed and analyzed the chemical elements before I start putting together this book. In other words, after I finished writing the chapters on the celestial bodies, I was tempted to use the insight I had on them to start writing the final book on the origin of the universe. But I resisted that temptation and first spent weeks studying and analyzing the literature available on subatomic particles and chemical elements.

My goal in some of the next few chapters is not to review all of the characteristics of the chemical elements, but to strategically pinpoint a few trends that can help me to explain the formation of chemical particles as components of the formation the whole universe. Because the variables or terms used to characterize the chemical elements are mostly different from those used to describe the celestial bodies, finding similarities and differences in their characteristics is challenging and caused me to implement or unleash my creative thinking skills. Although I studied more than 30 variables related to chemical elements, in this book, I emphasized some aspects of just 24 of them: atomic number, abundance, atomic mass, boiling point, conductivity, density, discovery year, electro negativity, electron affinity, energy of ionization, hardness, isotopes, location in the human body, melting point, modulus, number of neutrons, number of protons/electrons, occurrence, origin of name, radius, groups, special heat, state at STP (standard temperature and pressure: 273 K), and valence.

CHAPTER 8

WHAT CAN ELEMENTARY PARTICLES TEACH YOU THAT CAN QUICKLY START POINTING YOU TOWARD THE LONG-AWAITED PHYSICS BEYOND THE STANDARD MODEL?

Elementary particles are classified into 2 groups according to the value of their spin and their quantum statistics:

- Fermions: obey the Fermi-Dirac statistics and have a half-integer spin (1/2)
- Bosons: obey- the Bose-Einstein statistics and have a 0 spin or an integer spin

Before addressing the fermions and bosons, I would like to say that some literatures mentioned that "Ghost fields" are a type of elementary particle. However, due to the limited information on them, I did not address them in my writings. Like the previous chapter, this chapter is also a background chapter to help the non-chemists to catch up with some terminologies I will use later.

8.1. Fermions

The Fermi-Dirac statistics (which is the main characteristics of fermions) was independently discovered by the Italian physicist Enrico Fermi (1901-1954) and later by the British theoretical physicist Paul Dirac (1902-1984). In general, fermions are particles which are believed to obey what is called the Pauli exclusion principle which bans identical fermions from "occupying the same quantum state at the same time" (Raymond, 2006). The spin of fermions is half-integer (1/2). As of 2020, 12 variants of fermions have been observed and they are divided into 2 subgroups according to their interactions and what is called color charge:

- Leptons: elementary particles involved in the weak interaction and which have no color charge

- Quarks: particles involved in the strong interaction and which have a color charge

8.1.1. Leptons

Leptons are affected by the electromagnetism and the weak interaction. They have a half-integer spin, no color charge, and do not undergo strong interactions, but are involved in the weak interaction (Encyclopædia Britannica, 2010). Leptons are classified into 2 groups according to their electric charge:

- charged leptons (known as electron-like leptons), and
- neutral leptons (known as neutrinos).

According to their mass and electric charge, leptons are divided into 6 families:

- 3 charged leptons (known as electron-like leptons) with an electric charge of −1: electron (e−), muon (μ−), and tau (τ−)
- 3 neutral leptons (known as neutrinos) which electric charge is 0: Electron neutrino (v_e), muon neutrino (v_μ), and tau neutrino (v_τ)

These 6 leptons are grouped into 3 generations each containing 2 particles:

- First generation of leptons called *electronic leptons:* electron and electron neutrino
- Second generation of leptons called *muonic leptons:* muon and muon neutrino
- Third generation of leptons called *tauonic leptons:* tau and tau neutrino

While the charged leptons can combine with other particles to form various composite particles, neutrinos hardly interact with anything. Consequently, neutrinos are very rare.

The mass of the leptons varies between <0.0000022 MeV/c² and 1780 MeV/c². The smallest value was recorded with the electron neutrino. In contrast, the highest mass recorded for leptons was with tau. These values implied that the mass of the tau particle is:

- >809,090,909.1 times that of the electron neutrino
- >10,470.6 times that of the muon neutrino
- 3,483.4 times that of the electron
- >114.8 times that of the tau neutrino
- 16.8 times that of the muon

The mass of the muon is:

- >48045454.5 times that of the electron neutrino
- >621.8 times that of the muon neutrino
- 206.8 times that of the electron
- >6.8 times that of the tau neutrino

8.1.1.1. Electron (e−)

Postulated in the mid-19th century, the electron was discovered in 1897, and up until today, it is the most known lepton. After the discovery of the electron, it took almost 40 years before the second lepton, (muon) was discovered in 1936. As far as speed is concerned, it is commonly believed that electrons travel at a speed smaller than that of light, but they can be accelerated to reach speeds higher than that of light. The mass of the electron is about 9.11×10^{-31} kg which is equivalent to 0.511 MeV/c².

8.1.1.2. Muon (μ−)

Discovered in 1936 while scientists were studying cosmic radiation, muon is similar to the electron when the electric charge and spin are concerned, but its mass (105.66 MeV/c2) is about 207 times that of the electron. That is why sometimes the muon is called a heavy electron. Muon is heavier than any of the 3 neutrinos known as of 2025. This may also partially explain why muon can decay into electrons and neutrinos. Muons can be produced by a reaction between cosmic rays and particles in the atmosphere. In fact, muons naturally observed on the Earth's surface are postulated to be created indirectly as decayed products of collisions between cosmic rays and particles of the Earth's atmosphere (Demtröder, 2006). Unlike electron, which is the main lepton present in most atoms, muon does not appear in ordinary atoms, but in exotic atoms.

8.1.1.3. Tau (τ−)

Also called tau lepton, tau particle, or tauon, the tau was discovered between 1974 and 1977 (Perl et al., 1975). As the heaviest leptons, tau particles are heavier than muons, which, at their turn, are heavier than electrons. Indeed, the mass of the tau particle (1780 MeV/c²) is:

- 3483.4 times that of the electron
- 936.8 times that of the up quark
- 404.5 times that of the down quark
- 16.8 times that of the muon

The Tau particle has been shown to decay into other particles including composite particles (Tanabashi et al., 2018) such as the following that I studied in other sections:

- charged pion,
- electron antineutrino,
- electron,
- muon antineutrino,
- muon,
- neutral pion,
- tau neutrino, and
- even some hadrons (see the section on composite particles).

Because of the high mass of tau particles, it should not be surprising that it can decay into other particles. In the future, the tau particle may be one of the currently known elementary particles that will turn out to be classified as composite particle.

8.1.1.4. Neutrinos

Neutrinos are tiny particles that have no net electric charge. The name neutrino (Italian for "little neutral one") was given to these particles by the famous Italian physicist Enrico Fermi not only because they are electrically neutral, but also because their rest mass is so small (-ino) and was initially thought to be zero. With the advancement of scientific measurement, the neutrinos were proven not to be massless particles as originally thought. Although the absolute neutrino mass may still not be properly defined yet, the initial works that led to its estimation were so important in physics that the 2015 Nobel prize in Physics was awarded to the scientists who spearheaded the experimental discovery of neutrino oscillations, which proved that neutrinos have mass (Hut and Olive, 1979; Goobar et al., 2006). As of 2025, electrons were found to be a billionth time heavier than neutrinos. For instance, while the mass of the electron is estimated at 0.511 MeV/c^2, that of a neutrino is estimated to be < 2 eV/c^2 (Tanabashi et al., 2018). As of 2025, no elementary particle is lighter than the neutrinos.

The small mass of neutrino is one of the challenging arguments against the current standard model in physics (The T2K Collaboration, 2019). The detection of neutrinos is very difficult and sometimes requires sophisticated underground experiments. As of 2025, neutrinos are treated as point-like particles, without any width or volume. The small size of neutrinos and their electric charge (which is 0) make their detection very difficult because they are postulated not to ionize the materials they are passing through and they are said not to produce traceable radiations, which aid in detecting certain particles. The speed of a neutrino and the speed of light are considered not to be significantly different (Antonello et al., 2012). The small size of neutrinos may also favor their high speed. Because of its neutral charge, small size, and high speed, neutrinos are highly coveted particles with promising application in telecommunication, medicine, and physics.

As I was finishing this section, I felt like it may be useful to elaborate a little bit on the discovery of neutrino. Indeed, neutrino was discovered in Aiken (in the state of South Carolina) located about 15 miles from Augusta, Georgia, where I wrote the bulk of my books on the origin of the universe. In fact, during the summer of 2016, in the company of my family, I visited for the first time the city of Aiken, I was surprised as I was walking along the main street of Aiken and I saw a big sign mentioning that neutrino was discovered in that city, at the renowned radioactive research center called "Savannah River Site", a 310 square miles (~800 km²) US governmental site built in the 1950s to produce and/or refine nuclear defense materials mostly tritium (an isotope of hydrogen) and plutonian-239 used to fabricate nuclear bombs. The site sits alongside the Savannah River, which is near Plant Vogle (a nuclear power plant with 2 current reactors and 2 new reactors being built). The Savannah River itself opens to the

Atlantic Ocean. I later also learned that the Savannah River Site is the only site in the US that operates radiochemical separation, and also the only site that produces tritium in the US. As of 2019, the only site in the US to produce mixed oxide fuel (a fuel consisting of a mixture of plutonium and uranium) was said to be built at the same site (World Nuclear News, 2019) in order to convert weapon grade plutonium into fuels that can be used in commercial power reactors (Matthew, 2014). By the way, unlike normal hydrogen, which consists of a proton "orbited" by an electron, the tritium nucleus contains one proton and 2 neutrons, meaning that its nucleus is 3 times heavier than ordinary hydrogen nucleus. Additionally, it is very radioactive.

At the time I learned that neutrino was discovered in Aiken, I was already deeply reflecting on the origin and matter and I knew that, one day, I would talk about this experience in my writing. The environment around the site is highly contaminated by radioactive elements. Neighbors of that place witnessed that, at night for instance, fish in a lake near the site give light and the grasses glows as they are full of radioactive particles. Some who worked there have told stories about 2 headed reptiles and amphibians living out there. Many people who worked at that site have died of cancer and other unknown diseases. For instance, the rate of thyroid disease is reported to be very high for reasons claimed to be unknown in the surrounding area (southeast US). Some descendants of the people who worked at that site reported that many of their relatives who worked there died of cancer and that the lethality rate and sickness rate of the people who worked at that center and their relatives is so high that the center has been paying a huge lump sum amount to the affected families (Legoas, 2019). The US government has also made efforts to financially compensate some of those who were affected.

Yet, as of 2020, as I was writing the first draft of this current chapter, efforts are being made in Aiken and neighboring cities to win the confidence of the local population to dig plutonium pits to store plutonium and do some research with it. In fact, in 2019, near that city, a train was derailed and many containers leaked their materials out of which plutonium was one, suggesting that some classified work may still be going at that site. Seeing the danger for their community, the state of South Carolina and some activists in the local population have fought against the use of the land as dumpsters of unprocessed dangerous materials. This same site once shipped Plutonium on the highway all the way to Nevada (located thousands of miles away) and it was in the news that the state of Nevada sued Savannah River Site (SRS) because it said that the SRS put the people of Nevada at risks by transporting such material by highway. (Taylor, 2019) Most of the claims and lawsuits against the federal government failed even at the Supreme Court level.

During my reading, I later learned that the two scientists that discovered the neutrino at the Savannah River site received the 1995 Physics Nobel Prize. However, as I was reading to better know about these 2 famous scientists, I learned that one of them, Clyde Lorrain Cowan Jr (1919-1974), died of a sudden heart attack at the age of 54, while the other, Frederick Reines (1918-1998), who lived past his 80th birthday died after a long illness. Who can rule out that these scientists who devoted most of their life to studying neutrinos and other radioactive elements did not pay a price associated with their research? I brought this issue here just to show how the studies of some particles have been affecting the local population near the corresponding research centers. Who can even rule out the fact that several forms of thyroids diseases and cancer are predominant in the region can be connected to some radiation? A thyroid surgeon told me that, in the region surrounding the Savannah River Site, people are highly suffering from various forms of thyroid diseases! A few years, that doctor also died mysteriously. During my stay in Augusta, Georgia, I know people who developed nodules and cysts on their thyroid and had to have them drained and biopsied. I also know people who developed thyroid cancer and non-Hodgkin's lymphoma a few years after moving to the southeast.

Why am I bringing this issue up? I did not address problems here to expose anyone or any wrong doing, but to explain why those radioactive elements were not meant to be abundant on Earth where other abundant elements exist and suit the needs of human beings. Chemical elements were made according to the environmental conditions of their precursors; and the damages that some artificial radioactive elements cause to human health are an indirect testimony of the fact that, chemical elements were not just formed by chance. Even after their formation, they are still bound to the natural laws that formed them in such a way that they "react" or "respond" to human manipulations going against where they should be and how they should be "treated".

Anyways, neutrinos are divided into 3 types: electron neutrino, muon neutrino, and tau neutrino. Next, I will provide some additional information on each type of neutrino.

8.1.1.4.1. Electron neutrino (νe)
The electron neutrino has zero net electric charge and is a first generation of leptons just as the electron. Postulated in 1930 by Wolfgang Pauli, it was discovered in 1956 (Los Alamos Science, 1997).

8.1.1.4.2. Muon neutrino ($\nu \mu$)
Discovered about 6 years after the discovery of the electron neutrino, the muon neutrino is a second generation of leptons just as the muon. It was first hypothesized in the 1940s by several theorists but discovered in 1962 (Danby et al., 1962) by a team of researchers at the Brookhaven National Laboratory. That discovery was rewarded with a Nobel Prize in Physics in 1988. Although some retracted studies suggested that muon neutrinos could travel faster than light (CERN, 2012), others sceptic studies contradicted such a high speed (Anicin et al., 2012). As a result, an "apparent anomalous super-luminous propagation of neutrinos" was blamed to account for the explanation which raised the speed of the muon neutrino above that of light. Nevertheless, after the said correction, the

muon neutrinos appeared to travel with a speed similar to that of light (OPERA experiment reports anomaly in flight time of neutrinos from CERN to Gran Sasso (CERN, 2012). These kinds of experiments suggest that some of the characteristics measured of particles are just a reflection of the tools used to investigate them and are not even constant.

8.1.1.4.3. Tau neutrino ($\nu\tau$)
Also called tauon neutrino, the Tau neutrino was theorized in the mid-1970s, but discovered in 2000 (Physics Letters, 2001).

8.1.2. Quarks
Theorized in 1964, but discovered 4 years later (in 1968), quarks are subatomic particles that have never been directly observed or found in isolation, but in association with other quarks in composite particles called hadrons, which I detailed in another chapter below. The strong interaction is defined as the mechanism responsible for what is called the strong nuclear force. It is accepted that the strong interaction between quarks is facilitated by gluons, which are massless Gauge bosons (one of the 5 kinds of bosons that I explained in another section coming up). Quarks are also said to experience electromagnetism, strong interaction, and weak interaction.

The electric charge of quarks is not integer multiples of the elementary charge. Unlike the leptons, quarks have color charge. As of 2025, six types of quarks have been detected. The quarks are divided into 6 categories according to what is called "favor of quarks", which are subdivided into 3 generations:
- First generation of quark: up quark (u) and down quark (d)
- Second generation: charm quark (c) and strange quark (s)
- Third generation: top quark (t) and bottom quark (b)

The term up-type quarks and down-type quarks is also used to categorize the quarks:
- Up, charm, and top quarks are also referred to as up-type quarks and each have an electric charge of +2/3e, while the
- down, strange, and bottom quarks are also called down-type quarks, having a charge of -1/3e.

The mass of the quarks varies between 1.9 MeV/c^2 and 172,700 MeV/c^2. The highest mass is recorded with the top quark, whereas the smallest mass is with up quark followed by down quark (4.4 MeV/c^2):
- Up quark: 1.9 MeV/c^2
- Down quark: 4.4 MeV/c^2
- Strange quark: 87 MeV/c^2
- Charm quark: 1,320 MeV/c^2

- Bottom quark: 4,240 MeV/c²
- Top quark: 172,700 MeV/c²

These values implied that the mass of the top quark is:
- 90,894.7 times that of the up quark
- 39,250 times that of the down quark
- 1,985.1 times that of the strange quark
- 130.8 times that of the charm quark
- 40.7 times that of the bottom quark

The heaviest quarks are postulated to rapidly decay into up and down quarks through a process called particle decay, through which particles at a higher mass state are transformed into particles of a lower mass state. Consequently, up and down quarks are said to be the most abundant in nature, while the other 4 types of quarks (strange, charm, bottom, and top quarks) are rare and on Earth are said to only be produced in high energy collisions (e.g. involving cosmic rays and particle acceleration). A quark of one flavor can transform into a quark of another flavor through what is termed the weak interaction. For instance, by absorbing or emitting a W boson (a type of boson I explained in another section below), up-type quarks (i.e. up, charm, and top quarks) are said to be able to change into any down-type quark (i.e. down, strange, and bottom quarks), and vice versa.

8.2. Bosons

Unlike the fermions, the bosons obey the Bose-Einstein statistics, which was developed by Albert Einstein and Satyendra Nath Bose in 1924-1925. Indeed, Satyendra Nath Bose was a famous Indian physicist. To honor Bose, who was the first to work on the statistics of Bosons before Einstein joined hands with him to finalize the Bose-Einstein which was named after them, Paul Dirac named the bosons after Bose. Bosons are characterized by a 0 spin or an integer spin. As of 2025, twelve bosons have been observed and they are divided into 2 main groups according to their spin:

- gauge bosons: have a spin equal to 1 and are said to carry force, and
- scalar bosons (represented by the Higgs bosons, which as of 2025 is believed to be the only elementary particle which spin is zero) theorized not as a force carrier, but as the mediator of the mass of particles.

Gauge bosons have been classified into 4 groups:
- photon which is said to carry electromagnetism,
- W (W+ and W-) believed to carry the weak force,
- Z bosons believed to carry the weak force, and
- eight gluons (symbolized by g) suspected to carry the strong force.

In other words, elementary particles are believed to interact with one another through the mediation of gauge bosons, while they are believed to get their mass through the mediation of Higgs bosons. Although each of the bosons have been observed in nature, their characteristics, particularly, their function including the "force" or things they are believed to mediate may have been wrongly interpreted. For instance, it is important to underline that although the Higgs boson is postulated not as a mediator of interaction, but as the mediator of the mass of particles, the Higgs mechanism that would be the mechanism by which mass could have been given has not been observed.

Based on a report, at the range of 10^{-15} m, the "strong force is approximately 137 times as strong as electromagnetism, a million times as strong as the weak interaction, and 10^{38} times as strong as gravitation" (Strassler, 2013). According to the same source, the strong nuclear force is believed to "hold most ordinary matter together because it could be the force that is believed to confine quarks into hadron particles such as the proton and neutron". The same strong force is believed to bind neutrons and protons to form atomic nuclei, which are fundamental components of atoms, the main building block of every ordinary matter. Indeed, the strong force is believed to be mediated by two force carriers according to the scale of action or measurement:

- On a larger scale of about 10^{-15} m to 3×10^{-15} m (meaning 1 to 3 fm), the strong force is believed to be carried by mesons (composite bosons usually containing 2 quarks) that is believed to bind together nucleons (e.g. protons and neutrons) together to form the nucleus of atoms.
- On a smaller scale, less than 0.8 fm ($\sim 0.8 \times 10^{-15}$ m which is estimated as the radius of a nucleon), the strong force is said to be carried by gluons that are believed to hold together individual quarks forming protons, neutrons, and other hadron particles (Kolena, 2021). In other words, gluons are believed to be the carrier of the strong force, which is believed to maintain the cohesion of composite particles such as protons and neutrons.

Because the strength of that said force is stronger on the scale of an individual nucleon than on the scale of the entire atomic nucleus, when referring to that force which is also assumed to bind the entire nucleus together, the strong force is termed nuclear force or nucleon–nucleon interaction or residual strong force, for the theorists believed that it is the residuum of the strong force binding the individual nucleons that also binds the entire nucleus. After a distance of about the size of a hadron, the strength of the strong force is said to remain at about 10,000 newtons (N), no matter how much farther the distance between the quarks (About.com Education, 2017). To put the estimated value of the strong force into perspective, I would like to recall that 1 Newton (N), is the intensity of the force required to make a mass of one kilogram accelerated at a rate of one meter per second squared. At a distance smaller than 2.5 fm, the residual strong force is said to be much more powerful than the electrostatic force (that is believed to cause protons to repel each other) (Pfeffer and Nir, 2000).

Based on the above statement, it appeared to me that gluons may be mostly seen between individual nucleons as if they were squashed inside of them by the force or process that compressed these individual nucleons during the formation of the universe. On another hand, it sounds to me that mesons, which are said to carry the strong force, are like particles sandwiched between nucleons. The presence of gluons between quarks and of mesons between nucleons (e.g. neutrons and protons) may have been what caused the theorists to mistakenly believe that gluons and mesons are the mediator of the strong force, which could have compressed quarks into a different size of particles clustered into different groups.

Bosons differ from one another by their mass (Fig. 11) and electric charge. Besides the W bosons, no other observed boson to my knowledge has an antiparticle. Similarly, besides the W bosons, which electric charge is -e, all other bosons have a neutral electric charge.

The mass of the boson varies between 0 and 125,090 MeV/c². Photons and gluons are considered massless particles, meaning that their mass is assumed to be 0. In contrast, the highest mass was observed with Higgs boson followed by Z bosons (91187.5 MeV/c²) and W bosons (80385 MeV/c²) (Fig. 11). This means that the mass of Higgs boson, the particles postulated to be the mediator of mass is:

- 1.56 times that of the W boson, and
- 1.37 times that of the Z boson.

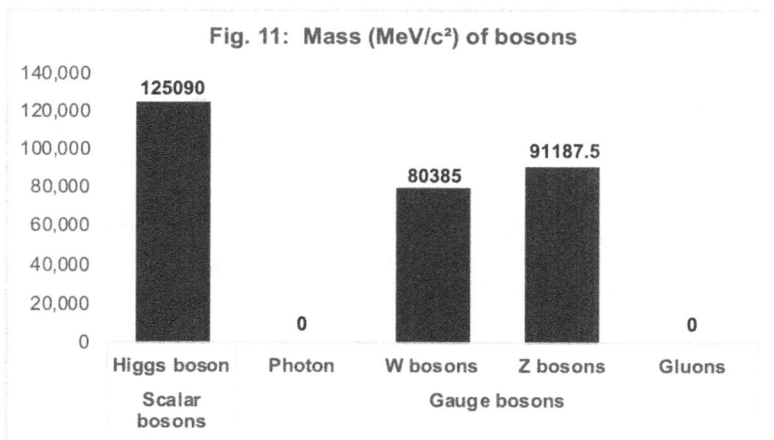

Fig. 11: Mass (MeV/c²) of bosons

Seeing how close the mass of the Higgs boson is to that of the W and Z bosons, I suspected that it would not take too much before other bosons (e.g. W and Z bosons) would be found and claimed to mediate mass. In the end, it will be accepted that particles are not the mediator of mass. Mass of particles was formed or gained because of how particles and other bodies in the universe were formed—I will explain later in this book. Next, I detailed the characteristics of the bosons.

8.2.1. Photons

Discovered by Albert Einstein, photons are considered as the quantum of light and of all other forms of electromagnetic radiation. Quanta is the plural of quantum, which is defined as the "minimum amount of any physical entity or property involved in an interaction". Saying that the photon is the light quantum means that it is a minimum amount of light or of any other type of electromagnetic radiation. In other words, photons are like the elementary particles of light and other types of electromagnetic radiations. According to Wikipedia (2022a), the photon is the "*quantum of the electromagnetic field including electromagnetic radiation such as light and radio waves, and the force carrier for the electromagnetic force*". Photons have no electric charge and are stable.

Photons are postulated as strictly massless particles (meaning they have no mass). Yet, a study done in 1971 suggested that the rest mass of photons could be set to a limit of $m \lesssim 10^{-14}$ eV/c2 if it is estimated using the Higgs mechanism, a mechanism theorized to give mass to particles (Williams et al., 1971). Other studies performed about 5 years after the previous one set the "upper bound on the photon mass" as $< 3 \times 10^{-27}$ eV/c^2 (Chibisov, 1976). Thirty-two (32) years later, meaning in 2008, a more recent research reset a "sharper upper limit" for photon mass as 1.07×10^{-27} eV/c^2, which is the equivalent of 10^{-36} Daltons (1 Dalton is about the mass of a proton or a neutron) (Amsler et al., 2008). In 2010, a study suggested that "photons inside superconductors develop a nonzero effective rest mass" (Wilczek, 2010).

Despite all of these studies and others that suggested that photons have mass, which scientists seem unable or unwilling to properly detect, may be because of its small value or a willful scientific blindness. Most theories have kept the mass of photons at 0 and its speed at a value that no other particle is supposed to surpass. Some assumptions about the photons have been set in such a manner that works which dare to prove that the speed of light is not a constant or that photons have a mass would be likely discarded much more than carefully reviewed by some scientific strongholds, which have been defining the so-called "universal" constants of nature based on shaking philosophical theories, which have been laying many scientific foundations on countless types of matters including photons, the particles without which no one could see! In other words, during my review of the literature, I felt like the efforts by some theorists to maintain the mass of photons as null stemmed from a "fear" that giving them a mass could throw many foundations of modern physics—such as speed of light, which could change and force a revision of certain physical laws (e.g. Coulomb's law, electromagnetic field, etc.)—into a big mess.

Photons have also been produced when particles are accelerated or when atoms lose energy or change their energy states. Photons are thought to be the most abundant particles in the universe. It was interesting that, although not a constituent of a normal atom, photons are believed to mediate the electromagnetic interaction, which some theories claim is present between electrons and protons. Furthermore, it may be useful to recall that, Albert Einstein demonstrated early in the 20th century that photons behave both like a wave and a particle. Around the year 2000, this wave-particle duality property was observed also with other particles (Arndt et al., 2000). Despite the rejection of creationism by several scientists, scientific data in quantum field theory evidenced that the interactions between particles are creation and extermination of quanta.

As of 2020, photons are generally considered as the fastest moving particles and in a vacuum, their speed is estimated at 299,792,458 m/s, known as the speed of light (denoted by the letter "c"). Although Albert Einstein is very famous for his work on photons, it is also important to give credit to the ground-breaking work of his predecessor Max Planck without whom Albert Einstein could not have quickly formulated the theory of photons.

Photon is sometimes symbolized by:

- photon energy, $h\nu$ (where h is Planck constant and the Greek letter ν (nu) is the photon's frequency), or
- hf, (where h is Planck constant and f is the photon's frequency).

Since the days of the German theoretical physicist Max Planck (1858-1947), it is accepted that the energy of any system that absorbs or emits electromagnetic radiation of frequency ν is an integer multiple of an energy quantum $E = h\nu$. This suggested that particles that transport energy may contain a special amount of it, leading the way to the quantization of energy. Photons of various energy levels are usually emitted during some natural processes and chemical reactions including the acceleration of charged particles, the transition of molecules, atoms, and nucleus to lower energy levels. Photons are known to be absorbed or emitted as a whole by some elementary particles, including electrons and systems of particles such as atomic nuclei, atoms, molecules, and much bigger chemicals.

I think that light is made of many types of particles that are unable to form a unified and large solid body, which constituents move all together just as some orbiting bodies failed to become a unified body. But because of their high speed and abundance, the different particles of light are present almost everywhere in nature, traveling at a very high speed. These particles can be divided according to their wavelength. The speed and amount of energy in them can hint at their characteristics, including their mass. Although it may still be impossible to differentiate between the speed of the various types of photons, the information related to their wavelength contain secrets that can allow to unravel fundamental truths about them. It should not be surprising that light was among the first matter that was formed from the original matter or precursors of all types of matter.

I also think that the photon is considered as the fastest particle in nature because, among other things, its mass is small and even considered to be zero. The day the mass of the photon will be properly defined and that more particles smaller than photons can be discovered, I believe that particles may be found with speeds higher than that of the photon, the particle of light.

8.2.2. W and Z bosons

One of the main characteristics of W and Z bosons is that they mediate the weak interaction. Two kinds of W bosons have been observed (W+ and W-).

8.2.3. Gluons

Theorized in 1962, but discovered 16 years later (meaning in 1978), the gluons are divided into eight types or eight colors of gluons. Although bosons are usually considered as particles which have no color charge, gluons are usually considered as carriers of color charge. Honestly, the mathematical characteristics of chemical particles seem to be confusing sometimes as many theoretical things apparently are mixed with reality, which is hard to explain. Gluons are postulated as the exchange particle (or gauge boson) for the strong force between quarks. In other words, not only are they postulated as the mediator of the strong interactions of quarks, but they are also said to participate in that strong interaction.

Just as the photons, gluons are also said to be massless particles. They have never been observed as free particles but in confinement within hadrons, which are composite particles of quarks. They are termed gluons because they are believed to "glue" quarks together, therefore postulated to play a crucial role in the formation of composite particles based on quarks such as hadrons that I detailed in the section below. In the literature, its mass is usually considered 0, or less than $1.3 \text{ meV}/c^2$ (Yao et al., 2006; Yndurain, 1995).

8.2.4. Higgs boson

Speculated in 1964, but discovered in 2012, the Higgs boson is a particle claimed to have no electric charge, no color charge and 0 spin. Approximated at 125 GeV (Taylor, 2014). the mass of the Higgs boson is about 72.4% that of the top quark (the heaviest elementary particle). This implies that although the Higgs Boson is theorized as the particle that mediates mass, itself is not the heaviest elementary particle.

While some studies predicted that the Higgs boson can decay into gluons, the Higgs boson has been observed to decay into other types of particles such as Bottom-antibottom pair, W bosons, Tau-antitau pair, Z bosons, photons, etc. (ATLAS collaboration, 2018; CMS collaboration, 2018).

The discovery of the Higgs boson was so celebrated in physics that, a year after the publication of its detection, the 2013 Nobel Prize in Physics was awarded to two of the physicists who spearheaded the theoretical research leading to its discovery. That particle was so praised by many theorists that some even called it "God particle" (Hill et al., 2013). However. as for me, when I put together all of the equations and details surrounding the matter in the universe, I felt like the Higgs boson is not the particle that mediates mass. I will demonstrate that later in this book. For instance, particles and matter have mass because of how they were amassed or aggregated into bodies. Unless one can explain how matter was formed, it would be impossible to explain how they get their properties, including mass.

8.3. Comparison of the characteristics of elementary particles

I wrote this section to allow a general comparative view of the elementary particles I explored so far. I would have presented Table 1 a little early, but I felt like some people may not understand its content if they don't have a little background on each particle.

Table 1: Diversity, classification, and characteristics of elementary particles

Types of elementary particles	Quantum statistics	Spin–statistics	Spin	Types of elementary particles	Types	Names (symbol)	Generations	Interaction mediated	Color charge	Mass (MeV/c²)	Electric charge (e)	Antiparticle
Fermions	Fermi–Dirac statistics	Half-integer	0.5	Fermions	Leptons	Electron (e⁻)	First generation	Electroweak interactions	No color charge	0.511	-1	Positron
						Electron neutrino (νₑ)				< 0.0000022	0	Electron antineutrino
						Muon (μ⁻)	Second generation			105.7	-1	Antimuon
						Muon neutrino (νμ)				< 0.170	0	Muon antineutrino
						Tau, also called tau lepton or tauon (τ⁻)	Third generation			1780	-1	Antitau
						Tau neutrino (ντ)				< 15.5	0	Tau antineutrino
					Quarks	Up quark (u)	First generation	Strong interactions	Have color charge	1.9	2/3	Up antiquark
						Down quark (d)				4.4	-1/3	Down antiquark
						Charm quark "c"	Second generation			1320	2/3	Charm antiquark
						Strange quark (s)				87	-1/3	Strange antiquark
						Top quark (t)	Third generation			172700	2/3	Top antiquark
						Bottom quark (b)				4240	-1/3	Bottom antiquark
Bosons	Bose–Einstein statistics	Integer	1	Bosons	Scalar bosons	Higgs boson (H⁰)		Mass of particles	No color charge	125090	0	Higgs boson
					Gauge bosons	Photon (γ)		Electromagnetic interaction		0	0	Photon
						W bosons (W⁺, W⁻)		Weak interaction		80385	-1	W bosons
						Z bosons (Z)				91187.5	0	Z bosons
						Gluons (g)		Strong interaction	charge	0	0	Gluon

8.3.1. Electric charge

The electric charge of the elementary particles varies between -e and +2/3e (Fig. 12). Three quarks recorded the highest electric charge (+2/3e):

- Up quark
- Top quark
- Charm quark

In contrast, the electric charge of 7 particles is 0, three of them are leptons and 4 are bosons:

- Leptons: electron neutrino, muon neutrino, and tau neutrino and
- Bosons: scalar bosons (Higgs boson) and gauge bosons (photon, Z bosons, and gluons)

Finally, the electric charge of 6 elementary particles is negative. That of 3 of them—3 leptons (Electron, Muon and Tau)—is equal to -e, while that of 3 is -1/3e. The 3 particles which electric charge is equal to -1/3e are all quarks:

- Down quark
- Strange quark
- Bottom quark

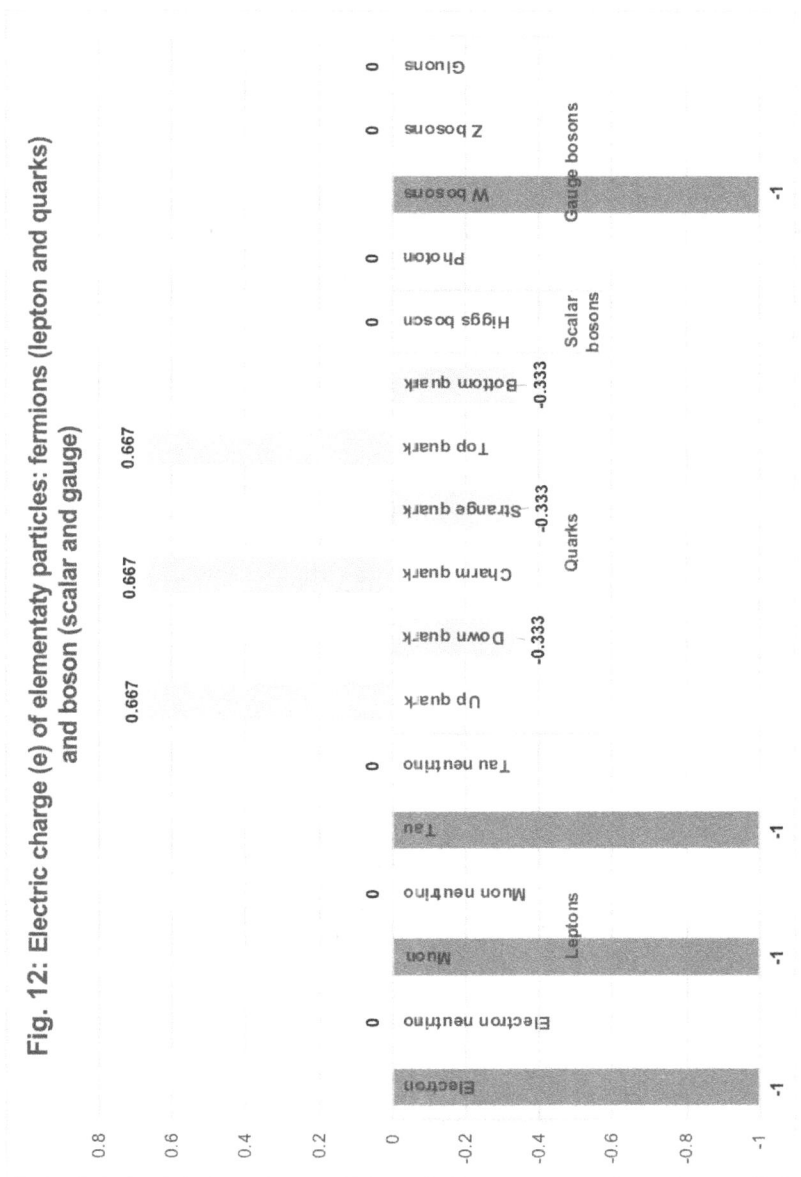

Fig. 12: Electric charge (e) of elementary particles: fermions (lepton and quarks) and boson (scalar and gauge)

8.3.2. Mass

The mass of the elementary particles varies between 0 and 172,700 MeV/c² (Fig. 13). No mass is reported for photons and gluons. The top quark is the heaviest elementary particles and its mass is:

- >78,500,000,000 times that of the electron neutrino,

- >1015882.4 times that of the muon neutrino,
- 337,964.8 times that of the electron,
- 90,894.7 times that of the up quark,
- 39,250 times that of the down quark,
- >11141.9 times that of the tau neutrino,
- 1,985.1 times that of the strange quark,
- 1,633.9 times that of the muon,
- 130.8 times that of the charm quark,
- 97 times that of the tau,
- 40.7 times that of the bottom quark,
- 2.1 times that of the W boson,
- 1.9 times that of the Z boson, and
- 1.4 times that of the Higgs boson.

I would like to recall that the tau particle is the heaviest lepton, thousands of times heavier than some other leptons. Yet, the mass of the top quark (the heaviest elementary particle) is 97 times heavier than the tau. This supports the idea that the mass of the particles that orbit the nucleus is very small. There is a reason for that and, in addition to the explanation I gave in the chapter on the system-additive variables, I will later elaborate in this book.

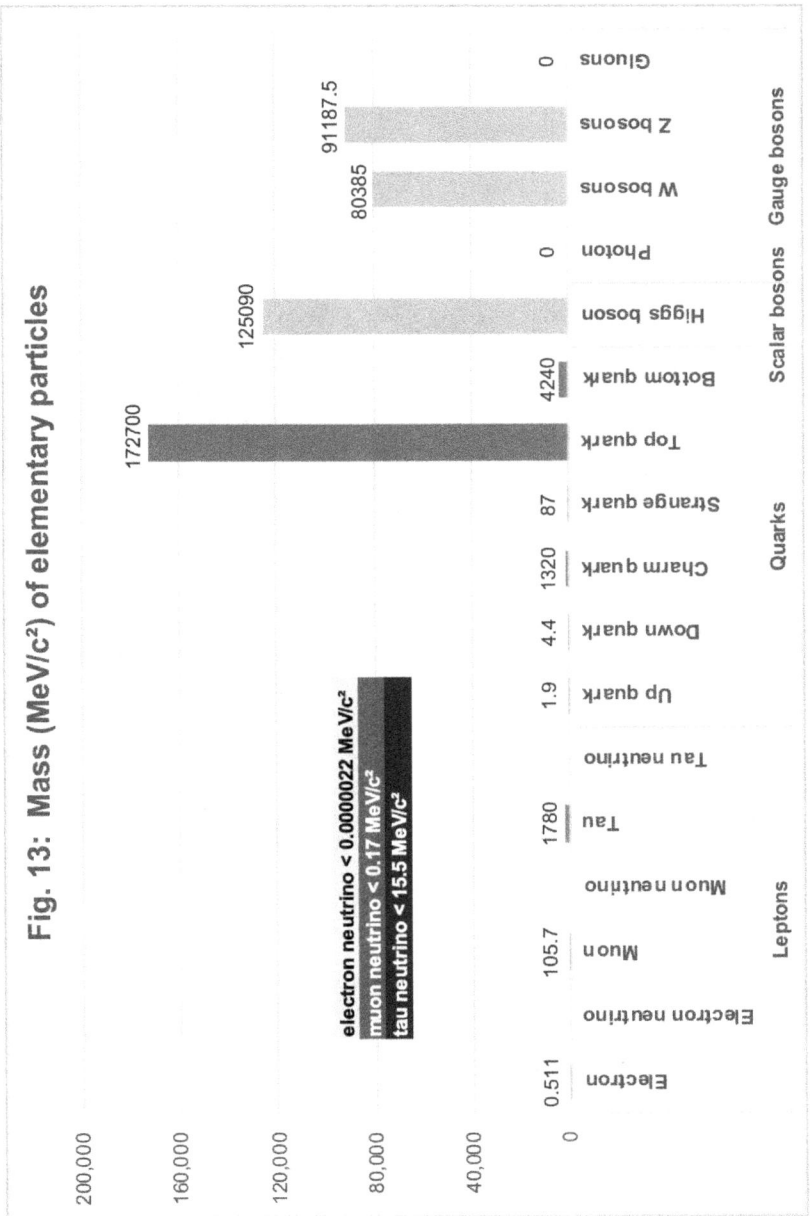

Fig. 13: Mass (MeV/c²) of elementary particles

In the next chapter, I will review how the composite particles are mixtures of elementary particles.

CHAPTER 9

WHY DON'T PEOPLE WHO ARE GIVING THE COMPOSITE PARTICLES A BAD INTERPRETATION LISTEN TO THIS EXPERT WHO AWAKENED THE WORLD TO A NEW SCIENTIFIC REALITY?

Composite particles consist of a blend of 2 or more elementary particles and are divided into 2 groups:

- Hadrons (quark-based particles made of two or more quarks postulated to be held together by the strong force) (e.g. baryon, meson, and exotic hadrons) and
- Others (e.g. atomic nuclei, atoms, superatoms, exotic atoms, and molecules)

Table 2 summarizes their characteristics.

9.1. Hadrons

Hadrons are usually divided into 3 categories:

- Baryons: composite fermions usually containing 3 quarks
- Mesons: boson composites habitually containing 2 quarks
- Exotic hadrons: consist of various combinations of particles sometimes higher than 3 quarks.

9.1.1. Baryons

The word baryon originates from the Greek word "barys", which means heavy. Therefore, heavy matters in the universe are thought to be made of baryons. As of 2025, baryons are defined as composite subatomic particles containing usually 3 valence quarks, which is why they are sometimes called triquarks. Some literatures defined them as having at least 3 quarks. However, as explained in another section

82

Nathanael-Israel Israel: Who Happens to be the World's #1 Authority on the Turbulent Origin of the Universe and Life

below, the tetraquark (a composite particle having 4 quarks) and the pentaquarks (which have 5 quarks) are not classified as baryons.

Table 2: Diversity and classification of composite particles

Composite particles	Name	Composite	Quark composition	Family	Examples	Constitution (e.g. quark content)	Other characteristics
Hadron	Baryon	Composite fermions	3	Nucleons (N)	Proton (uud)	two up and one down quark (uud)	Long-lived
					Neutron (udd)	two down and one up quark (ddu)	
					Other examples of nucleons include: N(1440), N(1520), N(1535) N(1650), N(1675), N(1680), N(1710), N(1720), N(1895) N(1900), N(2190), N(2220) N(2250)		
				Delta baryons (Δ)	Delta baryon (e.g.uuu, uud, ddd)	contain 3 up or down quarks	Short-lived and heavier than nucleons
				Lambda baryons (Λ)	Lambda baryon (e.g. uds, udc,udb	one up quark, one down quark and a third from a higher generation quark	
				Sigma baryons (Σ)	Sigma baryon (e.g.uus, uds, dds)	one up quark and/or one down quark and a third from a higher generation quark	
				Xi baryons (Ξ)	Xi baryon (e.g. uss, dss, usc)	contain one up or down quark and 2 more massive quarks (either strange, charm or bottom)	
				Omega baryons (Ω)	Omega baryon (e.g. sss, ssc, ssb)	contain no up or down quarks	
	Meson	Composite bosons	2	"Ordinary" mesons	B meson		
					Bottom eta meson		
					Charged rho meson		
					Charmed B meson		
					Charmed eta meson		
					D meson		
					Eta meson		
					Eta prime meson		
					J/Psi meson		
					Kaon		
					Neutral rho meson		
					Omega meson		
					Phi meson		
					Pion	up quark and/or down quark and their anti particles (anti-up quark or anti-down quark)	considered as the lightest meson
					Strange 3 meson		
					Strange D meson		
					Upsilon meson		
			4	Exotic meson	Tetraquark	four valence quarks	
	Exotic hadrons	Composite of quarks and gluons	5	Pentaquark		four valence quarks and one valence antiquark	
Others composite particles	Atomic nuclei	Nuclide				protons and neutrons	
	Atoms					nucleus surrounded by electrons	
	Superatoms					cluster or atoms	
	Exotic atoms	one or more subatomic particles are replaced by other particles of the same charge			Positronium	an electron bound to a positron	
					Muonium	electron bound with an antimuon	
					Tauonium		
					Onia		
					Hypernuclear atoms (e.g.uds, uus, dss,	e.g. hypernuclei (baryon containing strange quarks but no charm, bottom or top quark) orbited by electrons	
	Molecules					two or more atoms	

Baryons are divided into 6 families or groups according to their isospin (I) values (which is a kind of spin or orbital angular momentum), and their quark (q) content (or the number of each of the 6 types of quarks (up (u), down (d), strange (s), charm (c), bottom (b), and top (t)) they contain:

- Nucleons (N): combination of 3 u and/or d quarks and isospin (I) = 1/2
- Delta baryons (Δ) also called delta resonances: combination of 3 u and/or d quarks and isospin (I) = 3/2
- Lambda baryons (Λ): combination of two u and/or d quarks and isospin (I) = 0
- Sigma baryons (Σ): combination of two u and/or d quarks and isospin (I)

$= 1$

- Xi baryons (Ξ): containing one u or d quark and isospin (I) $=1/2$
- Omega baryons (Ω): containing no u or d quarks and isospin (I) $= 0$

To summarize, all nucleons and delta baryons have 3 u and/or d quarks, but differ by their isospin. Similarly, all Lambda baryons and Sigma baryons contain 2 u and/or d quarks, but differ by their isospin. Next (Table 3), I gave some examples of baryons, their quark content and mass based on the data reported by Beringer et al. (2012).

Table 3: Types, quark content, and mass of baryons

Family of baryon	Particle name	Quark content	Rest mass (MeV/c^2)
Nucleons (N)	Proton	uud	938.27
	Neutron	udd	939.57
Delta baryons (Δ)	Delta	uuu	1232
		uud	1232
		udd	1232
		ddd	1232
Lambda baryons (Λ)	Charmed Lambda	uds	1115.68
		udc	2286.46
	Bottom Lambda	udb	5619.4
Sigma baryons (Σ)	Sigma	uus	1189.37
		uds	1192.642
		dds	1197.449
	Charmed Sigma	uuc	2453.98
		udc	2452.9
		ddc	2453.74
	Bottom Sigma	uub	5811.3
		ddb	5815.5
Xi baryons (Ξ)	Xi	uss	1314.86
		dss	1321.71
	Charmed Xi	usc	2467.8
		dsc	2470.88
	Double charmed Xi	ucc	3621.4
	Bottom Xi	usb	5787.8
		dsb	5791.1
Omega baryons (Ω)	Charmed Omega	ssc	2695.2
	Bottom Omega	ssb	6071
	Omega	sss	1672.45

Designed by Dr. Nathanael-Israel Israel using the data reported by Beringer et al. (Particle Data Group) (2012).

Hyperons are baryons containing at least one strange quark, but no charm, bottom, or top quark (Greiner, 2001). They are also called strange baryons. A nucleus containing at least one hyperon is called hyper nucleus and is different from normal nuclei containing regular protons and neutrons. For instance, some Lambda and Sigma baryons are hyper nuclei (Table 4).

Table 4: Example of hyperons (baryons containing at least a strange quark)

Hyperon names	Quark content	Rest mass (MeV/c2)
Lambda	uds	1115.7
Omega	sss	1672.5
Sigma	dds	1197.4
	uds	1192.6
	uus	1189.4
Sigma resonance	dds	1387.2
	uds	1383.7
	uus	1382.8
Xi	dss	1321.3
	uss	1314.8
Xi resonance	dss	1535.0
	uss	1531.8

Because of the high mass of top quarks, baryons made of top quarks are postulated not to exist because of the top quark's short lifetime (Amsler et al., 2008). Although their existence in the vicinity of the Earth may be difficult, I personally believe that conditions of other planetary and stellar systems can allow these kinds of top quarks baryons to be dominant. For something which is impossible or rare on Earth and in the Solar System can be common and abundant in other stellar systems in the universe. Although little is known about most of the families of baryons, a lot of studies have been done on nucleons. Therefore, the next section provides some additional information about them.

Long-lived and not as heavy as the other types of baryons, nucleons are the main constituents of atomic nuclei and are the most known baryon family. Made of quark, protons and neutrons are the main nucleons. Although they both consist of 2 types of quarks (up quarks and down quarks), protons and neutrons differ by the number and types of quarks they contain (Table 5):

- Proton (annotated uud) consists of 2 up quarks (u) and 1 down quark (d)
- Neutron (annotated udd) consists of 1 up quark and 2 down quarks

Consequently, they differ by their electric charge, mass, magnetic moment, etc.

Table 5: Characteristics of protons and neutrons

Characteristics	Protons	Neutrons
Composite particles	Hadron	
Hadron type	Baryon (Composite fermions)	
Quark composition	3	
Family of baryon	Nucleon (N)	
Quark content	two up (u) and one down (d) quark	one up (u) and two down (d) quark
Symbol	uud	udd
Mass (kg)	1.6726E-27	1.6749E-27
Rest mass (MeV/c^2)	938.27	939.57
Electric charge (e)	1	0
Spin	0.5	0.5
Magnetic moment (nuclear magnetons: μN)	2.79	-1.91

Because the electric charge of an up quark is +2/3e whereas that of a down quark is -1/3e, the electric charge of the proton and neutron is claimed to be:

- Proton: 2*2/3 + (-1/3) = 3/3e= e
- Neutron: +2/3 + 2*(-1/3) = 2/3-2/3 = 0

The name neutron was derived from the convention that the electrical charge of a neutron is 0 (zero). The mass of a proton is estimated at 1.6726 x 10^{-27} kg (about 1,836 times that of an electron). In contrast, the mass of a neutron is 1.6749 x 10^{-27} kg. Neutrons are heavier than protons because down quarks (which are dominant in neutrons) are heavier than up quarks. Quarks in protons and neutrons are postulated to be linked together by the strong force, which is said to be mediated by gluons.

The magnetic moment of a proton (μ_p) is 2.79 nuclear magnetons (μ_N), whereas that of a neutron (μ_n) is -1.91 nuclear magnetons (μ_N).

Both protons and neutrons have a half-integer (1/2) spin. Atomic nuclei that have an even number of neutrons and protons are said not to have a spin (zero spin). But atomic nuclei that have an odd number of protons and neutrons have a spin (non-zero spin).

Besides protons and neutrons, all other hadrons are believed to be unstable and decay into other particles. Besides protons and neutrons, other kinds of nucleons have been postulated, some of which existence is said to be certain and which properties are said to be fairly well explored include: N(1440), N(1520), N(1535), N(1650), N(1675), N(1680), N(1710), N(1720), N(1895), N(1900), N(2190), N(2220), and N(2250).

9.1.2. Mesons

Unlike baryons, which consist of different combinations of 3 quarks and are made of fermions, most mesons consist of 2 quarks (1 quark and 1 antiquark) and are made of boson. Some literatures define mesons as composite bosons which have an even number of valence quarks, usually at least 2. Consequently, mesons have integer spin just like bosons (the elementary particles they are made of), while baryons have half-integer spin (just as the fermions they consist of). Because they are made of bosons (the elementary particles known to carry forces such as electromagnetic interaction, weak interaction, and strong interaction), bosons are postulated to act as a force mediating particles and are thought to play a role in nuclear interactions. They are also believed to intervene in interactions between particles including composite particles.

Most mesons are said to be unstable and composed of a quark-antiquark pair. Mesons are postulated to be also produced when cosmic rays interact with the atmosphere. For instance, when nucleons are artificially subdued to a higher energy source, they can interact with one another and produce mesons. Although mesons can be artificially produced, they are hard to manipulate because of their short life, suggesting that the conditions on earth are not favorable for the dominance of mesons. Many types of mesons have been discovered including:

- B meson
- Bottom eta meson
- Charged rho meson
- Charmed B meson
- Charmed eta meson
- D meson
- Eta meson
- Eta prime meson
- J/Psi meson
- Kaon
- Neutral rho meson
- Omega meson
- Phi meson

- Pion
- Strange B meson
- Strange D meson
- Upsilon meson

In Table 6, I presented the quark content and the rest mass of these mesons.

Table 6: Types, quark content, and mass of mesons

Meson names	Quark content	Rest mass (MeV/c^2)
B meson	ub	5279.26
	db	5279.58
Bottom eta meson	bb	9398
Charged rho meson	ud	775.11
Charmed B meson	cb	6275.6
Charmed eta meson	cc	2983.6
D meson	cd	1869.61
	cu	1864.84
Eta meson	based on a formula containing u, d, and s and their antiparticles	547.862
Eta prime meson	based on a formula containing u, d, and s and their antiparticles	957.78
J/Psi meson	cc	3096.916
Kaon	us	493.677
	ds	497.614
	us	891.66
	ds	895.81
Neutral rho meson	based on a formula containing u, d and their antiparticles	775.26
Omega meson	based on a formula containing u, d and their antiparticles	782.65
Phi meson	ss	1019.461
Pion	ud	139.5702
Strange B meson	sb	5366.77
Strange D meson	cs	1968.3
Upsilon meson	bb	9460.3
Designed by Dr. Nathanael-Israel Israel using the data reported by Olive et al. (Particle Data Group) (2014).		

9.1.3. Exotic hadrons

Unlike the "ordinary" hadrons previously discussed, and which contain just two quarks (mesons) or three quarks (baryons), exotic hadrons are composed of more than 3 quarks tied together by gluons. They can be either composite fermions or composite bosons. One of the most famous exotic hadrons is the pentaquark, which consists of 5 quarks (four valence quarks and one valence antiquark) and some gluons. The pentaquark is sometimes classified as an exotic baryon. It should not be surprising that in the future, other exotic hadrons may be found containing more than 5 quarks as already theorized:

- heptaquarks (5 quarks, 2 antiquarks),
- nonaquarks (6 quarks, 3 antiquarks), and
- many other candidates going through screening and characterization.

However, none of these hadrons which have more than 5 quarks have been seen as of 2020.

9.2. Other composite particles

Besides hadrons, other types of composite particles exist and have been observed in nature: atomic nuclei, superatoms, exotic atoms, atoms, and molecules.

9.2.1. Atomic nuclei, superatoms, and exotic atoms

Atomic nuclei are particles consisting of protons and neutrons found at the center or nucleus of an atom.

Superatoms are atom clusters that appear to display some properties of elemental atoms. Some of the most known superatoms are:

- Sodium (Na) atoms cooled from vapor.
- Aluminum (Al) clusters
- Lithium (Li) n clusters
- Platinum clusters
- Rubidium clusters
- Gold clusters also called Gold superatom complexes

In exotic atoms, one or more subatomic particles are replaced by other particles of the same charge. For instance, muonic atoms are exotic atoms in which electrons are replaced by the muon, a negatively charged particle but heavier than an electron.

Mesonic atoms are another kind of exotic atom in which, electrons are replaced by mesons (e.g. pion, kaon). For instance, in pionic atoms, electrons are replaced by pions whereas in kaonic atoms, electrons are replaced by kaons.

Most exotic atoms are very short-lived and are very unstable under normal conditions on Earth. They may be abundant and stable in other planetary systems or stellar systems. In other words, the conditions on Earth and in the Solar System may not have favored the formation and abundance of exotic atoms on Earth. These conditions can be also rolled back to those in the precursor of the Earth and of the Solar System.

9.2.2. Atoms

Ordinary atoms (or atoms in short) are the smallest constituent unit of ordinary matter that constitutes chemical elements. Atomic nuclei are also defined as atoms stripped of their electron shells. Many atomic nuclei are said to be present in space. For millennia, human beings have tried to understand the constitution of matter. Before the common era, people thought that atoms could not be cut into smaller particles. However, as explained in the section on subatomic particles, several particles smaller than atoms exist. Although particles that are not atoms exist in the universe, atoms are considered the basic units of matter. Symbolized by the letter Z, the atomic number of an atom is the number of protons in its nucleus. Atoms with an equal atomic number belong to the same chemical element. In other words, a chemical element is a species of atoms that have the same atomic number. In contrast, atoms that have an identical atomic number, but a different number of neutrons are isotopes of the same element. The total number of nucleons (protons and neutrons) of an atom is called the mass number. Lead is the heaviest stable atom and its mass number is 208 (Sills, 2003). More than 99% of the mass of atoms is said to be concentrated in the nucleus.

The total number of protons and neutrons is used to determine nuclides. About 339 of the nuclides found on earth occur naturally (Lindsay, 2000). According to the same source, 252 nuclides have not been detected to decay, and these are called "stable isotopes".

In 1869, the Russian chemist and inventor Dmitri Ivanovich Mendeleev (1834 –1907) published a periodic table, which organized chemical elements based on their properties. Later, the electron was discovered by the physicist J. J. Thomson in 1897. In 1909, the physicist Ernest Rutherford and his collaborators discovered that the mass of an atom is concentrated in a nucleus. Atoms are believed to consist of a dense nucleus surrounded by a cloud of electrons. The nucleus is believed to consist of protons and neutrons. The only atom that is said not to have a neutron is protium (also called hydrogen-1) and its positive hydrogen ion has no electron. Atoms are believed to be bound to one another by chemical bonds, and by doing so, they form molecules. The amount of energy required to remove or add an electron (electron binding energy) is said to be far less than the binding energy of nucleons. Every state of matter (solid, liquid, gas, and plasma) is composed of neutral atoms or ionized atoms. Ions are classified in two groups:

- Cations, which have lost at least one more electron and
- Anions, which have gained at least one electron.

Unlike conventional atoms consisting of protons, neutrons, and electrons, exotic atoms have other types of subatomic particles. For instance, in the muonic atom, the electron is replaced by the muon. Similarly, in the hadronic atom, the electron is replaced by a hadron such as a meson (e.g. kaon, pion).

Until today, scientists are unable to properly define and see with the naked eye the constitution and dimension of atoms. Estimations published in the literature are theoretical. They suggest that the radius of a nucleon (proton or neutron) is about 0.8 fm (fm = fento metre 10^{-15} meter). The size of most atoms is believed to be about 100 picometers (10^{-10} m).

Electrons are believed to be attracted to protons (later in this book, we will see if that is true) in a nucleus by the electromagnetic force which is believed to be inversely proportional to the distance separating the electrons from the nucleus. Although the position of the electrons has been deemed impossible to properly know, electrons are thought to behave as both a particle and a wave. At one point in history, scientists believed that electrons orbit the nucleus in a precise trajectory as satellites orbiting a planet. However, that idea has been discarded. As of today, electrons are believed to move like a wave in a cloud-like region called atomic orbital, which position is not precisely known, but is based on a mathematical function characterizing the probability that an electron can be found at a particular location (Mulliken, 1967). Electrons are thought to occupy specific energy levels, from which they can move as they emit or absorb a photon. Orbitals around a nucleus are believed to exist in only a discrete or quantized set (Brucat, 2008). Orbitals are said to vary in size, shape, and orientation and are said to have one or more ring or node structures (Manthey, 2001).

These characteristics seem to point at a similitude between the satellite systems around planets on one hand and on the other hand a similitude between the planetary systems and asteroids systems around the Sun in regards to the turbulence I decoded that took place during the formation of the universe. For the form, shape, and orbital inclination of satellites, planets, and asteroids are not the same, and sometimes, rings are present around some primary celestial bodies. These observations I am making here are not an endorsement of previous theories that tried to explain the movement of electrons around a nucleus as that of satellites around planets, but they suggest that some similitude exists and, if well studied, they can help explain the origin and formation of both the smaller bodies in the universe (e.g. atoms and subatomic particles) as well as the biggest celestial bodies (.e.g. Sun, planets, asteroids, and satellites), and that of much larger clusters of bodies such as galaxies and the universe as a whole.

Scientists believe that electrons that are the farthest from the nucleus can be transferred to other neighboring atoms or shared between atoms, a mechanism that is believed to allow atoms to bond to molecules and other types of chemical compounds such as ionic and covalent network crystals (Smirnov, 2003). In a laboratory setting, scientists can increase the number of protons and neutrons in a nucleus through processes including a nuclear fusion, which joins multiple atomic particles to form a heavier nucleus. Fusion reactions can form other particles such as gamma rays and beta particles (Shultis et al., 2002). In contrast to nuclear fusion, which increases the size of the nucleus, nuclear fission split a nucleus into two smaller nuclei. Furthermore, the size of a nucleus can be changed by bombarding it with high energy subatomic particles or photons. The subsequent modification of the nucleus can form a different chemical element if the number of protons change (Staff, 2007; Makhijani and Saleska, 2001).

Atoms can go through a radioactive decay, a process by which a nucleus can emit particles or electromagnetic radiation. There are 3 common types of radioactive decay (L'Annunziata, 2003; Firestone, 2000).

- Alpha decay, when a nucleus emits an alpha particle (e.g. helium nucleus consisting of two protons and two neutrons).

- Beta decay (also called electron capture) when a neutron is transformed into a proton, and vice versa. When a neutron is transformed into a proton, an electron and an antineutrino are emitted. In contrast, when a proton is transformed into a neutron, a positron (also called an antielectron (e+)) and a neutrino are emitted or an electron is absorbed or captured. Electron capture (i.e. an absorption of an electron by a nucleus) is said to be more common than positron emission. Beta particles are electrons or positrons emitted by a beta decay.
- Gamma decay (i.e. decrease of the energy level of a nucleus) followed by the emission of electromagnetic radiation is usually a consequence of an alpha or beta decay.

9.2.3. Molecules

Usually defined as an electrically neutral group of at least two atoms held together by chemical bonds, the term molecule can also apply to polyatomic ions, which are groups of ionized or charged atoms. Because noble gases (also called inert gases or aerogens) are not connected by chemical bonds (because of their low chemical reactivity), they are considered monatomic molecules. In other words, the presence of chemical bonds between groups of atoms is not always required before they can be called molecules. Molecules constituted of one chemical element are called homonuclear and a well-known example is oxygen (O_2) without which no human and most other living organisms can live. Molecules consisting of more than one element are called heteronuclear and a good example is carbon dioxide (CO_2).

Two types of chemical bonds are usually found in molecules:

- covalent bonding: termed as a "sharing of electron pairs (also called shared pairs or bonding pairs) between atoms", and
- ionic bonding: believed to be an "electrostatic attraction between oppositely charged ions"

Ions are classified in two groups:

- cations, which have lost at least one more electron and
- anions, which have gained at least one electron.

The length (Fig. 14) and energy (Fig. 15) of covalent bonds depend on their types and their atoms (Chemistry Libre Texts, 2019). In general, the highest length of chemical bonds and the lowest energy between bonds were found with halogens. The highest energy was found between carbon and oxygen. I think that if the noble gases could have been considered, they could have held the record of the highest chemical bond length and lowest bond energy.

When molecules composed of atoms from more than one chemical element are bound together by chemical bonds, they form chemical compounds. Compounds are generally divided into 4 groups depending on the bonds of their constituent atoms:

- intermetallic compounds connected by metallic bonds
- ionic compounds connected by ionic bonds
- molecules connected by covalent bonds
- other complexes connected by coordinate covalent bonds

Before I close this section on molecules, I would like to mention that, in living organisms, molecules are found in various, complex forms, most of which are called macromolecules, including nucleic acids (DNA, RNA), proteins, carbohydrates, etc. I did not elaborate on these macromolecules in this book, but in my book on the origin of life.

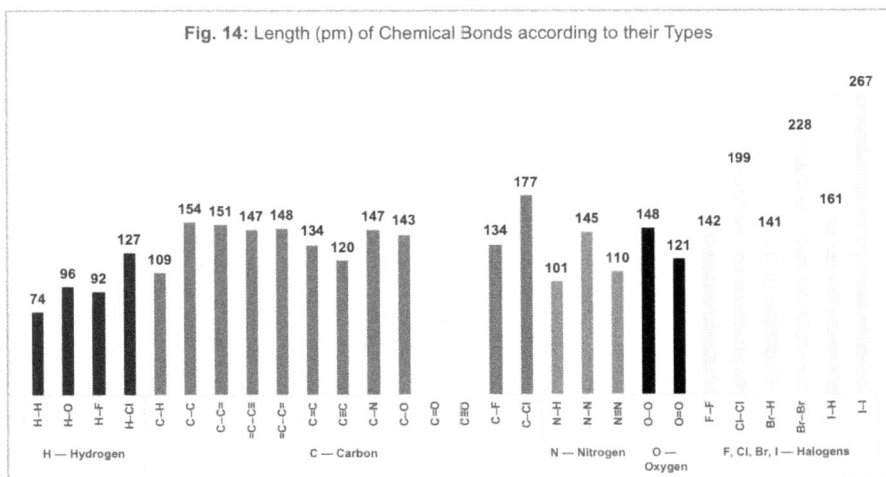

Fig. 14: Length (pm) of Chemical Bonds according to their Types

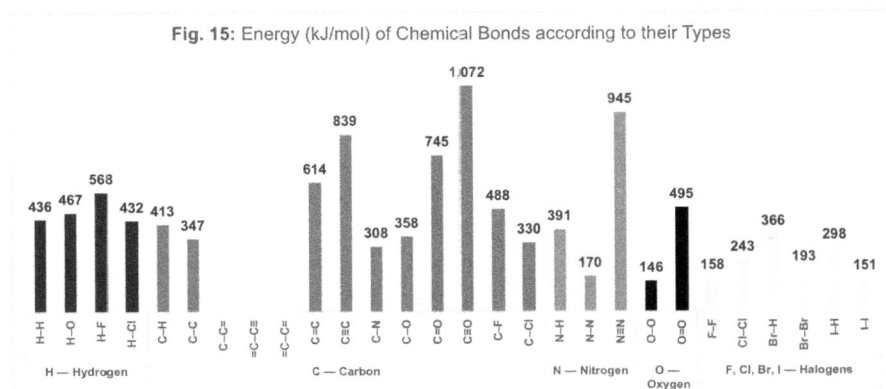

Fig. 15: Energy (kJ/mol) of Chemical Bonds according to their Types

9.3. Quasiparticles and other types of composite particles

In addition to the elementary and composite particles mentioned above, other particles called quasiparticles also exist. In other words, quasiparticles are not classified as elementary particles, composite particles, or as the conventional and condensed forms of matters, but as excited and vibrational fields. According to Wikipedia, the main types of quasiparticles include (Wikipedia 2020b):

- Phonons (vibrational modes in a crystal lattice)
- Excitons (bound states of an electron and a hole)
- Plasmons (excitations of a plasma)
- Polaritons (photons mixed with other quasi-particles)
- Polarons (particles surrounded by ions in a moving and charged material)
- Magnons (excitations of electron spins in a material)

9.4. Minerals and rocks

Efforts have been made to explore the crust of the Earth and even that of the Moon. A human expedition to the Moon provided some materials that allowed scientists to have a glimpse at the composition of the lunar surface. More recently, a robot was sent to Mars to study its chemical composition. Spectroscopy is one of the indirect measurements used to try to determine the composition of the crust and atmosphere of celestial bodies. The interior of the Earth is not well studied yet. Worse, the interior of other celestial bodies has not been experimentally analyzed. This implies that everything said about the core of celestial bodies is theoretical. Although the interior of the Earth has not been fully explored, geologists believe that the interior of the Earth is filled with magma or larva, which erupt during volcanic activities.

Minerals are aggregate of one or more compounds consisting of atoms held together by bonds such as chemical bonds. The International Mineralogical Association defined a mineral as "*an element or chemical compound that is normally crystalline and that has been formed as a result of geological processes*". Earth's minerals are dominated by 8 chemical elements, which in order of decreasing abundance are: oxygen, silicon, aluminum, iron, magnesium, calcium, sodium, and potassium. According to the International Mineralogical Association, more than 5,000 minerals exist on Earth. Minerals are classified based on their chemical composition, structure, hardness, color, cleavage, density, and other criteria. Minerals are divided into two categories: silicates and non-silicates depending on whether they contain silica (silicon and oxygen) or not.

Also called stones, rocks are solid aggregates of one or more minerals or mineraloids. Mineraloids are like minerals but do not form crystal or solid. Rocks are parts of the solid material found on Earth. Other materials are unconsolidated. Rocks are dominated by certain major minerals called rock-forming minerals, most of which are silicates except calcites: amphiboles, calcites, feldspars, micas, olivines, pyroxenes, and quartz (SiO_2).

For instance, limestone is a rock made of the mineral calcite or aragonite ($CaCO_3$). A granite is a rock composed of quartz, alkali feldspar, and plagioclase feldspar. Considered as a sedimentary rock composed primarily of organically-derived carbon, coal is an example of rock composed of non-mineral material.

Rocks have been classified into 3 groups: igneous, metamorphic, and

94

sedimentary. Believed to make more than 60% of the Earth's crust, igneous rocks are thought to come from the cooling and solidification of magma. Igneous rocks were divided into 2 groups according to the place they were formed: plutonic rocks and volcanic rocks. Plutonic rocks are thought to form when "magma cools and crystallizes slowly within the Earth's crust", whereas volcanic rocks are formed when magma comes out of the interior of the Earth and cools at the surface. Some of the key minerals found in igneous rocks are amphiboles, feldspars, feldspathoids, micas, olivines, pyroxenes, and quartz.

I will give more details about minerals and rocks in the chapter where I addressed their formation. Finally, it is important to recall that, although they are very astronomical, celestial bodies (galaxies, stars, stellar systems, planets, planetary systems, satellite systems, satellites, asteroids, and their clusters) are also called particles. Although some people may deny that celestial bodies are not particles, a closer look at their characteristics made me to understand that they behave like atoms and subatomic particles, but just on different scales and with different proprieties. In *"Turbulent Origin of the Universe"*, I devoted hundreds of pages to them. Therefore, I did not deepen them in this book.

CHAPTER 10

WHY THE CRITICS OF THE STANDARD MODEL OF PARTICLE PHYSICS AND THE PROTON RADIUS PUZZLE ARE NO JOKE IN AN ERA WHEN HYPOTHETICAL PARTICLES CAN'T HELP?

It would have been unthoughtful to report the classification of chemical particles without addressing some of the ongoing challenging problems and debates related to them and to other hypothetical particles that have never been observed, but which are highly postulated and considered in many theories. In this chapter, I will discuss some of the ongoing challenges and debates related to:

- Standard Model or particle physics (the theory that tries to connect most of the known particles)
- Hypothetical particles
- Conversion of particles into one another
- Proton radius puzzle and the EMC debate (concerning the discrepancy of the radius of the proton depending on the leptons orbiting it)

Because I will be addressing all of those issues as I tackle the origin of chemical particles later in this book, I feel like, to ensure everybody understands their meaning, I need to explain them using simple terms that the non-chemists can comprehend. However, if you properly understand those concepts, you can skip this chapter and go to the next.

10.1. Standard Model and its critics

Can dark energy and dark matter really help explain the universe-origin without totally eclipsing themselves for a new physics?

Indeed, based on the knowledge available on elementary particles, physicists have developed a theory called "Standard Model", which aims at explaining most of the observed elementary particles as a whole and the interactions between them, energy, and matter. Although the Standard Model of physics attempts to explain the particles and the fundamental interactions between them, it fails to elucidate gravity and many other key physical facts. As of 2012 that the Higgs boson has been confirmed, all of the particles in the Standard Model have been experimentally discovered. In short, as of 2025, the elementary particles of the Standard Model are classified as (Cottingham and Greenwood, 2007):

- Six "flavors" of quarks (i.e. up, down, strange, charm, bottom, and top)
- Six types of leptons: electron, electron neutrino, muon, muon neutrino, tau, and tau neutrino;
- Twelve gauge bosons (postulated to carry forces):
  - photon (believed to carry electromagnetism),
  - three W and Z bosons (believed to carry the weak force), and
  - eight gluons (believed to carry the strong force); and
- The Higgs boson.

The Standard Model is based on particles and antiparticles including:

- 36 Quarks
- 12 Leptons
- 8 Gluons
- 1 Photon
- 1 Z Boson
- 2 W Bosons
- 1 Higgs

Below, I summarized some of the main hypotheses of the Standard Model. Indeed, atoms having an equal number of protons and electrons are said to be neutral. As they interact with other atoms, atoms are believed to be able to gain or lose electrons. An atom that has a different number of protons and electrons is charged and is called an ion. An ion which has more electrons than protons is negatively charged, while an ion that has more protons than electrons is positively charged. It is postulated that charged particles having opposite charge tend to attract one another, while they tend to repulse one another if they have the same charge. In other words, positive ions would repulse one another and negative ions would also repulse one another. In contrast, positive ions and negative ions would attract one another. It is postulated that the interaction (e.g. attraction and repulsion) between charged particles are mediated by the electromagnetic force, which is believed to cause electric fields, magnetic fields, light, and other types of radiations. Likewise, electrons are thought to be attracted by protons via the electromagnetic force. Inside a nucleus, protons are said to be connected to one another and to neutrons by the nuclear force, called a strong force because its strength is very strong and considered as the strongest interaction force in nature. The strong force is estimated at more than 100 times the electromagnetic force, and is said to be mediated by gluons, which is believed to unite quarks to form the nucleons. At the atomic level, the Standard Model maintains that the electromagnetic force mediates the interactions between electrons and the nucleus, and without the electromagnetic force, proponents of the Standard Model think that atoms could lose their cohesion. Similarly, electrons are believed to be moving around the nucleus in orbitals or in shells, but their position cannot be precisely determined, facts that led to the elaboration of the uncertainty principle, which states that it is impossible to know the exact position of electrons but the probability of its location. The Standard Model postulated that nucleons are relatively immobile in the nucleus. After reading these conclusions of the Standard Model, many questions came into my mind:

- If the electromagnetic force connects electrons to the nucleons, why do electrons and nucleons not go as fast as photons, which are particles believed to mediate the electromagnetism?
- If electrons are really attracted by protons, why is the distance between protons and electrons in atoms not changing but is considered a constant?

As early as 2013, which is 10 years before I published my findings on the origin of the universe, these kinds of questions have caused me to believe that existing theories would have misinterpreted the origin and interactions between particles in nature. It should not be surprising that the scientific community itself believed that the Standard Model and most of its ramifications are incomplete because they do not explain everything in the universe, including facts like gravity (considered the fourth fundamental force) or the so-called "dark matter", "dark energy", and exotic hadrons, which, unlike the hadrons described in the Standard Model, contain more than 3 quarks. Many hypotheses in theoretical particle physics (including string theory on which many people had put their hope to fully describe the universe) have been evoked, but none of them properly explained the key unanswered questions in physics. Attempting to incorporate gravity into a revised version of the Standard Model, some physicists predicted that a hypothetical elementary particle called graviton (theorized by Albert Einstein in 1916) and others exist in nature. Yet, none of them have been discovered as of 2025. Similarly, all efforts to construct a unified theory, which some scientists planned to call "Theory of Everything", because they thought it could describe quantum mechanics and general relativity, failed.

Unfortunately, most of the efforts to find a new physics to explain things in nature are focused on colliding or accelerating particles in collider experiments as if breaking matter into pieces could reveal its true origin, constitution, and function. Many countries have been investing billions of dollars in particles physics research. In general, quantum mechanics is the branch of physics devoted to microscopic and subatomic particles. For instance, particle physics (also called high energy physics) focuses on the smallest particles (e.g. electron, quark). Nuclear physics studies atomic nuclei, their constituents (e.g. protons and neutrons) and interactions. Finally, atomic physics studies the structure and properties of atoms as a whole (including their electrons) without detailing the nuclear aspects. The Standard Model and its derivatives theories continue to speculate that the universe is mostly made of quarks, the main component of baryonic matter. However, as I will explain very soon, I personally believe that under other planetary systems or stellar systems, spatial and environmental conditions could have allowed the prevalence or dominance of other types of particles that are not the quarks known on Earth. Because the composition of bodies in the universe depends on the environmental conditions that shaped their precursors, it will be unthoughtful to assume that matters in other stellar systems are the same as those in the Solar System. Furthermore, I believe that, most physical theories have failed to properly clarify the origin of mass and consequently, they continue to think that mass is a property mediated by certain types of particles just as how they continue to wrongly believe that gravity or gravitation is mediated by a particle. As billions of dollars have been invested into particle research and several active efforts are still being made to break into pieces matter of various size by colliding them with one another, it should be expected that the number and characterization of particles would keep changing and some of the terminologies used today will be rendered useless, archaic, or inappropriate to cope with the massive growing data sometimes collected under wrong and fictitious assumptions that have nothing to do with reality.

10.2. Hypothetical particles

Before I addressed the hypothetical particles, I would like to mention that, many high energy particles called "cosmic rays" are believed to exist and traveling through space. Defined as "high-energy protons and atomic nuclei which move through space at nearly the speed of light", cosmic rays can be separated into two types (Wikipedia, 2020c):

- Solar energetic particles believed to be emitted by the Sun, and
- galactic cosmic rays and extragalactic cosmic rays believed to come from sources outside of the Solar System.

Many of those particles have been observed:

- solitary electrons,
- atomic nuclei (e.g. hydrogen nuclei or proton, and nuclei of heavier elements),
- alpha particles (e.g. helium nuclei), and
- many types of elementary and composite particles already discussed in previous sections.

Some of these particles such as the so-called the "Oh-My-God particle" claimed to be observed in 1991 have "energies similar to the kinetic energy of a baseball moving at a speed of 90 km/hour (equivalent to 56 miles / hour) (Anchordoqui et al., 2003). Many theories exist concerning how the collisions of cosmic rays with other particles in space can create other particles. Many cosmic ray particles are still unknown, and I personally believe that if the space outside of the Earth's atmosphere and outside of the Solar System could be fully explored, new particles would be found. Nevertheless, based on my discoveries, collisions between particles cannot explain the origin and the diversity of all the particles in the universe.

As I am finishing talking about the types of elementary and composites observed in the universe, it is important to notice that many hypothetical particles have been postulated, but none of them has been discovered. Most non-baryonic particles (i.e. particles that are not baryonic) are generally hypothetical particles. Some of them are called dark matter because they could not be seen, and their mass is postulated to be zero. The belief in the existence of the hypothetical particles seem to "answer" the aspirations and interrogations of some scientists who are blocked in their theories that keep requiring more input, more unknown variables to try to justify errors they don't want to acknowledge or fix. Although the postulation of some hypothetical particles seems to help some theorists to "balance" some of their theories, the problems they create is outrageous for they complicate the path that should lead to the truth! For instance, I don't think that the hypothetical particle called graviton, which some people think mediates gravity or gravitation, exist. To avoid building my discoveries on hypothetical particles, I didn't focus on the unobserved particle in my discussion. Furthermore, based on my findings, I think that the scientific efforts that cling to the existence of hypothetical particles is one of the mistakes that have been preventing some scientists from properly understanding and explaining gravity and other issues connected to the origin of the universe. In other words, I did not address the hypothetical particles in my writings, for I think it is a waste of time and resources to focus on fictitious particles, while experimentally-detected particles still need serious attention. Sometimes, I am surprised that most theories have been trying to explain the existence of forces and interactions between particles by the existence and functions of other particles. Based on the data I studied, I realized that forces and interactions between particles and celestial bodies are not mediated by particles, but are a consequence of how the constituents of the universe were formed and interact with one another.

10.3. Conversion of some particles into others

In this section, I will explain how some particles can be converted into others, and how some particles known as elementary today may become composite in the future. Indeed, subatomic particles can really be converted into others under certain conditions. For instance, the nucleus of an atom can emit an alpha particle through a process called "alpha decay". A neutron can be transformed into a proton through a process called "beta decay" and this can emit also a neutrino. A nucleus can emit electromagnetic radiation through a process called "gamma decay", which usually follows alpha or beta decay. Some types of radioactive decay are said to eject neutrons, protons, and beta particles from a nucleus. I already provided details in the segments of atoms in the previous chapter. These reactions suggested to me that subatomic particles could contain others particles of different size and which have undergone different transformations. In other words, because the radioactive disintegrations of some atoms can produce electrons, neutrons, and protons, it is possible that some atoms, isotopes, and other particles found in nature could have been produced from the disintegration or gathering together of others. These reactions also suggested to me that, under different conditions, a similar particle could have been converted into the particles currently found in nature.

Prior to 1960, most of the particles known today as composite particles were considered elementary because, in those days, their composition was unknown. This suggests that, with the advancement of science, as scientific equipment and methodology improve and more microscopic details are understood, some current elementary particles can become also composite particles. For as time passes by, new details are discovered and scientific theories change. For instance, it was shown that an electron dropping to a lower orbit can emit a photon. This suggests that it may be possible that one day, scientists may find that electron is a complex, composite particle consisting of photon-like particles. Otherwise, how can an electron emit a photon if itself does not contain something like photons? Some models predict that high-energy photons can transform into electrons. Although early models suggested that electrons in atoms move on orbit, advances in quantum mechanics instead suggested that the position of an electron in an atom cannot be precisely defined by an orbit, but rather by an orbital, which is a region where there is about a 95% chance of finding the electron. Therefore, to some extent, it can be inappropriate to view the electrons in an atom as orbiting atomic nuclei. The Schrödinger equation describes how electron waves are propagated. Chemical bonds are believed to be mediated by the exchange or sharing of electrons between atoms. Scientists believe that the speed of electrons depends on the material they are part of and also on the electrical current they may be in. In some cases, the speed of electrons can near that of light, but in other cases, electrons can almost be immobilized. Like light, the speed of an electron can depend on the medium it is traveling through. This suggests electrons around atoms may not likely be traveling all at the same speed. As I unravel the formation of the chemical particles, you will better understand the significance of the things I said above.

10.4. The proton radius puzzle

As I finish the description of elementary and composite particles, I would like to introduce another key fact that has been puzzling the scientific community for more than a decade: the proton radius puzzle. Indeed, as explained in the previous chapter, the proton is a key particle in nature and its characteristics are used to describe matter. Proton is found in the nucleus and is usually "orbited" by electrons.

Using electron-proton scattering and atomic spectroscopy, proton radius was defined as 0.88 ± 0.01 femtometers (fm = 10^{-15} m). As a reminder, the muon is said to be similar to the electron in all respects except its mass, which is said to be about 200 times heavier than that of an electron. Attempting to improve the accuracy of the measurement of the proton radius, a group of scientists have placed a muon in orbit around a proton (forming a muonic hydrogen), and surprisingly, they found that the radius of the proton was 0.84 ± 0.001 fm, which is about 4% smaller (Pohl et al., 2010; Antognini et al., 2013). In another study, when a muon was placed around a proton, the muon is said to "orbit" a proton about 200 times closer than the electron does (Carlson, 2015). In other words, the radius of a proton orbited by an electron is greater than that of a proton orbited by muon. To state the problem in a different way, when researchers measured the proton radius using electrons and muons, they got incompatibly different results, therefore puzzling researchers who are still unable to properly explain them. As for me, the "proton radius puzzle" is a proof that current models of physics are not correct.

The ordinary hydrogen has no neutron in its nucleus, but just one proton. However, the hydrogen isotope called deuterium has a neutron and a proton in its nucleus, which is "orbited" by one electron. The nucleus of a deuterium atom is called a deuteron, meaning consisting of a proton and a neutron. When a deuteron is orbited by a muon, it forms an exotic atom called a muonic deuterium. When a deuteron is orbited by an electron, it forms an atom called an electronic deuterium. When a deuteron is orbited by a muon, it forms an atom called an muonic deuterium. To confirm the "proton radius puzzle", researchers in another experiment formed a muonic deuterium, and then measured the radius of the proton. They thought that the presence of a neutron by the proton could have changed the proton radius. They found that the radius of the proton of a muonic deuterium is similar to that of the muonic hydrogen, which is smaller than the radius of a proton orbited by an electron (Pohl et al., 2016), therefore confirming that the proton radius puzzle is real. As of 2025, several studies have confirmed the proton radius puzzle, but until my discoveries, it was still an unresolved problem in physics. I explained the underlying cause of the proton radius puzzle in the chapter on the formation of subatomic particles and atoms.

CHAPTER 11

WHY ARE SCIENTIFIC LIES AND CONFUSION ABOUT THE GROUPS OF CHEMICAL PARTICLES STILL TOLERATED?

11.1 Generalities

When I started describing the chemical elements, the first variable I studied was their classification into groups as I explained in this chapter. I initially felt like the grouping of the chemical elements may contain similarities and differences that could allow me to get a quick insight into the variation of their data. The 118 chemical elements known as of 2025 are classified into 11 groups:

- Actinoid,
- Alkali metal,
- Alkaline earth metal,
- Halogen,
- Lanthanoid,
- Metal,
- Metalloid,
- Noble gas,
- Non-metal,
- Transition metal, and
- Unknown.

These groups are also called series or classification. These chemical groups were defined according to criteria that include patterns in their electronic configuration, trends in behavior, etc. As explained in the following sections, because the proprieties of the chemical elements overlap a lot, the boundaries of the chemical groups are not absolute. Unlike the typology of celestial bodies (e.g.

stars, planets, satellites, and asteroids) which means specific things, chemical groups do not always mean anything specific. This taught me that chemical elements are "like" different planetary systems belonging to different stellar systems or like different stellar systems belonging to different galaxies. Before I explained that statement, let's first review these groups.

11.2. Actinoids

Also called actinides, actinoids are 15 metallic chemical elements with atomic numbers from 89 (actinium) through 103 (lawrencium). They are all radioactive and have very large atomic and ionic radii. While 6 of them occur naturally, the other 9 are synthetic:

- Actinium (Ac, Z=89), occurs naturally and known as one of the most dangerous radioactive poisons
- Thorium (Th, Z=90), occurs naturally
- Protactinium (Pa, Z=91), occurs naturally
- Uranium (U, Z=92), occurs naturally
- Neptunium (Np, Z=93), occurs naturally
- Plutonium (Pu, Z=94), occurs naturally
- Americium (Am, Z=95)
- Curium (Cm, Z=96)
- Berkelium (Bk, Z=97)
- Californium (Cf, Z=98)
- Einsteinium (Es, Z=99)
- Fermium (Fm, Z=100)
- Mendelevium (Md, Z=101)
- Nobelium (No, Z=102)
- Lawrencium (Lr, Z=103)

Except for actinium or lawrencium, all actinides are said to be f-block elements, which is a term used to describe the filling of what is term of the 5f electron shell, which is about the configuration of some electrons around the nucleus of those elements. Actinides are believed to release a significant amount of energy when they go through radioactive decay. Most actinides are believed to be rare. Highly used in nuclear reactors and nuclear weapons, Uranium, thorium, and plutonium are the most abundant actinides on Earth. Thorium and uranium are believed to be the main naturally-occurring actinides that are abundant in nature, while Plutonium (naturally present in trace) is usually synthetized. The radioactive decay of some of the actinides produced other chemical elements including other actinides. Many synthetic actinides and even natural actinides were discovered, isolated, and synthesized by bombarding some atoms (e.g. Uranium) by other smaller particles (e.g. neutrons, deuterons, neutrons, α-particles, nitrogen, neon, oxygen, etc.) (Greenwood and Earnshaw, 1997):

- Neptunium: bombardment of ^{238}U by neutrons (in 1940)
- Plutonium: bombardment of ^{238}U by deuterons (in 1941)
- Americium: bombardment of ^{239}Pu by neutrons (in 1944)
- Curium: bombardment of ^{239}Pu by α-particles (in 1944)
- Berkelium: bombardment of ^{241}Am by α-particles (in 1949)
- Californium: bombardment of ^{242}Cm by α-particles (in 1950)
- Einsteinium: As a product of nuclear explosion (in 1952)
- Fermium: As a product of nuclear explosion (in 1952)
- Mendelevium bombardment of ^{253}Es by α-particles (in 1955)
- Nobelium: bombardment of ^{243}Am by ^{15}N, or ^{238}U with ^{22}Ne (in 1965)
- Lawrencium: bombardment of ^{252}Cf by ^{10}B or ^{11}B, and ^{243}Am with ^{18}O (1961-1971)

Actinides are said to be pyrophoric, meaning they "*spontaneously ignite upon reacting with air at or below 55 °C*" (Greenwood and Earnshaw, 1997). I think that the ability of actinides to detonate or explode in the air can be a consequence of how their constitutive particles were wrapped or spiraled around the nucleus. For because these actinides are among the heaviest atoms, have a small radius, and are less dense atoms (except Uranium), they release their content, while in contact with the air, for the atmospheric conditions may not fit those which maintain their internal structure.

Uranium was the first actinide discovered. I used to wonder why that chemical element is named after the planet Uranus until I finally learned that Uranium was discovered by the German chemist Martin Heinrich Klaproth in 1789, who named it after the planet Uranus, because that planet had been discovered only 8 years earlier, and therefore, it was probably still a breaking news in those days. Actinides have many isotopes. For instance, Uranium only has about 25 isotopes among which 3 (i.e. ^{234}U, ^{235}U, and ^{238}U) are said to be present in significant quantities in nature. Because of its long half-life believed to be thousands of years, ^{238}U is generally used in most chemical studies.

11.3. Alkali metals

These are:

- Lithium (Li, Z=3)
- Sodium (Na, Z=11), the most abundant
- Potassium (K, Z=19)
- Rubidium (Rb, Z=37)
- Cesium (Cs, Z=55), the most reactive metal
- Francium (Fr, Z=87), very rare and extremely radioactive

In addition to hydrogen (which is not an alkali metal, but a nonmetal), all alkali metals belong to group 1 and they are classified in the s-block of the periodic table. The alkali metals are also called the lithium family after the name of the leading element of that group. According to Wikipedia (2016a), the alkali metals have their outermost electron in an s-orbital and they easily and quickly lose their outermost electron to become cations with a charge +1. Alkali metals are also said to be soft (due to the weak metallic bonding that connects them), highly reactive at standard temperature and pressure. For instance, Caesium is said to be the most reactive of all metals. All alkali metals react with water and some of them even react so strongly that when they are put into water, they detonate like a bomb. According to Wikipedia (2016a), the alkali metals are never found as free elements, but usually in salts (Krebs, 2006). They are said to have a very low density, melting point, and boiling point (Royal Society of Chemistry, 2012a). They are also said to have a low heat of sublimation, vaporization, and dissociation (Greenwood and Earnshaw, 1997). According to the previous source, alkali metals have what is called the ns^1 configuration, which is said to cause them to have a very large atomic and ionic radius, as well as very high thermal and electrical conductivity. In the periodic table where they are present, alkali metals are said to have the lowest first ionization energies (Lide, 2003), which was postulated to be due to their low effective nuclear charge and ability to easily lose their outermost electron (Royal Society of Chemistry, 2012a). When moving down the table, meaning going down the groups, alkali metals are said to have increasing atomic radius, increasing densities (except for potassium, which is less dense than sodium) (Clark, 2005), increasing reactivity (Royal Society of Chemistry, 2012a), but decreasing electronegativity, decreasing melting point, and boiling points (Clark, 2005). The ionic radii of alkali metals is said to be much smaller than their atomic radii (Royal Society of Chemistry, 2012a).

I believe that the outermost electron of the alkali metals may be located far away from the electrons just before it, hence when it is lost, the radius of the ions is much smaller. For the satellite systems of the planets, I remember that some outermost satellites are far away from the satellites just upstream of them. This suggested to me that in the case of the satellites, an event could have projected or ejected some outermost satellites further away than the regular distance increment separating consecutive satellites. Similarly, in the case of the alkali metals, I felt like during their formation, the precursor of the outermost electron was ejected further away, which in the end caused it to be easily lost. The positioning of the outermost electron could also explain why the first ionization energy of alkali metals is very small, for not much effort could be needed before removing that lone electron.

As I started carefully looking at the data of the chemical elements, it easily appeared to me that chemistry is just physics done on the scale of a lab or on the microscopic level, while physics itself is like chemistry done on the scale of space or at a gigantic level. If chemists can also expand their knowledge in physics and physicists expand their knowledge in chemistry, a lot of progress could be made in science. For most scientific breakthroughs are rooted in chemistry and physics. For instance, without the breakthroughs to discover and characterize chemical elements, it could be nearly impossible to have much advancement in most biological fields. And similarly, without using some physical laws to characterize the dynamics, cinematics, and interactions of bodies in nature, not much scientific progress could have been made. Therefore, it is very sad that chemists and physicists seem to be compartmentalized into different scientific entities without interacting much with one another.

11.4. Alkaline earth metals

They are:

- Beryllium (Be, Z=4)
- Magnesium (Mg, Z=12)
- Calcium (Ca, Z=20)
- Strontium (Sr, Z=38)
- Barium (Ba, Z=56)
- Radium (Ra, Z=88), very radioactive

The alkaline earth metals are slightly reactive at standard temperature and pressure (Royal Society of Chemistry, 2012b). According to the same author and others, together with helium, alkaline earth metals have in common a full outer s-orbital. In other words, these authors postulated that the orbital of these metals contains its full complement of two electrons, which can be easily lost to form cations with a charge +2, and an oxidation state of +2 (Greenwood and Earnshaw, 1997). The alkaline earth metals are said to be silver-colored and soft. They have relatively low densities, melting points, and boiling points. Except beryllium, all of the alkaline earth metals react with water and the heavier alkaline earth metals are said to react more vigorously than the lighter ones (Royal Society of Chemistry, 2012b).

11.5. Halogens

Here, I will discuss:

- Fluorine (F, Z=9)
- Chlorine (Cl, Z=17)
- Bromine (Br, Z=35)
- Iodine (I, Z=53)
- Astatine (At, Z=85)

Some people believe that the synthetic element Tennessine (Ts, Z=117) may be a halogen. Also known as group 17, halogens are "salt-producing" chemical elements, meaning that their reaction with other chemical elements such as metals yields different kinds of salt (e.g. sodium chloride known as the common table salt). When they are bonded to Hydrogen, they form acids (e.g. hydrochloric acid). Halogens are extremely reactive, which is believed to be due to the high electronegativity of their atoms. They are believed to have 7 valence electrons in their outermost energy level, and by reacting with atoms of other chemical elements, they are believed to be able to gain an electron.

An emphasis is usually put on the high reactivity of fluorine as it is one of the most reactive chemical elements, even more electronegative than oxygen. It can even react with noble gas to form compounds. Fluorine and chlorine gases are extremely toxic and lethal. On the periodic table, halogens are located just at the left of the noble gases.

11.6. Lanthanoids

They are:

- Lanthanum (La, Z=57)
- Cerium (Ce, Z=58)
- Praseodymium (Pr, Z=59)
- Neodymium (Nd, Z=60)
- Promethium (Pm, Z=61)
- Samarium (Sm, Z=62)
- Europium (Eu, Z=63)
- Gadolinium (Gd, Z=64)
- Terbium (Tb, Z=65)
- Dysprosium (Dy, Z=66)
- Holmium (Ho, Z=67)
- Erbium (Er, Z=68)
- Thulium (Tm, Z=69)
- Ytterbium (Yb, Z=70)
- Lutetium (Lu, Z=71)

Also called lanthanides, lanthanoids are 15 metallic chemical elements which atomic number (number of protons) ranges from 57 (for lanthanum) through 71 (lutetium). Since 1985, the term "lanthanoid" rather than "lanthanide" has been recommended by the International Union of Pure and Applied Chemistry. Together with chemical elements scandium and yttrium, lanthanoids are habitually acknowledged as the "rare earth elements" or "rare earths".

The term rare elements have misled some people to think that those elements were rare or not abundant on Earth. Indeed, deprecated by International Union of Pure and Applied Chemistry (IUPAC), the term rare earth elements is mistakenly implying that those chemical elements are the rarest elements on Earth or are found only on Earth, but that it is very difficult to separate out each of the individual lanthanide elements and their propensity of "hiding" behind each other in minerals. For instance, although called a rare element, Cerium (Ce, Z=58) is said to be the 26th most abundant element in the Earth's crust, even more abundant than copper (Gray, 2007). Neodymium (Nd, Z=60), another lanthanoid is said to be more abundant than gold (Aspinall, 2001). Besides lanthanum or lutetium (usually considered as d-block elements), all lanthanoids are said to be f-block elements, a nomenclature used to describe the filling of what is called the 4f electron shell. They all form trivalent cations, but they can also form +2 complexes in solution (MacDonald et al., 2013). Their ionic radius is postulated to decrease progressively from lanthanum (the lightest lanthanoid) to lutetium (the heaviest lanthanoid). The chemistry of lanthanides is said to be similar to that of lanthanum, the leading element in their groups.

The lanthanoids are said to be chemically very similar and often occur together in nature with Yttrium (Y, Z=39, a transition metal) commonly present by their side (Gray, 2007). According to the same source, 3 to all of the 15 lanthanoids occur in minerals such as samarskite, monazite, and many others. Furthermore, the same source reported that most of the rare earths were discovered at the same time in Ytterby, Sweden, and four of them are named (yttrium, ytterbium, erbium, terbium) after that city. The proximity of the localization of these lanthanoids in nature suggests that they may have been formed in a similar environment, close to one another, hence they have been mixed together to form compounds containing most of them.

11.7. Metals

Their names are:

- Aluminum (Al, Z=13)
- Gallium (Ga, Z=31)
- Indium (In, Z=49)
- Tin (Sn, Z=50)
- Thallium (Tl, Z=81)
- Lead (Pb, Z=82)
- Bismuth (Bi, Z=83)
- Polonium (Po, Z=84)

Metals conduct electricity and heat fairly well. Physicists generally define metals as any substance that can conduct electricity at a temperature of absolute zero (Yonezawa, 2017). According to that source, Sir Nevill Mott (1905-1996) wrote a letter to a fellow physicist, Prof. Peter P. Edwards, in which he noted *"I have thought a lot about 'What is a metal?' and I think one can only answer the question at T =0 (the absolute zero of temperature) at which a metal conducts and a nonmetal doesn't"*.

In fact, several chemical elements and compounds that are not typically defined as metals become metallic under high pressures. For instance, *"the nonmetal iodine gradually becomes a metal at a* pressure of between 40 and 170 thousand times atmospheric pressure" (Wikipedia, 2020d). Likewise, certain materials viewed as metals can become nonmetals under some conditions. This is a reason chemists and physicists seem to define metal differently. For instance, chemists prefer calling certain metals (e.g. arsenic and antimony) as metalloids, while physicists could call them as metals. Astrophysics went further to widely define metals as any chemical element in a star that is heavier than hydrogen and helium. In other words, according to the field of study, the nomenclature, definition, or boundaries of some chemical groups can differ. These struggles of the characterization of the elements and the change they can go through suggests that they could have been shaped by different conditions, which, to some extent, may be overturned or affected by subduing them to other environmental conditions. As of 2025, about 95 of the 118 known chemical elements are like metals, although most of them are not called metals. In fact, although some chemical elements are called nonmetals and metalloids, some of them behave like metals and the distinction between the groups vary and no strict and universal definition exists to clearly cut the chemical elements into specific categories which properties do not overlap. In other words, although efforts have been made to group chemical elements, the similarity between properties of elements belonging to different groups is a challenge suggesting that the origin and characterization of some chemical elements by scientists may be arbitrary and misunderstood. Anyways, as of 2025, according to their physical and chemical properties, metals are divided into categories such as:

- Actinoides or actinides
- Alkali metals
- Alkaline earth metals
- Lanthanoids or lanthanides
- Transition and post-transition metal
- Elements that are possibly metals (meitnerium, darmstadtium, roentgenium, nihonium, flerovium, moscovium, livermorium, tennessine, and oganesson)
- Elements that are sometimes considered metals (germanium, arsenic, selenium, antimony, tellurium, and astatine)

I studied some of the groups of metals in the section labeled according to their names. The ability of metals to easily lose their outer shell electrons has been postulated as the reason of their solid or liquid state. The atomic structure of metals has been "visualized as a collection of atoms embedded in a cloud of relatively mobile electrons, forming a type of interaction called a metallic bond" (Mortimer, 1975). Metals are generally denser than nonmetals and most of them react with oxygen to form oxides. Nevertheless, many types of metals do not react with oxygen.

11.8. Metalloids

The six chemical elements commonly recognized as metalloids are:

- Boron (B, Z=5)
- Silicon (Si, Z=14)
- Germanium (Ge, Z=32)
- Arsenic (As, Z=33), known as very poisonous
- Antimony (Sb, Z=51)
- Tellurium (Te, Z=52)

CHAPTER 11: HIDDEN MESSAGE ABOUT GROUPS OF CHEMICAL ELEMENTS

Metalloids are chemical elements having properties between those of metals and nonmetals, or properties of both metals and nonmetals. They are a mixture of metals and nonmetals and generally have intermediate or hybrid properties between metals and nonmetals. The terminology metalloid is much more used in chemistry. Although the above-listed 6 chemical elements are those commonly recognized as metalloids, 5 other elements are sometimes classified as metalloids: carbon, aluminum, selenium, polonium, and astatine. However, to align my writing with the definition in chemical literature, I did not consider these 5 chemical elements as metalloids. Metalloids are poor conductors of electricity, and, from a chemical standpoint, they are said to behave generally as nonmetals. They are highly used as semiconductors and in electronic devices. However, it is important to mention that the electrical conductivity of antimony and arsenic is near that of metals.

Silicon is a metalloid agreeing with conditions favoring life on Earth. The Earth's crust is dominated by silicon and oxygen. The abundant silicon in the Earth can be a shield preventing a massive leak of the energy and heat stored at the center of the Earth, and which is sometimes released during volcanic activities. Similarly, because oxygen is a nonmetal, its presence and abundance by the side of silicon can help protect the surface of the Earth from receiving much of the electricity contained in the Earth's interior.

11.9. Noble gases

The noble gases are:

- Helium (He, Z=2)
- Neon (Ne, Z=10)
- Argon (Ar, Z=18)
- Krypton (Kr, Z=36)
- Xenon (Xe, Z=54)
- Radon (Rn, Z=86)

Although the recently synthetic gas Oganesson (Og) is projected to be a noble gas, its chemistry has not yet been explored. Also called inert gases or aerogens, noble gases are chemical elements monatomic gases, meaning their atoms are not bound to each other. These gases are highly unreactive, meaning that they have a very low chemical reactivity. Because of their inertness, noble gases are highly used in systems where reactions are undesired. For instance, helium is something used to study turbulence, implying that the presence of helium in turbulent reactions can prevent the atoms present in the environment from interacting with others as they would if helium is not blocking them. In other words, the use of helium and other noble gases in turbulence experiments and others can bias the results.

Noble gases are believed to be inert because of their atomic structure, which has an outer shell of valence electrons said to be "full", implying that these gases have a small propensity to partake in chemical reactions. According to Wikipedia

(2016b), their melting and boiling points are similar, differing by less than 10 °C. Under some limited and small temperature range, they can be in liquid form, but it general, they stay as gases. They are said to have a weak interatomic force, and subsequently a very low melting and boiling point.

Besides radon, which has no stable isotope, all of the other noble gases have multiple stable isotopes. In each period of the periodic table, noble gases are said to have the highest ionization potential. They are believed to be unable to accept an electron to form stable anions, and which is postulated to be translated into their negative electron affinity (Wheeler, 1997). Their macroscopic physical properties are dominated by the weak van der Waals forces between the atoms (Wikipedia, 2016b). The bigger the atoms, the higher the so-called "attractive force" between the atoms, and this is thought to explain the positive correlation between their density and atomic mass (Greenwood and Earnshaw, 1997). I will later explain the cause of this relationship.

Although noble gases are less reactive, their atoms are sometimes combined with atoms of other chemical elements to form compounds, (Encyclopædia Britannica, 2008), most of which are based on xenon (e.g., xenon tetrafluoride, xenon hexafluoride, xenon tetroxide, sodium perxenate) and krypton difluoride. Because of the low reactivity of noble gases, naturally-occurring noble gas compounds may be due to these gases being trapped into other compounds rather than they interacting with others or forming chemical bonds with other atoms. That can explain why the most used mean to produce large-scale helium is its extraction by fractional "distillation from natural gas, which can contain up to 7% helium" (Winter, 2008). In addition to fractional distillation, liquefaction of ambient air is also used to obtain other noble gases such as neon, argon, krypton, and xenon. I believe that most of the helium in natural gases found underground were trapped after their formation during the genesis of the Earth. Some of them may have made their way by passing through rocks until they condensed into natural gases. Some noble gases in the Earth's atmosphere could have escaped the ground or the crust to station in the atmosphere because of their inability to interact with others. Some of those gases can be abundant in the upper atmosphere due to their inability to sustain the higher pressure and turbulence near the surface of the Earth. Because pressure and turbulence may be smaller in the upper atmosphere, noble gases may prefer that location because it may reduce their interactions with other atoms.

It is very difficult to convert noble gases into solid forms. For instance, to convert helium into a solid by cooling, the reaction must occur at pressure of 25 standard atmospheres (2,500 kPa; 370 psi) at a temperature of 0.95 K (-272.200 °C; -457.960 °F) (University of Alberta, 2008). Considering what I just said above about efforts to solidify helium by cooling, noble gases may have been formed in environments that have mimicked those of turbulence Zone 5 (at the atomic levels), hence their small radius, low reactivity, and interaction with other atoms. The temperature, pressure, and energy available in the space where noble gases were formed could have been much higher than those that human beings can submit them to in lab settings. Because the harsh conditions that prevailed during their formation were unable to bend or force their precursors not to become gas, lab efforts to cool or compress them cannot disregard their low reactivity and force them to stick together under standard atmospheric conditions. I personally believe that chemical elements were formed under very hot conditions. Therefore, efforts to solidify noble gases by cooling them and by using high pressure are against the conditions under which they were naturally formed. The slowing down of their temperature can slow down the movement of their subatomic particles and also their atoms.

11.10. Non-metals

According to their ability to form chemical compounds with other chemical elements, nonmetals are divided into 2 categories:

- reactive nonmetals and
- noble gases

I already expounded on the noble gases in a previous section. Also called reactive nonmetals, the common and clearly-accepted nonmetals are:

- Hydrogen (H, Z=1),
- Carbon (C, Z=6),
- Nitrogen (N, Z=7),
- Oxygen (O, Z=8),
- Phosphorus (P, Z=15),
- Sulfur (S, Z=16), and
- Selenium (Se, Z=34).

In addition to the 7 nonmetals listed above and the 6 common noble gases listed in the previous section, four other chemical elements are sometimes classified as nonmetal: fluorine, chlorine, bromine, and iodine. Therefore, the total number of chemical elements that are generally classified as nonmetal is 17. Some literatures even lengthen that list by adding some metalloids such as boron, silicon, and germanium. They are more inclined to have a relatively low melting point, boiling point, and density. Their thermal conductivity and electrical conductivity are usually poor, while their ionization energy, electron affinity, and electronegativity are usually high. When reacting with other chemical elements or compounds, they are said to gain or share some of their electrons. Unlike the noble gases (known not to interact much with other chemical elements), reactive metals react with others and form chemical compounds. Carbon and sulfur are considered the least electronegative nonmetal and are frequently said to have weak to moderately strong nonmetallic properties and tendency to form covalent compounds with metals. In contrast, the most electronegative nonmetals are said to be oxygen and fluorine and are considered to have stronger nonmetallic properties and a propensity to mainly form ionic compounds with metals.

Unlike metals, nonmetals are poor conductors of electricity and heat. Their melting point, boiling point, and density are usually smaller than that of metals. Ionization energy, electron affinity (e.g. nitrogen and the noble gases have negative electron affinity) and electronegativity are frequently used by some authors to clearly define nonmetals (Yoder et al., 1975). When nonmetals react with oxygen, they form acidic oxides. In contrast, the oxides formed by the reaction of metals and oxygen are usually basic.

As far as the chemical elements discovered as of 2025 are concerned, although there are less nonmetals than metals, the nonmetals constitute the majority of chemical elements found in nature. For instance, some scientists believe that the nonmetals—hydrogen and helium—are believed to make up over 99% of the chemicals in the observable universe (Sukys, 1999). However, I think that hydrogen and helium as known on Earth may not be the exact same in all other stellar systems or galaxies. Even in the Solar System, some data suggests that hydrogen in the outer planetary systems have other isotopes less abundant than that on Earth (see data on atmosphere of planets). Oxygen (a nonmetal) is believed to make up nearly half of the Earth's crust, oceans, and atmosphere (Bettelheim et al., 2016). More chemical compounds discovered so far are made of nonmetals than metals (Steurer, 2007). Finally, nonmetals (e.g. carbon, hydrogen, oxygen, and nitrogen) are also the dominant and most abundant chemical elements in living organisms (Schulze-Makuch and Irwin, 2008). It is not by chance that the most abundant chemical elements in living organisms are some of the key nonmetals (C, H, O, and N), which poorly conduct heat and electricity. For the properties are compatible with life as, for instance, they help conserve the internal temperature of living organisms. Else, life would have been very difficult and even impossible, for the energy needed for biological reactions and activities could have been dissipated or transmitted to the environment instead of being kept mostly inside those organisms. The ability of these key nonmetals to preserve heat and to conduct it to some extent, greatly contribute to allowing the compartmentalization of biological organelles and other entities indispensable for biological reactions required to produce and regulate biological compounds essential for living organisms. If the chemical elements constituting living organisms could not to some extent allow the flow of energy, life could have been impossible. In other words, if the nonmetals are 100% unable to conduct some electricity to some extent, life could have been impossible, for some exchange between cells and organelles could have been difficult.

11.11. Transition metals

The transition metals are:

- Scandium (Sc, Z=21)
- Titanium (Ti, Z=22)
- Vanadium (V, Z=23)
- Chromium (Cr, Z=24)

- Manganese (Mn, Z=25)
- Iron (Fe, Z=26)
- Cobalt (Co, Z=27)
- Nickel (Ni, Z=28)
- Copper (Cu, Z=29)
- Zinc (Zn, Z=30)
- Yttrium (Y, Z=39)
- Zirconium (Zr, Z=40)
- Niobium (Nb, Z=41)
- Molybdenum (Mo, Z=42)
- Technetium (Tc, Z=43)
- Ruthenium (Ru, Z=44)
- Rhodium (Rh, Z=45)
- Palladium (Pd, Z=46)
- Silver (Ag, Z=47)
- Cadmium (Cd, Z=48)
- Hafnium (Hf, Z=72)
- Tantalum (Ta, Z=73)
- Tungsten (W, Z=74)
- Rhenium (Re, Z=75)
- Osmium (Os, Z=76)
- Iridium (Ir, Z=77)
- Platinum (Pt, Z=78)
- Gold (Au, Z=79)
- Mercury (Hg, Z=80)
- Rutherfordium (Rf, Z=104)
- Dubnium (Db, Z=105)
- Seaborgium (Sg, Z=106)
- Bohrium (Bh, Z=107)
- Hassium (Hs, Z=108)
- Meitnerium (Mt, Z=109)
- Darmstadtium (Ds, Z=110)
- Roentgenium (Rg, Z=111)
- Copernicium (Cn, Z=112)

Also called transition elements, the transition metals are not clearly defined. Although some classifications considered them as elements belonging to the d-block (a term used to characterize atoms having up to 10 d electrons) of the periodic table, others suggested that as lanthanoids and actinoids can also be considered transition metals and older references called them "inner transition metals". Transition metals form compounds which color is commonly believed to be caused by their d–d electronic transitions. These compounds are also said to be in various oxidation states, and they hold the record of the highest density, melting point, and boiling point. They also form various paramagnetic compounds (Figgis and Lewis, 1960) due to what is termed the presence of unpaired d electrons. Paramagnetism is a *"form of magnetism whereby some materials are weakly attracted by an externally applied magnetic field, and form internal, induced magnetic fields in the direction of the applied magnetic field"* (Miessler and Tarr, 2010). Although it is believed that paramagnetism is caused by the presence of unpaired electrons (Figgis and Lewis, 1960), this special form of magnetism suggested to me that the precursors of these atoms may have gone through movements that aided in creating a special magnetic field. Among the transition metals, 3 are notably known as the only elements that produce a magnetic field: Iron (Fe, $Z=26$), Cobalt (Co, $Z=27$), and Nickel (Ni, $Z=28$). The fact that these chemical elements are near one another as far the number of their protons is concerned may have something to do with some of their proprieties.

11.12. Unknowns
Examples of chemical elements labeled as unknowns are:
- Nihonium (Nh, $Z=113$)
- Flerovium (Fl, $Z=114$)
- Moscovium (Mc, $Z=115$)
- Livermorium (Lv, $Z=116$)
- Tennessine (Ts, $Z=117$)
- Oganesson (Og, $Z=118$)

The unknown chemical elements are some of the most recently discovered synthetic chemical elements. As more studies are done on them, their properties will be better known and they will be consequently classified.

Nathanael-Israel Israel: Member of the American Association for the Advancement of Science

'Science180 Academy' Success Strategy:
SCIENCE180 SERVICES AND PRODUCTS YOU WILL LOVE

Because you are reading this book, you are probably very interested in answering your questions about the origin of the universe, of life, and of chemicals. Imagine you want to be trained by Dr. Nathanael-Israel Israel and his team so you can benefit from their outstanding expertise to empower yourself or your team. Or you want him to give a keynote speech, a seminar, or any other kind of talk or conference at your organization. Or you want him to mentor you or some people or team at your organization. Maybe you have critical origin-related questions that you need his help to accurately answer. You want a true expert to talk with you about the customized program or game plan that fits your needs. You want him to tailor his advice, expert feedback, and proven shortcuts to the stage of life you are in and help you get to where you want to be in your desire to properly understand the origin of the universe, life, and chemicals and harness the benefits that come with it. Perhaps you don't know how to properly get any of these important tasks done according to your specific needs or the needs and demands of your organization. That is what Science180 Academy is all about. Visit Science180.com/services for more details about how to benefit from the services that Science180 provides.

Maybe you are a leader that wants to hire Dr. Nathanael-Israel Israel and his team to train some departments at your organization. Or you want to refer them to other companies like a good dish passed around the dinner table, and you want to explore how Nathanael-Israel Israel can pay you something for that referral. Maybe you attended Nathanael-Israel Israel's speaking program, for which, without going into details, he accurately raised your awareness about how the universe, life, and chemicals were formed. Or maybe you attended his training, in which he detailed and showed you how he decoded the scientific data using various tools and certain thinking strategies that helped him and which transferred some skills to you; and now, you are interested in a long term one-on-one consulting, or mentoring program with him, so that, he delves into more details about how to use proven techniques to decode the universe (strategies for data collection, data analysis, data presentation, writing, and even tips for future research) and change your behavior on a long term basis. If you related to any of the points mentioned above, Science180 Academy is the right fit for you!

Other customizable services that Science180 provides include: Assessments, Books and other products, Book publishing, Conferences, Consulting, Online courses, Podcasting, Retreats, Seminars, Speaking engagements, Survey and research tools, Training, Video programs

Here are other reasons why you should choose to work with or hire Nathanael-Israel Israel and the team at Science180:

- Accurately understand chemicals-origin. Be happy forever!
- Breaking cosmological and traditional chemicals-origin nonsenses
- Chemicals-origin problems final bus stop
- Chemicals-origin theory that helps you fight wasteful programs
- Complex chemicals-origin questions solved accurately in a simple language
- Current hot chemicals-origin solutions for adults and kids
- Customizable chemicals-origin trainings with unique materials
- Discover the key variables needed to decode the chemicals-origin
- Easily understand complex chemicals-origin equations in minutes
- Enter the realm of true knowledge about the chemicals-origin
- Impressively accurate and impossibly easy-to-understand chemicals-origin theory
- Improve your understanding of chemicals-origin with new, accurate products and services
- Nonconformist, rule-breaker, and accurate demonstrator of chemicals-origin
- Personalized chemicals-origin decoding package
- Properly understand turbulence. Only on Science180
- Receive an exceptional chemicals-origin decoding perspective
- Simplified and generalizable explanations of the real origin of chemicals
- Source of unconventional wisdom and knowledge on the origin of chemicals
- State-of-the-art decoding experience of the origin of chemicals
- The accurate and most trusted chemicals-origin decoder
- The go-to source for valuable chemicals-origin information
- The most accurate, reliable, safest, best explanation of the chemicals-origin ever
- Think chemicals-origin differently
- Where chemicals-origin is accurately decoded, full stop

CHAPTER 12

WON'T YOU REFRESH YOUR MIND WITH THIS SUMMARY ABOUT PROTONS, ELECTRONS, NEUTRONS, ISOTOPES, MASS, AND THE DISCOVERY YEAR OF CHEMICAL ELEMENTS?

12.1. Atomic number or number of protons, number of electrons

The atomic number is the number of protons (symbol Z) in the nucleus of an atom. All atoms which have the same atomic number belong to the same chemical element. Because neutrons have a neutral electrical charge, the atomic number is also an indicator of the charge number of the nucleus of an atom. When an atom is not electrically charged, its atomic number and the number of electrons is the same. The atomic number of the chemical elements known as of 2020 varies between 1 and 118 (Fig. 16 to Fig. 26). The number of electrons and that of protons are said to be the same. Therefore, as I deal with the number of protons below, I also mean the number of electrons.

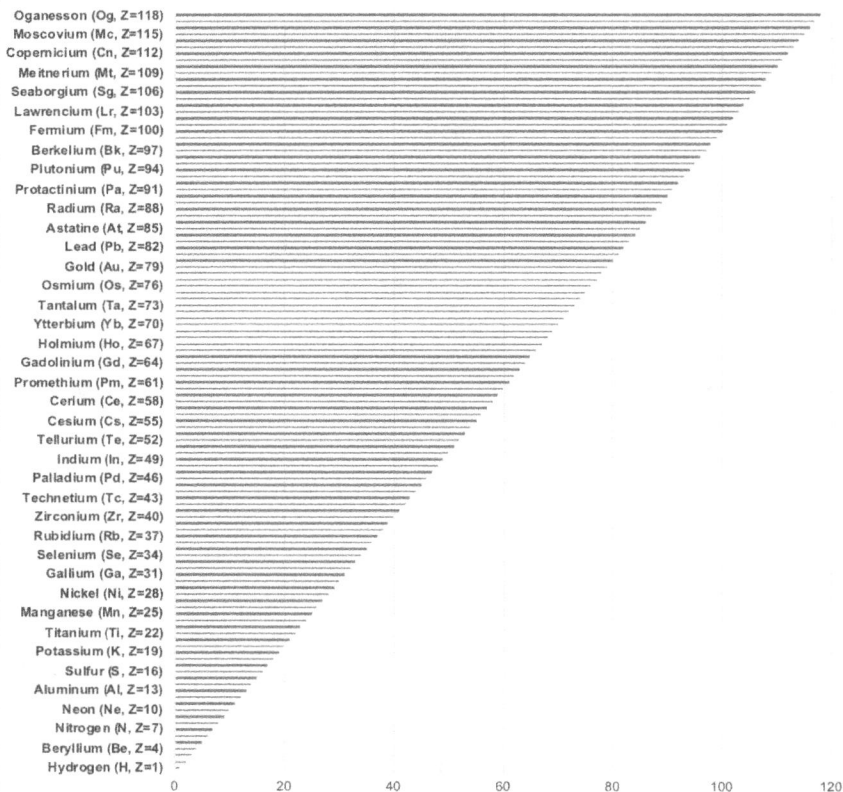

Fig. 16: Atomic Number of the 118 chemical elements known as of 2020

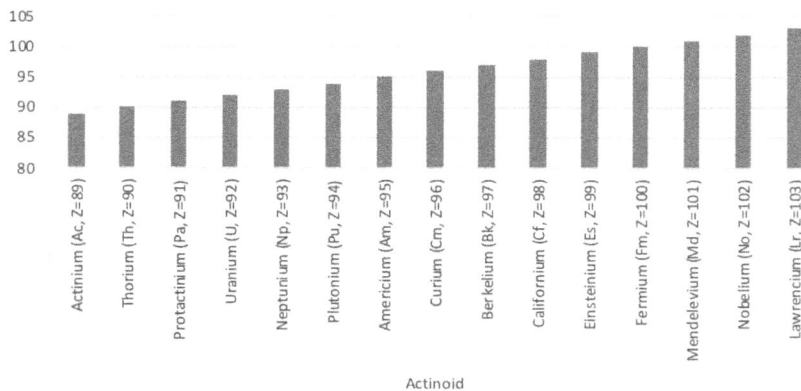

Fig. 17: Atomic Number of Actinoids

Fig. 18: Atomic Number of Alkali metals and Alkaline earth metals

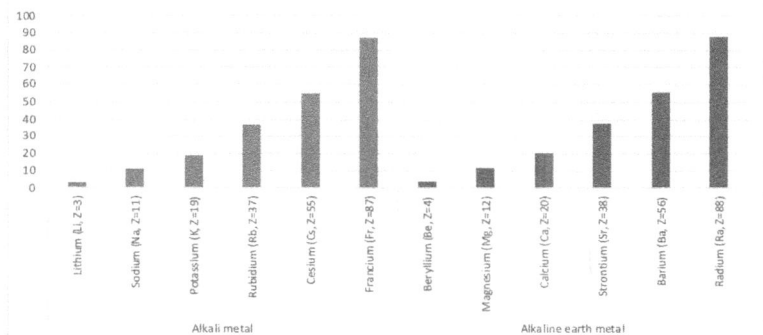

Fig. 19: Atomic Number of Halogens

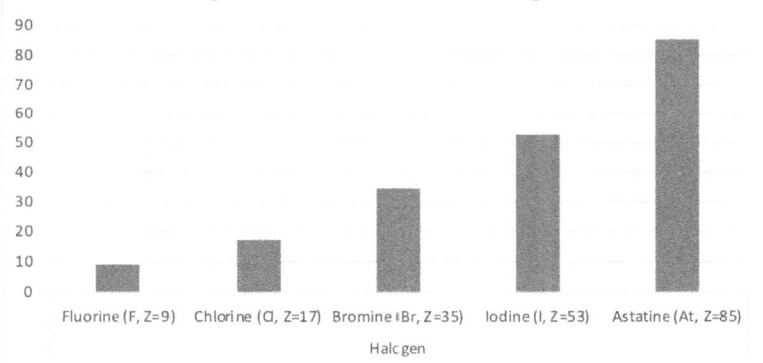

Fig. 20: Atomic Number of Lanthanoids

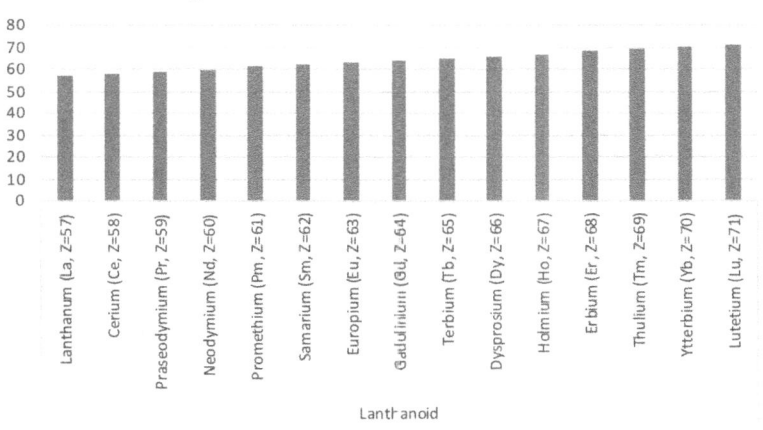

Science180: Complex Chemicals-Origin Questions Solved Accurately in a Simple Language

Fig. 21: Atomic Number of Metals

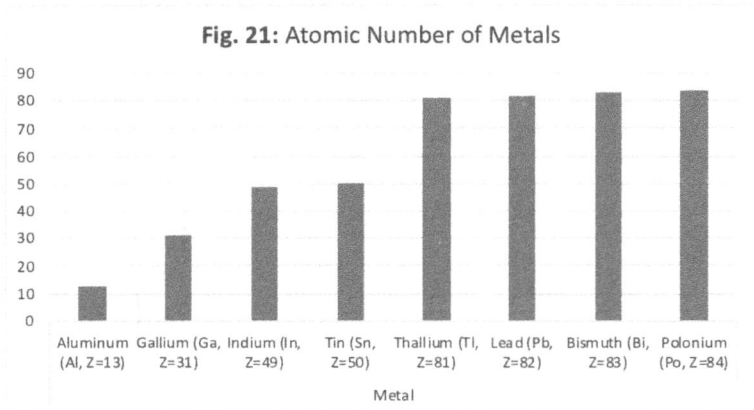

Fig. 22: Atomic Number of Metalloids

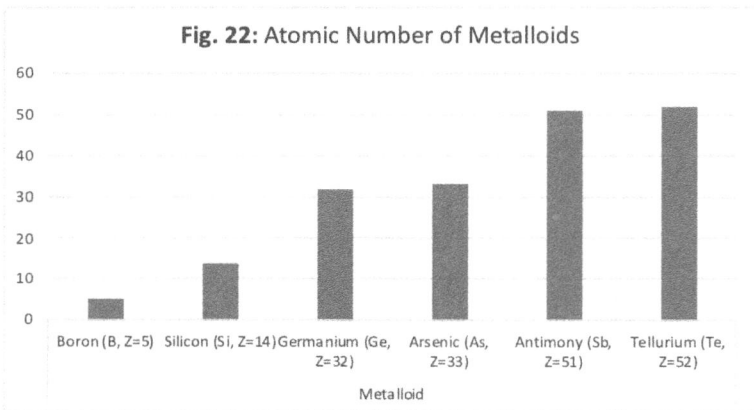

Fig. 23: Atomic Number of Noble gases

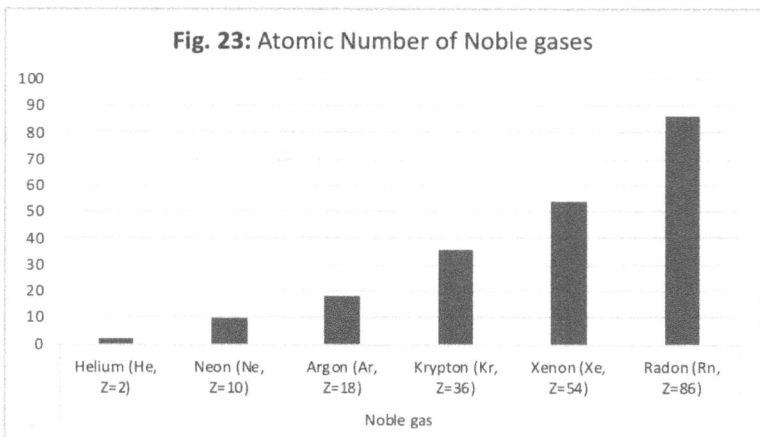

Fig. 24: Atomic Number of Non-metals

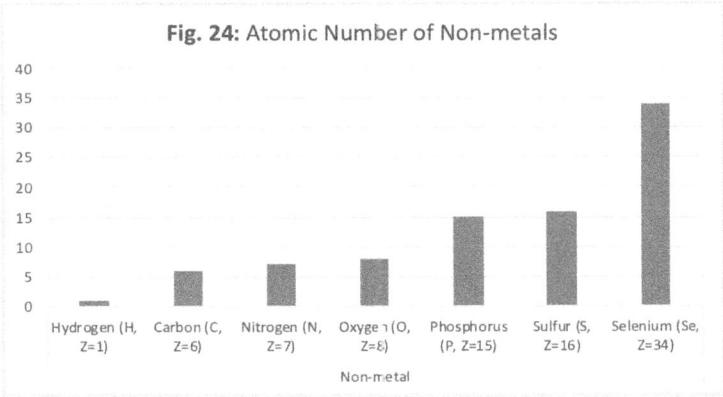

Fig. 25: Atomic Number of Transition metals

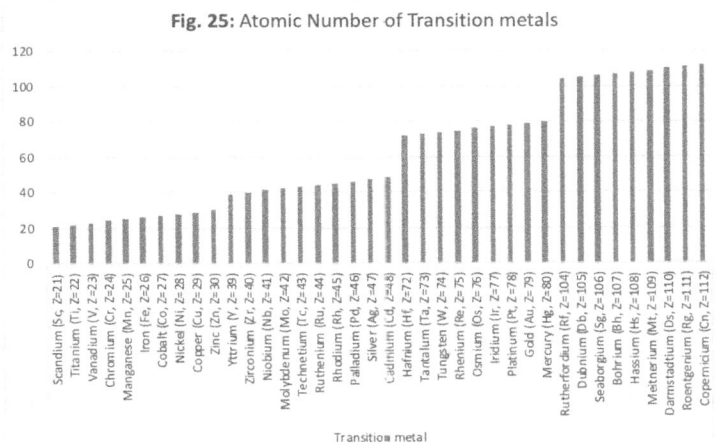

Fig. 26: Atomic Number of Unknowns

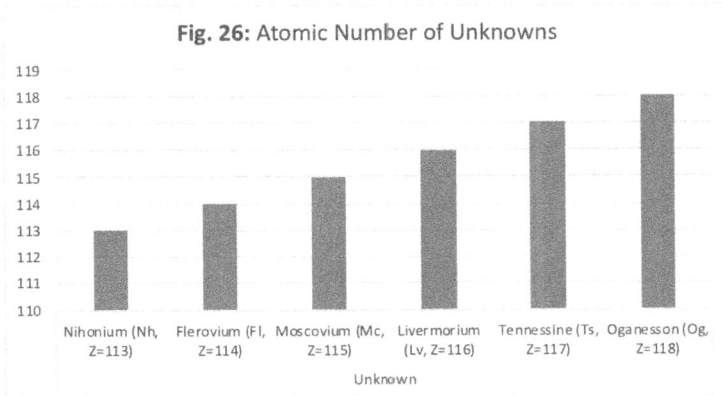

Science180: Complex Chemicals-Origin Questions Solved Accurately in a Simple Language

12.2. Number of neutrons

The number of neutrons of the chemical elements varies between 0 and 165. The ordinary hydrogen (H, Z=1) is the only chemical element which has no neutron. However, it is important to mention that hydrogen has 3 isotopes, 2 of which have neutrons: deuterium (which has one neutron), and tritium (which has 2 neutrons). Copernicium (Cn, Z=112) is the chemical element that has the highest number of neutrons. As more data is available on the chemical elements which atomic number is higher than 112, it is possible that other elements may have more than 165 neutrons which was recorded with Copernicium (Cn, Z=112).

Unlike the number of protons which is the same for all the atoms belonging of the same chemical elements, many chemical elements have the same number of neutrons although they differ by their number of protons. For instance, Boron (B, Z=5) and Carbon (C, Z=6) all have 6 neutrons each. Similarly, 10 neutrons are found with Fluorine (F, Z=9) and Neon (Ne, Z=10). Aluminum (Al, Z=13) and Silicon (Si, Z=14) each have 14 neutrons. Radon (Rn, Z=86) and Francium (Fr, Z=87) each have 136 neutrons. Based on the data I reviewed, all of the following synthetic elements have 157 neutrons:

- Fermium (Fm, Z=100)
- Mendelevium (Md, Z=101)
- Nobelium (No, Z=102)
- Rutherfordium (Rf, Z=104)
- Dubnium (Db, Z=105)
- Seaborgium (Sg, Z=106)
- Hassium (Hs, Z=108)
- Meitnerium (Mt, Z=109)

The actinoids followed by the lanthanoids have the smallest coefficient of variation of the number of neutrons. While the actinoids have at least 138 neutrons, the lanthanoids have at least 82. In general, the number of neutrons increased as the number of protons increased. However, many chemical elements have more protons than others and yet, the number of their neutrons is smaller. In other words, the fact that a chemical element has more protons than another one does not guarantee that the number of neutrons of the former is higher than that of the latter. For instance, Argon (Ar, Z=18) has 18 protons and 22 neutrons, while Potassium (K, Z=19) and Calcium (Ca, Z=20) which both have more protons than Argon have 20 neutrons meaning 2 neutrons less than Argon. Similarly, Tellurium (Te, Z=52) has 76 neutrons, while Iodine (I, Z=53), which has one proton more than it, has 74 neutrons.

12.3. Isotopes

Isotopes are atoms of the same chemical element but which differ by their number of neutrons. While some isotopes are stable, most of them are unstable. Some isotopes are radioactive while others are not. Isotopes that are radioactive are called radioisotopes. Based on the data I collected on isotopes in 2016, at least 110 chemical elements have an isotope. Based on that data, some chemical elements have up to 23 isotopes. Some of the chemical elements which have the highest number of isotopes include:

- Tellurium (Te, Z=52), a metalloid: 23
- Xenon (Xe, Z=54), a noble gas: 21
- Tin (Sn, Z=50), a metal: 20

Among the 8 chemical elements that have no reported isotope or no stable isotope, Phosphorus (P, Z=15) is the only natural one. All of the other 7 chemical elements are synthetic:

- Roentgenium (Rg, Z=111)
- Nihonium (Nh, Z=113)
- Flerovium (Fl, Z=114)
- Moscovium (Mc, Z=115)
- Livermorium (Lv, Z=116)
- Tennessine (Ts, Z=117)
- Oganesson (Og, Z=118)

As I was searching the literature, I realized that Phosphorus has 23 isotopes, but only one, ^{31}P, is stable. Consequently, phosphorus is considered a monoisotopic element meaning a chemical element that has only a single stable isotope. I was very interested in learning more about phosphorous, a fundamental chemical element constituting the backbone of nucleic acids (e.g. DNA and RNA), the fundamental macromolecules without which life on Earth may not be possible as it is today. I believe that the stability of phosphorous was one of the reasons it was chosen to be part of the elements of DNA. The number of isotopes of the other 4 fundamental chemical elements abundant in most living organism are:

- Hydrogen (H, Z=1): 3
- Carbon (C, Z=6): 3
- Nitrogen (N, Z=7): 4
- Oxygen (O, Z=8): 4

In other words, despite its small size, Hydrogen has 3 isotopes:

- protium (^{1}H), which has no neutron,
- deuterium (^{2}H), which has one neutron, and
- tritium (^{3}H), with two neutrons.

Protium is the ordinary hydrogen and mostly represents the hydrogen element. It is the main hydrogen isotope. Deuterium is also stable but tritium is very radioactive and is used in the manufacturing of nuclear bombs. The half-life of tritium is estimated at 12.3 years. Other hydrogen isotopes having more than 2 neutrons were artificially produced, but they all are unstable and likely very radioactive. Each lanthanoid has 4 to 13 isotopes. Based on the data I had, metalloids have the maximum number of isotopes per chemical element. The smallest coefficient of variation of the number of isotopes was found with lanthanoids (29.1%).

12.4. Atomic Mass

Expressed in atomic mass units (amu), or Dalton, or gram per mol, the atomic mass is the mass of an atom. The atomic mass of the 118 chemical elements known as of 2020 varies between 1.00794 and 294 grams per mol (Fig. 27). The lightest atom is that of Hydrogen whereas the heaviest is that of Oganesson (Og, Z=118), the last chemical element on the periodic table. The heaviest natural element is Plutonium (Pu, Z=94) and its mass is 244 g/mol. In contrast, the mass of the synthetic elements varies between 243 g/mol (which is for Americium (Am, Z=95)) and 294 g/mol. As of 2020, Plutonium (Pu, Z=94) is the only natural chemical element which is heavier than a synthetic element (i.e. Americium (Am, Z=95)). In other words, most of the synthetic chemical elements are heavier than the natural ones.

In general, the average mass of the nonmetal is smaller than that of any other group, suggesting that the environment of their formation could have not allowed them to gain much mass. In contrast, the highest mean mass was obtained by the elements which group is classified as unknown, followed by the actinoids. The minimum mass was obtained by a nonmetal while the heaviest elements found were the unknown, the transition metals and the actinoids. The second lightest element is a noble gas (Helium). All of the groups contained chemical elements heavier than that of the heaviest nonmetal (Selenium (Se, Z=34, mass is 78.96 g/mol).

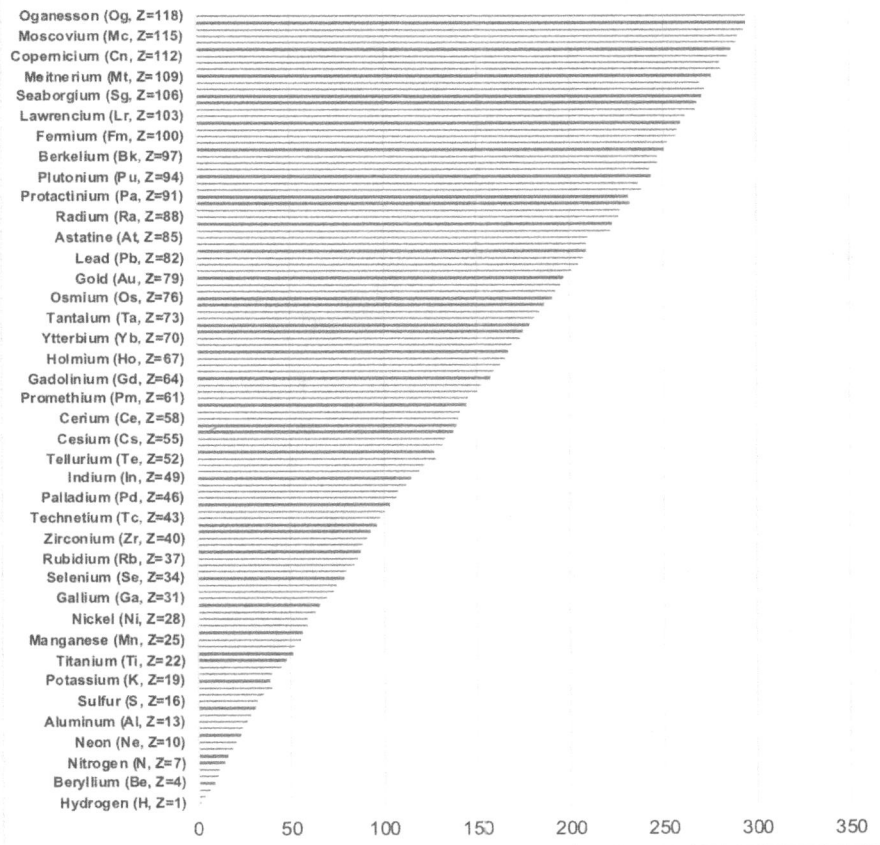

Fig. 27: Atomic Mass (g/mol) of the 118 chemical elements known as of 2020

12.5. Discovery year

Among the 118 chemical elements known as of 2025, nine were known before common era:

- Sulfur (S, Z=16)
- Iron (Fe, Z=26)
- Copper (Cu, Z=29)
- Silver (Ag, Z=47)
- Tin (Sn, Z=50)
- Antimony (Sb, Z=51)
- Gold (Au, Z=79)
- Mercury (Hg, Z=80)

- Lead (Pb, Z=82)

Besides those 9 chemical elements known for more than 2000 years, the discovery year of the other elements I studied ranges from 1400 to 1996. But some synthetic chemical elements which properties are still under scrutiny were more recently discovered. Besides the 9 elements mentioned above, only 4 others were discovered by the end of the 17th century:

- Bismuth (Bi, Z=83): 1400
- Zinc (Zn, Z=30): 1500
- Phosphorus (P, Z=15): 1669
- Carbon (C, Z=6): 1694

Then, in the 18th century, 19 chemical elements were discovered. In the 19th century alone, meaning from 1801 to 1899, 50 additional chemical elements were discovered. In the 20th century, 30 elements were discovered.

'Science180 Academy' Success Strategy
SCIENCE180 SEMINARS

People whose awareness is raised by Science180 usually ask me to go deeper or they wonder "what's else?". That is one of the reasons Science180 trains them through strategic work sessions (during seminars or training sessions) that transfer customizable skills and solutions to them. Science180 Seminars are client-centered and tailored to strongly engage the clients so they maximize the discovery of and the tapping into new opportunities, and exponentially outperform their expectations. Science180 offers customizable seminars that can be labeled as a colloquy, conference, consultation, discussion, forum, keynote speech, lecture, lesson, meeting, symposium, summit, study group, tutorial, workshop or working section accordingly on any topic related to:

- Universe-origin for scientists and mathematicians, philosophers, laypeople, and the general public
- Universe-origin or universe creation for believers
- Life-origin for life scientists, for all other scientists, and for believers
- Chemical-origin for scientists
- Universe-origin seminars for children
- Universe and life-origin for pseudepigraphic believers

As you contact us with your needs, we can customize your program accordingly. Learn more at Science180Seminars.com.

CHAPTER 13

CAN CONDUCTIVITY AND SPECIAL HEAT HELP YOU TO DEBUNK THE ORIGIN OF CHEMICAL PARTICLES?

Two types of conductivity are typically used to characterize chemical elements: electric conductivity and thermal conductivity. Occasionally, some literatures address other types of conductivity such as ionic conductivity, which is a measure of the tendency of a chemical substance to conduct ions. Hydraulic conductivity is also used to express the ability of a substance to let a fluid pass through it. Furthermore, the inverse of the electric conductivity, also called electrical resistivity (also known as specific electrical resistance or volume resistivity), is a measure of how materials resist the flow of electric current. The unit of electrical resistivity is the ohm-meter ($\Omega \cdot m$).

To get the content of this chapter visit
www.Science180.com/ConductivityAndSpecialHeat

'Science180 Academy' Success Strategy

SCIENCE180 CONSULTING

Because Science180's trainings, seminars, or strategic work sessions (through which it transfers skills and training solutions) are great, some customers want to go even deeper on a long-term, sustainable basis. That is where Science180 Consulting, one-on-one consulting, and mentoring (that some people may prefer calling coaching programs) comes in. That is where Science180 can truly change people's behavior on a long-term basis according to their specific needs. With Science180 Consulting, you will discover and understand the deep secrets of the formation of the universe, life, and chemicals around you. Hear Dr. Nathanael-Israel Israel's personal selection and teaching on key topics that will help you break the code of the universe formation and functioning. All strategically designed to enlighten you, guide you to navigate and filter the massive data collected on the universe and its content so you know how to answer the world's most challenging origin questions, remove any scientific and philosophical cataracts that may be blocking you, and help bring you many steps closer to your best life today and forever. Science180 Consulting will train you, transfer unconventional skills to you and change your behavior so you go deeper. To get started today or to learn more, go to Science180Consulting.com.

CHAPTER 14

WHY DON'T PEOPLE PAY ATTENTION TO THE DENSITY OF CHEMICAL ELEMENTS?

Two kinds of density are usually used to characterize chemical elements:
- Liquid density, and
- Density at 293 K.

I addressed each of them in the following sections.

14.1. Liquid density

Reported for no synthetic chemical elements but only for 68 natural elements, the liquid density varies between 512 kg/m³ and 20,000 kg/m³ (Fig. 31). Most of these 68 elements are metals and only 3 are nonmetals. Indeed, as of 2025, the only nonmetals which liquid density are reported are:

- One halogen, Bromine (Br, Z=35): 3120 kg/m³
- 2 nonmetals:
  - Sulfur (S, Z=16): 1819 kg/m³
  - Selenium (Se, Z=34): 3990 kg/m³

The liquid density is not reported for any noble gas. The smallest liquid density was recorded with Lithium (Li, Z=3), while the highest value was with Osmium (Os, Z=76), a transition metal. The density of the following chemical elements is higher than 86.5% that of Osmium:

- Platinum (Pt, Z=78): 19770 kg/m³, which is 98.85% that of Osmium
- Iridium (Ir, Z=77): 19000 kg/m³, which is 95% that of Osmium
- Rhenium (Re, Z=75): 18900 kg/m³, which is 94.5% that of Osmium
- Tungsten (W, Z=74): 17600 kg/m³, which is 88% that of Osmium
- Gold (Au, Z=79): 17310 kg/m³, which is 86.55% that of Osmium
- Uranium (U, Z=92): 17300 kg/m³, which is 86.5% that of Osmium

The highest liquid density is generally found with transition metals. The smallest CV (16.7%) was obtained with lanthanoids.

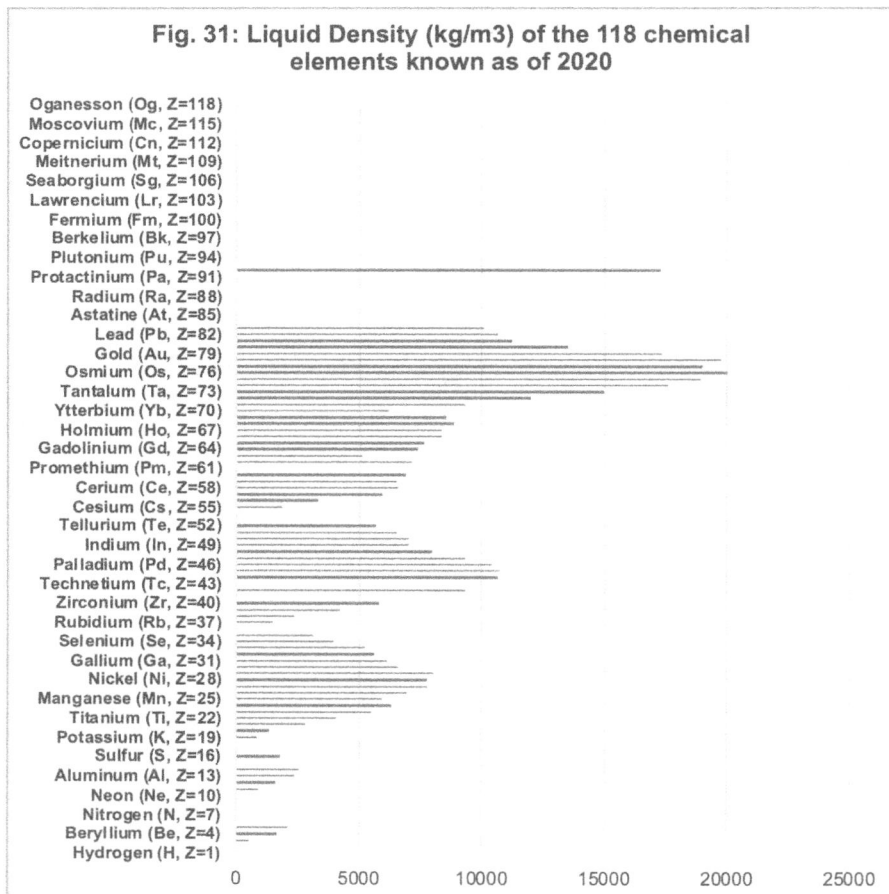

Fig. 31: Liquid Density (kg/m3) of the 118 chemical elements known as of 2020

14.2. Density at 293 K

The density at 293 K is available for all of the 118 chemical elements known as of 2025. It varies between 0.09 kg/m³ and 40700 kg/m³ (Fig. 32). The smallest density was observed with hydrogen. In contrast, the highest density at 293 K was recorded with Hassium (Hs, Z=108), a synthetic transition metal. The densest natural chemical element is Osmium (Os, Z=76): 22,610 kg/m³ followed by:

- Iridium (Ir, Z=77) 22,560 kg/m³
- Platinum (Pt, Z=78) 21,460 kg/m³
- Rhenium (Re, Z=75) 21,020 kg/m³

In general, the highest densities at 293 K were recorded with transition metals, while the smallest densities were recorded with the 3 groups of chemicals classified as nonmetals (halogens, noble gases, and nonmetals). Furthermore, the density of the halogens and noble gases is positively correlated with their atomic number. If it was not because of the high value of the density of carbon (C, Z=6), 2267 kg/m³, the correlation would also have been positive for the other nonmetals. Among the chemical elements commonly put into the category of nonmetals, only 4 have a density at 293 K higher than that of carbon and 3 of them are halogens:

- Bromine (Br, Z=35): 3122 kg/m³
- Selenium (Se, Z=34), a nonmetal: 4809 kg/m³
- Iodine (I, Z=53): 4930 kg/m³
- Astatine (At, Z=85): 7000 kg/m³

Lanthanoids recorded the smallest coefficient of variation.

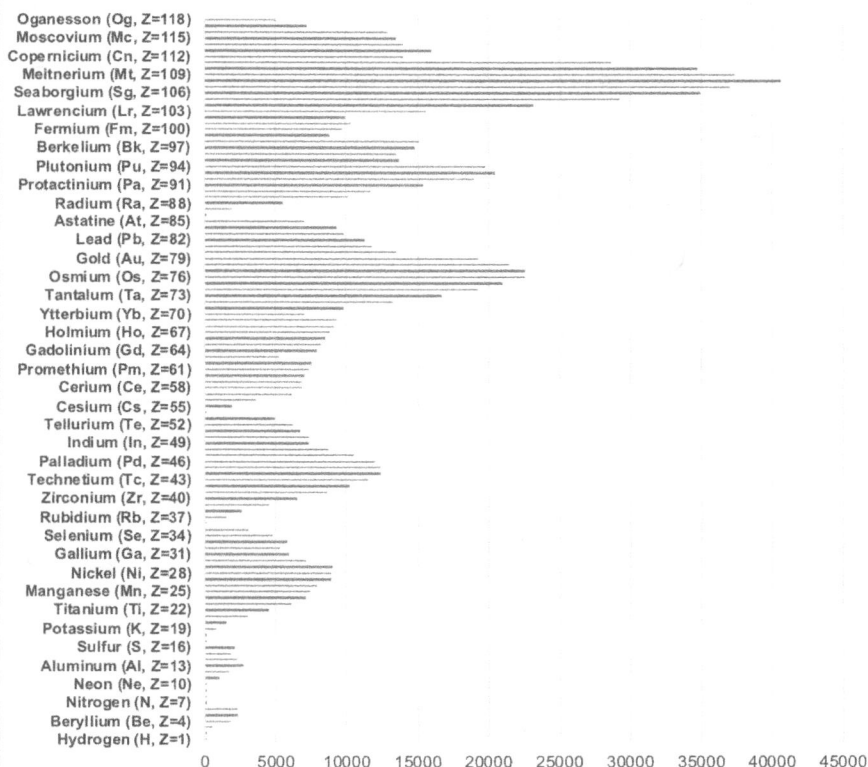

Fig. 32: Density at 293 K (kg/m3) of the 118 chemical elements known as of 2020

Another Book by Nathanael-Israel Israel:
FROM SCIENCE TO BIBLE'S CONCLUSIONS

THE # 1 UNIVERSE-ORIGIN MASTERPIECE OF ALL TIME … AND THE MOST ACCURATE SCIENTIFIC FORMULA THAT STOOD AND WILL STAND THE TEST OF TIME AND OF MATHEMATICS

The real reason scientists have been struggling to accurately understand the universe-formation is because they have spent centuries collecting expensive, complicated, and massive amounts of data, but learned very little, if not nothing, about how to unconventionally step back to properly analyze it to decode the universe. Consequently, people learned to collect all kinds of data everywhere to build models and imaginary concepts that betray their discernment, but they never learned to unlearn wrong theories, nor learned how to stop trashing great raw data hidden in theories they dislike or misunderstand, never knew where to find and how to properly combine the fundamental variables without which it is impossible to ever clear the way so their data can properly work for and precisely lead them to the real origin of the universe. How can people abandon the dangerous theories they think are correct because they don't know any better ones?

Lucky you, that is where Dr. Nathanael-Israel Israel, the founder of Science180 (Science180.com) came in to properly reanalyze and put under control these costly, underrated data to provide the accurate and simple solution people have been looking for throughout the ages, but that they have ignored.

In *"From Science to Bible's Conclusions"*, you will:

- Get a world class explanation of the 4 fundamental variables without which it is unquestionably impossible to ever decode the universe-formation scientifically
- Save time and money, and enjoy a life filled with the wonderful peace that the accurate understanding of the universe-origin can create
- Discover the errors in the scientific theories and religious belief systems about the universe-formation that are putting you at risk, and learn how to take control over cosmological threats lurking at the edge of your rational mind, faith, disbelief, or doubt
- Unlock the accurate scientific formula to rationally test the existence of God in a historic way that uncompromisingly satisfies both believers and skeptics (Science180.com/public)

- Get all you need to become a knowledgeable person who will never again need anybody else to explain to you the origin of the universe, for, you will fully understand and articulate it yourself and rationally know whether science is really at war with religion
- Receive deep insights that even those who went to university for years were not able to decrypt by themselves, so you can equip yourself to eliminate all forms of scientific and religious universe-origin prejudices
- Discover whether the scientific data finally confirms that the formation of the Earth was completed on the 3rd day, while that of the Moon and the Sun was on the 4th day of creation like the Bible says, or whether the data proves that it took billions of years to progressively form the universe
- Understand the celebrated scientific formula that rationally puts to rest all debates about the relationship between science, faith, and all theories about the universe-origin so you can properly develop yourself, expand your network, and shape your future

Quickly grab and read this scientifically verifiable, bestselling book to finally get the accurate, jaw-dropping answer that has been rationally shaking both believers, skeptics, and all freethinkers. Don't wait!

Dr. Nathanael-Israel Israel has had the honor to be acknowledged as the #1 universe-origin, life-origin, and chemicals-origin expert. He is the author of *"Turbulent Origin of the Universe"*, *"Reconciling Science and Creation Accurately"*, *"Turbulent Origin of Chemical Particles"*, *"Turbulent Origin of Life"*, *"How Baby Universe Was Born"*, *"Science180 Accurate Scientific Proof of God"*. Visit Israel120.com to learn more about this world's most trusted expert that helps scientists and laypeople to properly decode the origin and formation of the universe, life, and chemicals so people can live more effectively nonstop.

CHAPTER 15

ELECTRONEGATIVITY AND ELECTRON AFFINITY OF CHEMICAL ELEMENTS

Electronegativity is used to measure "the tendency of an atom to attract a shared pair of electrons or electron density towards itself" (IUPAC, 2006). This means that a positive relationship exists between the electronegativity number and the ability of an element to "attract" electrons towards itself. Electro positivity is the opposite of electronegativity and is a measure of the ability of an element to donate electrons. Based on the energy between chemical bonds, the electronegativity scale mostly used today is after the pioneering work of the Nobel prize recipient Linus Pauling in 1932. Usually referred to as the Pauling scale, the electronegativity scale is a dimensionless quantity. Unlike most of the variables and parameters related to the chemical elements, and which were calculated, electronegativity values were not based on a mathematical formula or a measurement, but instead were based on a pragmatic value ranging between the 2 extremes: the highest possible electronegativity given to fluorine and the lowest possible electronegativity given to Francium. Pauling did not view electronegativity as a property of an atom alone, but as a property of an atom in a molecule (Pauling, 1960). In this perspective, some of the properties used to characterize a free atom are ionization energy and electron affinity.

For the sake of space, I put the content of this chapter online and you can get it at www.Science180.com/ElectronegativityElectronAffinity

Nathanael-Israel Israel: Member of the American Society of Biochemistry and Molecular Biology

CHAPTER 16

HARDNESS OF CHEMICAL ELEMENTS: A LITTLE VARIABLE YOU CANNOT IGNORE WHILE DEMONSTRATING THE ORIGIN OF CHEMICAL PARTICLES

16.1. Generalities

Hardness of a chemical element measures its resistance to localized plastic deformation caused by mechanical indentation or abrasion. Harness is usually measured in 3 ways: indentation, rebound, and scratch. Scratch hardness assesses how resistant a material is to fracture or to permanent plastic deformation due to friction from a sharp object (Wredenberg and Larsson, 2009). Harder materials can scratch softer ones. Indentation hardness measures the resistance of a sample to material deformation due to a constant compression load from a sharp object. Examples of indentation hardness scales include: Brinell hardness, Rockwell hardness, Shore hardness, and Vickers hardness. Related to elasticity, and also called dynamic hardness, the rebound hardness is defined as a measure of the "*height of the "bounce" of a diamond-tipped hammer dropped from a fixed height onto a material*" (Allen, 2006).

Also called stiffness sometimes, the hardness is related to what is called modulus, which I addressed in a different section. When a force is applied to them, solids generally respond by:

- exhibiting elasticity, which is an ability to temporarily change shape but return to the original shape when the pressure is removed.
- displaying plasticity, an ability to permanently change shape (in response to a force) but remain in one piece. When undergoing plastic deformation, some materials can display elasticity and viscosity at the same time and in that case, it is called viscoelasticity.
- fracturing, which is a split into at least two pieces.

137

While characterizing chemical element groups, some terms like brittle and ductile are used. Now is a good time to clarify them a little bit. Indeed, the strength of a material measures the extent of its elastic range, or elastic and plastic ranges together (Wikipedia, 2020f). Opposed to ductility, brittleness is the tendency of a material to fracture with very little or no detectable plastic deformation beforehand. In other terms, it is the tendency to fracture under a small amount of force. Finally, toughness is a term used to measure the maximum amount of energy a material can absorb before fracturing. Therefore, brittle materials have a small toughness. In the following section, I focused on 3 types of hardness: Mohs hardness, Brinell hardness, and Vickers hardness.

16.2. Mohs hardness

Reported for 55 chemical elements, all of which are naturally occurring elements, the Mohs hardness is a scale varying between 0.2 and 9.3 (Fig. 35). The smallest Mohs hardness was obtained with Cesium (Cs, Z=55), an alkali metal, while the highest value was with Boron (B, Z=5), a metalloid. Most of the highest Mohs hardness were recorded on metalloids and transition metals. Most of the available data on Mohs hardness are for metallic elements except for 3 nonmetals. In other words, of all kinds of nonmetals (noble gases, halogens, and nonmetals), Mohs hardness was reported for only 3, none of which is a noble gas or a halogen:

- Carbon (C, Z=6): 0.5
- Sulfur (S, Z=16): 2
- Selenium (Se, Z=34): 2

Because of the limited number of elements which hardness was measured, I was not able to properly study it.

Fig. 35: Mohs Hardness of the 118 chemical elements known as of 2020

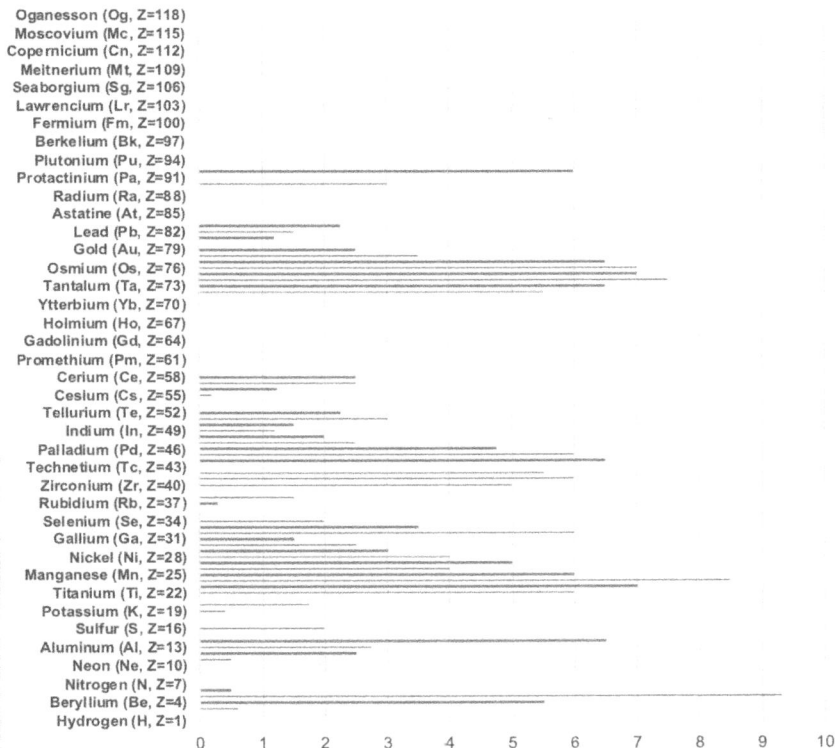

16.3. Brinell hardness

The Brinell hardness is measured for 59 natural chemical elements. No data is reported for the synthetic chemical elements. The Brinell hardness varies between 0.14 MPa and 3920 Mpa (Fig. 36). Cesium (Cs, Z=55), an alkali metal, recorded the smallest Brinell hardness just as it did for the Mohs hardness. In contrast, Osmium (Os, Z=76), a transition metal, recorded the highest value. While Boron (B, Z=5) has the highest Mohs hardness, no data is reported for its Brinell hardness. The second and third highest Brinell hardness were recorded with:

- Tungsten (W, Z=74): 2570 MPa and
- Uranium (U, Z=92): 2400 MPa.

In general, the highest values were recorded with transition metals and actinoids. Of the 3 kinds of elements classified as nonmetallic, Brinell hardness was not reported for any halogen or noble gas, but only for 1 nonmetal, Selenium (Se, Z=34): 736 MPa. Brinell hardness is not reported for any of the key elements forming the building block of livings organisms:

- Hydrogen (H, Z=1)

- Carbon (C, Z=6)
- Nitrogen (N, Z=7)
- Oxygen (O, Z=8)
- Phosphorus (P, Z=15)

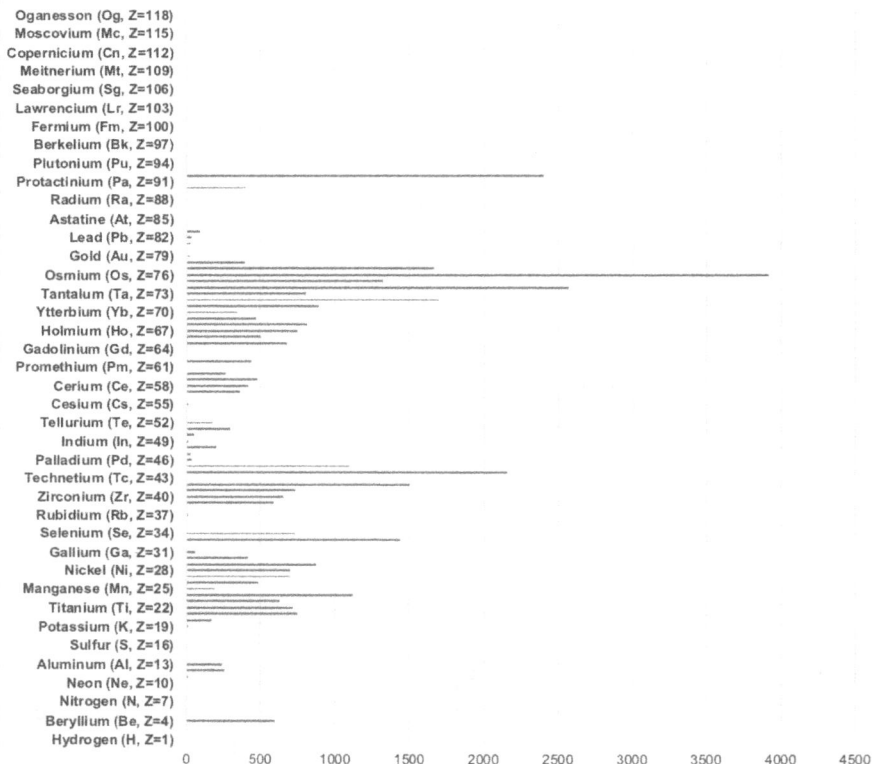

Fig. 36: Brinell Hardness (MPa) of the 118 chemical elements known as of 2020

16.4. Vickers hardness

The Vickers hardness is reported for 39 natural chemical elements, 87.2% of which are either lanthanoids or transition metals. It ranges from 167 MPa to 49000 Mpa (Fig. 37). The smallest values were recorded with Aluminum (Al, Z=13) and Europium (Eu, Z=63). In contrast, the highest Vickers hardness was obtained with Boron (B, Z=5), a metalloid which also recorded the highest Mohs hardness. The Vickers hardness of Boron is at least 14.3 times that of any other chemical element. For instance, Vickers hardness of Iron (Fe, Z=26), that people refer to as a very hard metal), 608 MPa, is 1.24% that of Boron. The Vickers hardness of the next 7 hardest elements after Boron are:

- Tungsten (W, Z=74): 3430 MPa 7% that of Boron

140

- Rhenium (Re, Z=75): 2450 MPa 5% that of Boron
- Uranium (U, Z=92): 1960 MPa 4% that of Boron
- Hafnium (Hf, Z=72): 1760 MPa 3.59% that of Boron
- Iridium (Ir, Z=77): 1760 MPa 3.59% that of Boron
- Beryllium (Be, Z=4): 1670 MPa 3.41% that of Boron
- Molybdenum (Mo, Z=42): 1530 MPa 3.12% that of Boron

Vickers hardness is not reported for any halogen, noble gas, or nonmetal.

Fig. 37: Vickers Hardness (MPa) of the 118 chemical elements known as of 2020

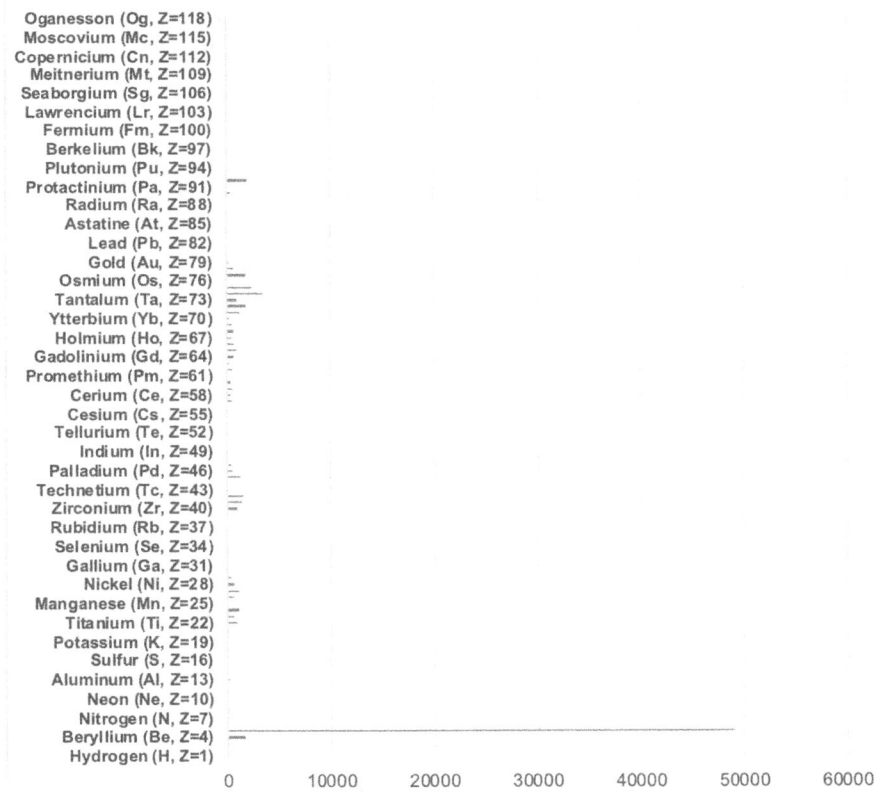

CHAPTER 17

WHAT IS THE MODULUS OF CHEMICAL ELEMENTS ALL ABOUT?

Modulus measures the stiffness of materials. When addressing chemical elements, substances, or materials, 3 kinds of modulus are usually mentioned:

- Bulk modulus: measure the resistance to compression and is also defined by the "*ratio of the infinitesimal pressure increases to the resulting relative decrease of the volume*" (Wikipedia, 2020g).
- Shear modulus: also called modulus of rigidity, describes the response to shear, and defined by the IUPAC as the ratio of shear stress to the shear strain, and
- Young's modulus which describes the response to linear stress.

Moduli are expressed in Pascal, but according to their intensity, they can also be in megapascal (MPa) or gigapascals (GPa).

17.1. Bulk modulus

The bulk modulus is usually considered a thermodynamic quantity and is temperature dependent. It estimates the ability of a material to change its volume under its pressure. It is believed to be directly connected to the interatomic potential and volume per atoms. I studied the bulk modulus for 69 natural chemical elements, and it varies between 1.1 and 380 Mpa (Fig. 38). The smallest bulk modulus was obtained with Chlorine (Cl, Z=17) followed by:

- Cesium (Cs, Z=55), an alkali metal: 1.6 MPa
- Bromine (Br, Z=35), a halogen: 1.9 MPa

Most of the smallest values were recorded on halogens, nonmetals, and alkali metals. In contrast, the highest bulk modulus was recorded with Rhodium (Rh, Z=45), a transition metal, followed by:

- Rhenium (Re, Z=75): 370 MPa

- Boron (B, Z=5), a metalloid: 320 MPa
- Iridium (Ir, Z=77): 320 MPa
- Tungsten (W, Z=74): 310 Mpa

Most of the highest bulk moduli were recorded with transition metals and metalloids. Some of the chemical elements that have the highest hardness also have the highest bulk modulus. The 15th highest bulk modulus (170 MPa) was recorded with Iron (Fe, Z=26) while the 8th highest value (220 MPa) was with Gold (Au, Z=79). While bulk modulus is not reported for any noble gas, it is reported for 3 halogens and 4 nonmetals. These 3 halogens are:

- Iodine (I, Z=53): 7.7 MPa
- Bromine (Br, Z=35): 1.9 MPa
- Chlorine (Cl, Z=17): 1.1 MPa

The 4 nonmetals are:
- Carbon (C, Z=6): 33 MPa
- Phosphorus (P, Z=15): 11 MPa
- Selenium (Se, Z=34): 8.3 MPa
- Sulfur (S, Z=16): 7.7 MPa

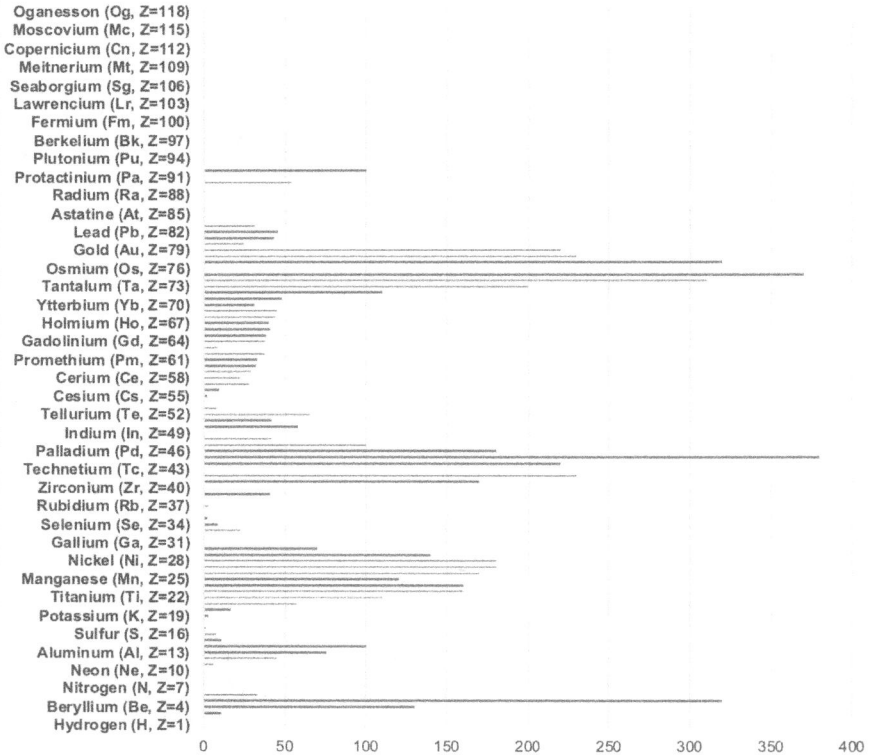

Fig. 38: Bulk Modulus (GPa) of the 118 chemical elements known as of 2020

17.2. Shear modulus

Investigated for 60 natural chemical elements and no synthetic elements, the shear modulus varies between 1.3 GPa and 222 GPa (Fig. 39). Potassium (K, Z=19) has the smallest shear modulus while Osmium (Os, Z=76) recorded the highest value. The shear modulus of Iron (Fe, Z=26) is 82 GPa. Besides Osmium, the other chemical elements which shear modulus is higher than Iron are:

- Iridium (Ir, Z=77): 210 GPa
- Rhenium (Re, Z=75): 178 GPa
- Ruthenium (Ru, Z=44): 173 GPa
- Tungsten (W, Z=74): 161 GPa
- Rhodium (Rh, Z=45): 150 GPa
- Beryllium (Be, Z=4): 132 GPa
- Chromium (Cr, Z=24): 115 GPa
- Uranium (U, Z=92): 111 GPa

Some of these elements also recorded some of the highest hardness (e.g.

Rhenium (Re, Z=75), Tungsten (W, Z=74), and Uranium (U, Z=92)). Most of the highest shear moduli were obtained with transition metals. The only nonmetal which shear modulus was reported is Selenium (Se, Z=34): 3.7. No data exists for halogens and noble gases.

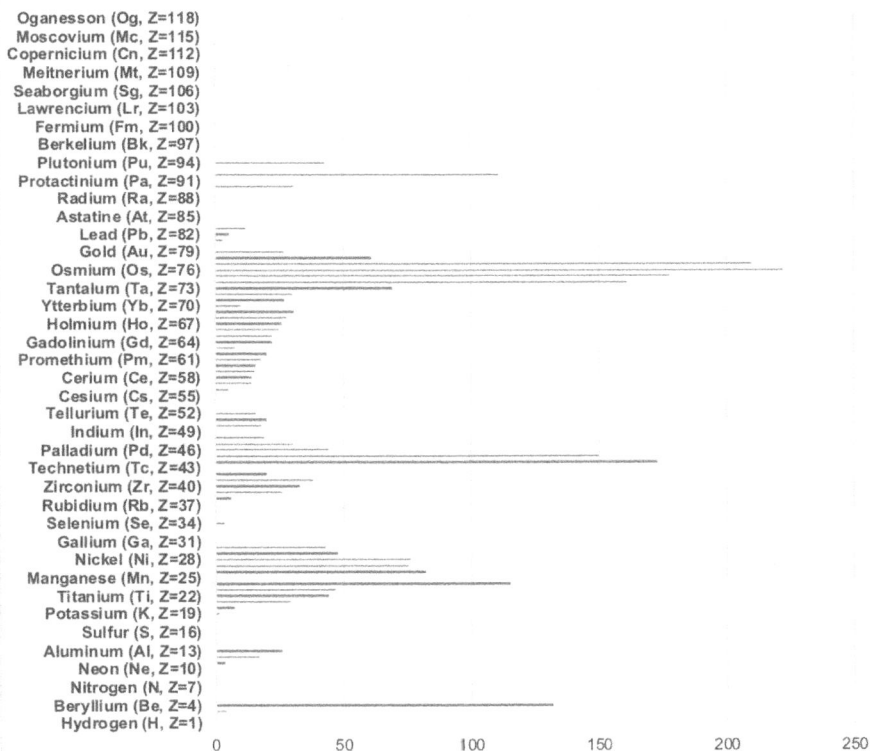

Fig. 39: Shear Modulus (GPa) of the 118 chemical elements known as of 2020

17.3. Young's modulus

I studied the Young's modulus of 63 natural chemical elements. It varies between 1.7 GPa and 528 GPa (Fig. 40). Cesium (Cs, Z=55) has the smallest Young's modulus, whereas Iridium (Ir, Z=77) has the highest value. Besides the value recorded on Uranium (U, Z=92), an actinoid, and Beryllium (Be, Z=4), an alkaline earth metal, all of the other 19 highest Young's moduli higher than 100 were recorded with transition metals. Young's modulus is not reported for any halogen or noble gas. The only report for nonmetals, is for Selenium (Se, Z=34): 10 GPa. Based on the data that I studied, only 8 chemical elements have a Young's modulus higher than that of Iron (Fe, Z=26): 211 GPa, and besides Iridium (Ir, Z=77), which has the highest value, these elements are:

- Rhenium (Re, Z=75): 463 GPa

- Ruthenium (Ru, Z=44): 447 GPa
- Tungsten (W, Z=74): 411 GPa
- Molybdenum (Mo, Z=42): 329 GPa
- Beryllium (Be, Z=4): 287 GPa
- Chromium (Cr, Z=24): 279 GPa
- Rhodium (Rh, Z=45): 275 GPa

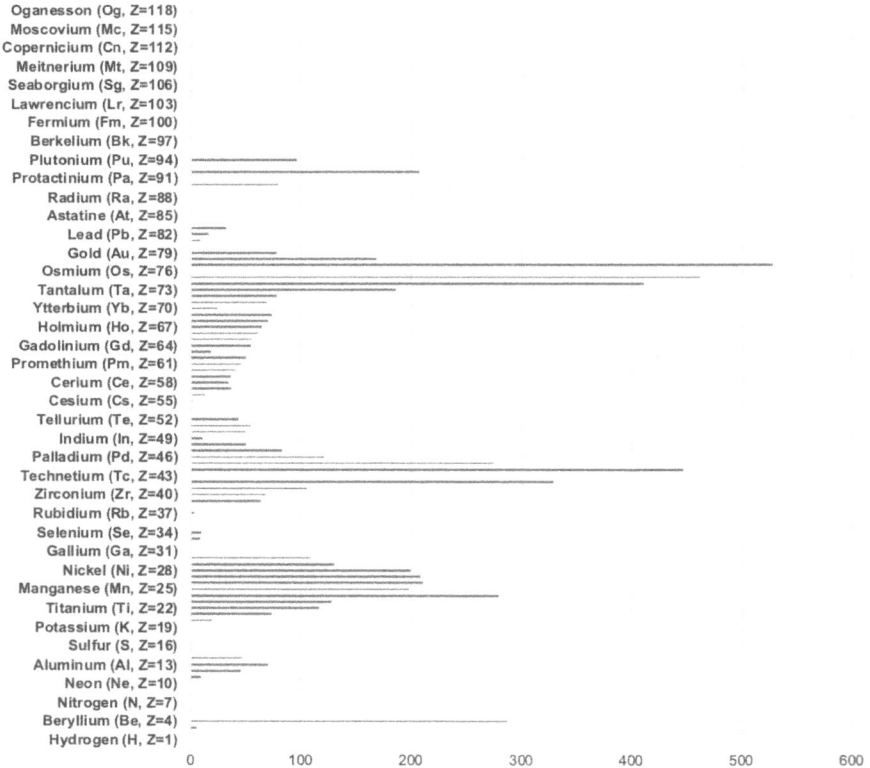

Fig. 40: Young's Modulus (GPa) of the 118 chemical elements known as of 2020

17.4. Relationship between the types of moduli

A positive relationship exists between the types of moduli. The highest relationship was found between Shear modulus and Young's modulus (Fig. 41).

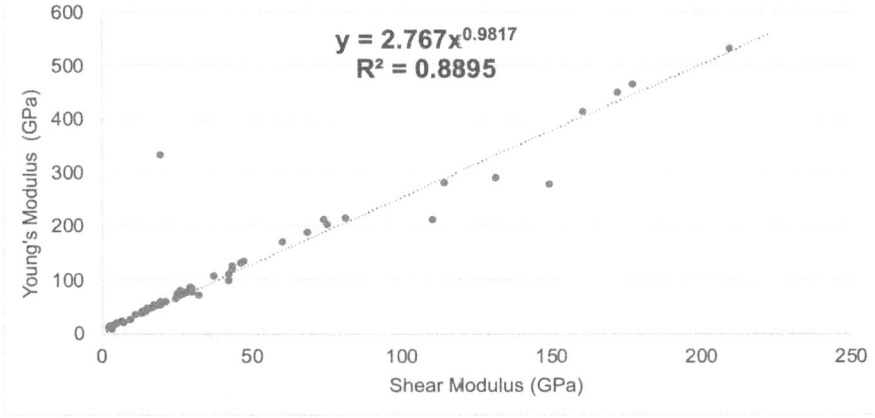

Fig. 41: Relationship between Shear Modulus and Young's Modulus

$y = 2.767x^{0.9817}$
$R^2 = 0.8895$

CHAPTER 18

THE UNEXPLAINABLE TRUTH, JOY, AND ACCURACY IN DISCOVERING THE PRICELESS SECRETS CONCEALED IN THE IONIZATION ENERGY OF CHEMICAL ELEMENTS

18.1. General trends of the ionization energy

The ionization energy is the minimum amount of energy needed to remove the most loosely bound electron, also called the valence electron, of an atom or molecule. In other words, the ionization energy is the energy required to make a free atom or molecule to lose an electron. It measures the strength of electron bonds to an atom or molecule. To put it another way, it shows how difficult it is to remove electrons from an atom. Expressed in kJ/mol or kcal/mol, the molar ionization energy or enthalpy is the amount of energy needed for the atoms in a mole of a chemical to lose one electron each.

Because most atoms have many electrons, there are many levels of ionization energy according to the number of electrons removed. For instance, the first ionization energy is the amount of energy required to remove an electron from a neutral atom. The second, third, etc., ionization energy is the energy required to remove the second, third, and so on and so forth, electron from a singly, doubly, etc. charged ion. As a general rule, the nth ionization energy is the amount of energy necessary to remove an electron from an atom or molecule having a charge of $(n-1)$. I studied 30 ionization energy, meaning from the 1st ionization energy until the 30th ionization energy, of the chemical elements (Hoffman et al., 2006; Cotton, 2006; Fricke, 1975) as reporter by Wikipedia (2021b).

The ionization energy of the chemical elements varies between 375.7 kJ/mol and 1,116,105 kJ/mol. The smallest value was the 1st ionization energy of Cesium (Cs, Z=55), an alkali metal. In contrast, the highest ionization energy was the 29th ionization energy of Copper (Cu, Z=29), a transition metal. In general, most of the smallest values were recorded with the 1st ionization, while most of the highest values were with the 29th ionization energy. The two 30th ionization energies reported are much smaller than the maximum of the ionization energy ranging from the 16th to the 29th ionization energy. Fig. 42 and Fig. 43 illustrate the mean ionization energy and the maximum ionization energy of the chemical elements respectively.

The number of chemical elements which nth ionization energy was determined decreased as the number of electrons removed increased. For instance, while the 1st ionization energy is reported for 112 chemical elements, the 30th ionization energy is reported for just 2 chemical elements. The lack of complete data on the ionization energies of all of the chemical elements may be due either to the difficulty in measuring them or to some complications or complexities related to such measurements beyond the 30th electrons.

Fig. 42: Mean ionization energy (kJ/mol) of the chemical elements

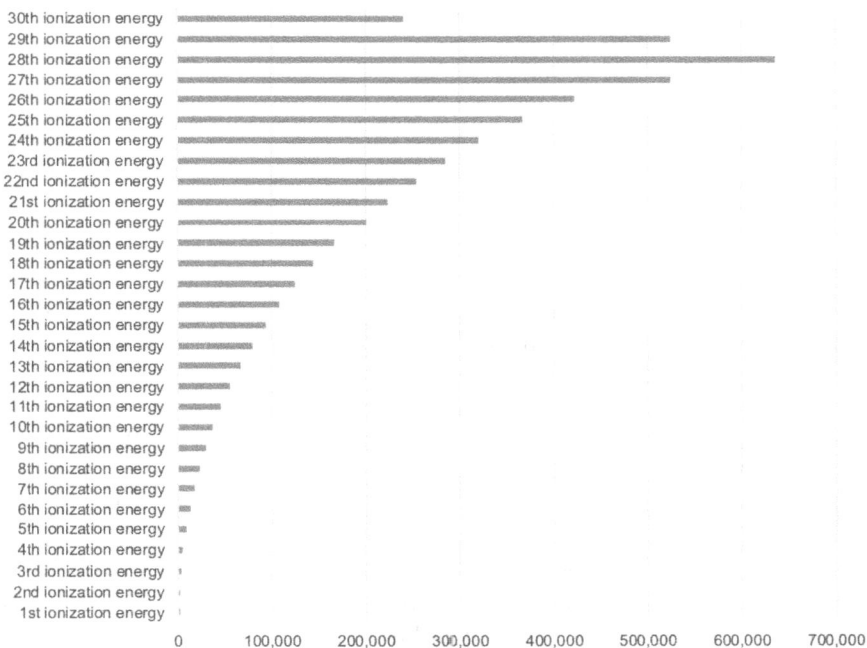

Fig. 43: Maximum ionization energy (kJ/mol) of the chemical elements

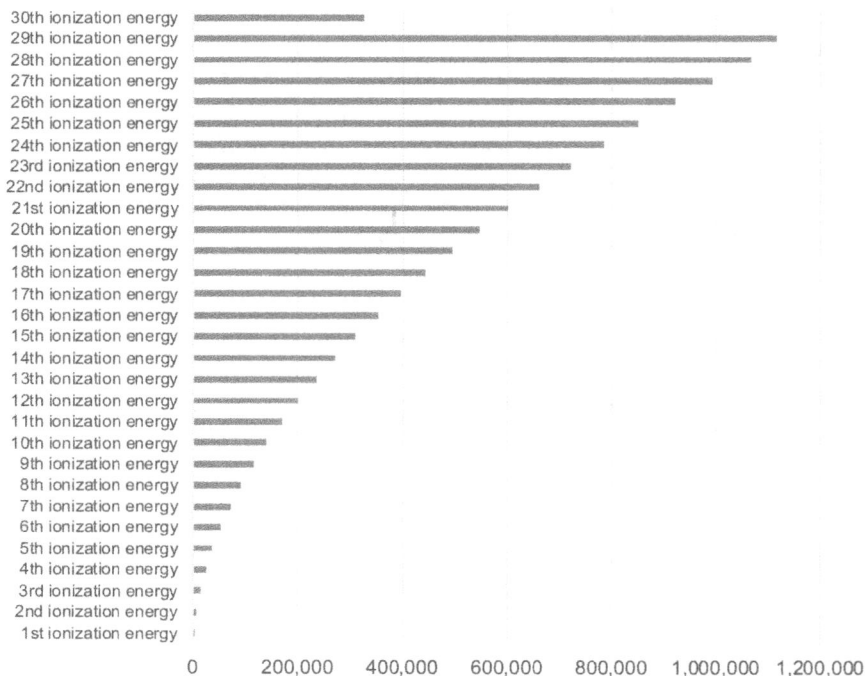

For each chemical element, the ionization energy of the outermost electrons is smaller than that of the innermost electrons. By innermost electrons, I mean the electrons that are closer to the nucleus. Because the mass of the electron is considered as a constant for all electrons, an increase in ionization energy can also be interpreted as an increase in the speed or energy of those electrons. For in general, energy is associated with speed and just as planets are orbiting their primary stars, and just as satellites are orbiting their primary planets, so electrons are also "orbiting" the nucleus of the atoms they belong to. Therefore, the increase in the ionization energy from the outermost electron to the innermost electrons can be a consequence of the increase of the orbital speed of the electrons (around the nucleus of the atoms they belong to) as they are located deep and deeper inside the atoms. On the other hand, considering the trends I obtained for the speed of the systems of celestial bodies I studied, I personally think that the mass and speed of electrons are not constant. Speed and mass of electrons may have been set as constant because of ignorance, or a lack of precision of scientific equipment, or the limits imposed on the details scientists can see. In other words, I expect that someday, scientific measurements will confirm that the speed, size, and mass of electrons are not always the same not even within the same chemical element. Based on trends I found with other systems of bodies, I think some outermost and innermost electrons would be smaller than some between them. I also think that the innermost electrons would be faster than the outermost electrons. The trends of the dimension and speed of the electrons could result from how the precursors of electrons in each atom were split into the various electrons. To some extent, the physics that explains the formation of celestial bodies (e.g. clusters of galaxies, stellar systems, planetary systems, and satellite systems) can also explain the formation of the atoms including their nucleus and electrons.

The variation of the ionization energy according to the number of electrons of the chemicals and according to the level of ionization suggested to me that the size and energy of the electrons are not the same. Just as in the planetary systems, most of the energy of the satellites is concentrated in Zone 3 (where the largest secondary bodies are found), it is possible that most of the energy of the electrons may be concentrated not at the outer region but somewhere toward the middle. If the ionization energy of all electrons around atoms can be known for all atoms, it may be possible, using also other variables, to delimitate electrons into turbulence zones, just as I did for satellites. Here, because the mass of the electrons is considered the same, which I think is a mistake, the turbulence zones of the electrons around an atomic nucleus can be divided according to their ionization energy levels. The zone where the energy may be concentrated can give a better insight into the density and distribution of electrons around the nucleus. The lessons I learned by deeply studying celestial bodies taught me that the organization of electrons around the nucleus and the variation of their characteristics contributed to conferring diverse characteristics to chemical elements and the compounds made with them.

In general, the ionization energy increases from the outermost electron to the innermost electron just as the orbital speed of the satellites increases from the outermost satellite to the innermost ones or just as the speed of planets and asteroids increases from the outermost to the innermost ones. In nature, younger offspring or those born last are usually (but not always) smaller than the older ones. The energy and resources of children depend on their position of birth. Those who are farther from their parents or born when their parents were less vigorous inherit less energy. Outermost bodies are like those who are farther from the position of their parents while the innermost bodies are closer to the position of their parents. The movement of living and non-living things differs just by some few factors related to their ability to cause changes to themselves on top of the changes that their nature can confer to them.

To sum it up, ionization energy increases as the level of ionization increases because the energy and/or orbital speed of the electron increases as their distance from the nucleus decreases. Even if the mass of the electron is assumed the same, the kinetic energy of the electron should increase as they are located toward the innermost part of the atoms. Because the energy of the electron increases as their distance from the nucleus decreases, it requires also more energy to pull electrons as the level of ionization increases. I was not interested in studying the 30 ionization energies of the chemical elements, but to explore how they related among the chemical elements. I learned that as the mass of atoms increases, the ionization energy of their outermost electrons become increasingly similar. This implies that the outermost electrons of the larger atoms are bound to their atoms by an amount of energy in the same order of magnitude. Below, I illustrated the ionization energy of the chemical elements one by one (Fig. 44 to Fig. 62).

18.2. Ionization energy according to the chemical elements

Below, I will show trend following trends:

- From Hydrogen (H, Z=1) to Argon (Ar, Z=18) (Fig. 44)
- From Nitrogen (N, Z=7) to Magnesium (Mg, Z=12) (Fig. 45)
- From Aluminum (Al, Z=13) to Argon (Ar, Z=18) (Fig. 46)

Because of the number of their electrons, elements ranging from Hydrogen (H, Z=1) to Carbon (C, Z=6) have no more than 6th ionization level.

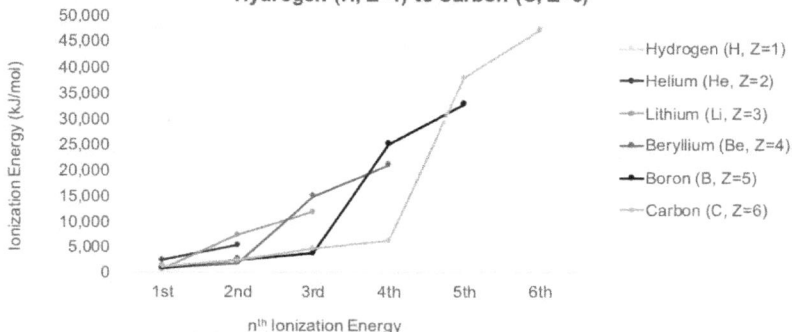

Fig. 44: 1st to 6th ionization energy (kJ/mol) of elements ranging from Hydrogen (H, Z=1) to Carbon (C, Z=6)

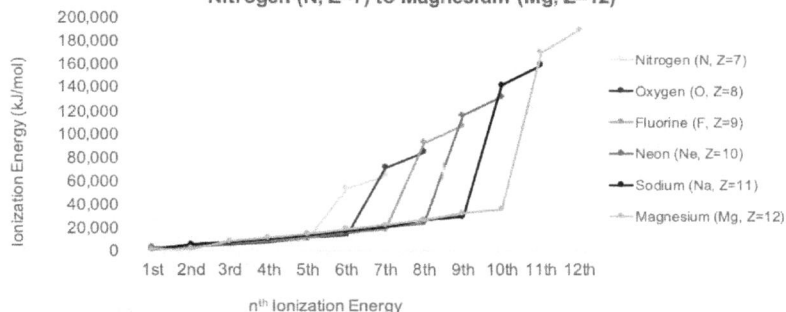

Fig. 45: 1st to 12th ionization energy (kJ/mol) of elements ranging from Nitrogen (N, Z=7) to Magnesium (Mg, Z=12)

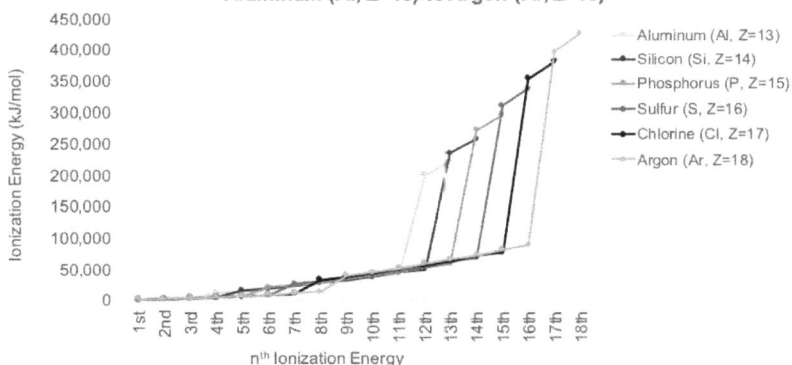

Fig. 46: 1st to 18th ionization energy (kJ/mol) of elements ranging from Aluminum (Al, Z=13) to Argon (Ar, Z=18)

The next trends that I will show are:

- From Potassium (K, Z=19) to Chromium (Cr, Z=24) (Fig. 47)
- From Manganese (Mn, Z=25) to Zinc (Zn, Z=30) (Fig. 48)

153

- From Gallium (Ga, Z=31) to Krypton (Kr, Z=36) (Fig. 49)

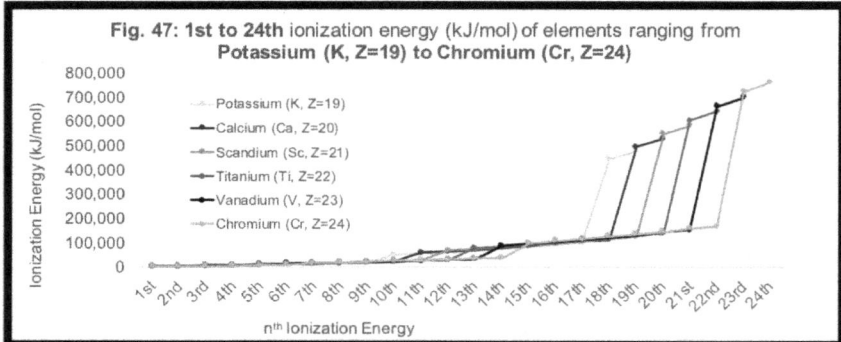

Fig. 47: 1st to 24th ionization energy (kJ/mol) of elements ranging from Potassium (K, Z=19) to Chromium (Cr, Z=24)

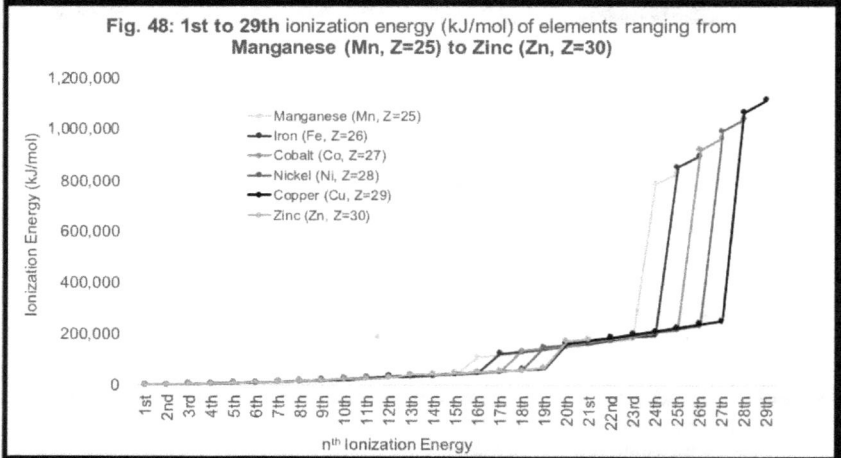

Fig. 48: 1st to 29th ionization energy (kJ/mol) of elements ranging from Manganese (Mn, Z=25) to Zinc (Zn, Z=30)

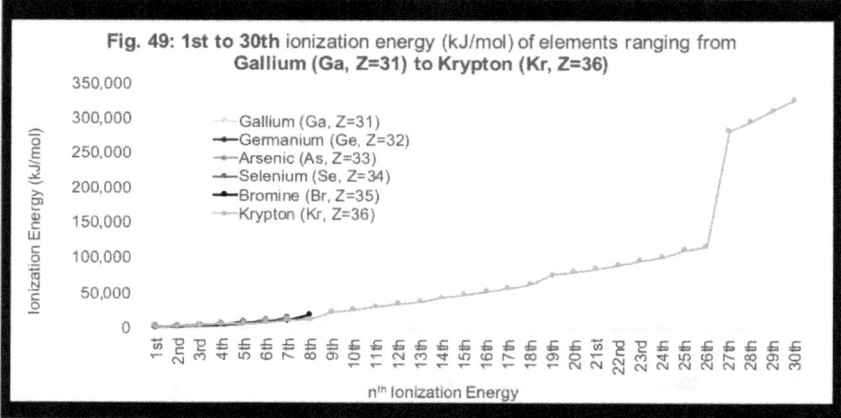

Fig. 49: 1st to 30th ionization energy (kJ/mol) of elements ranging from Gallium (Ga, Z=31) to Krypton (Kr, Z=36)

The next trends are:

- From Rubidium (Rb, Z=37) to Molybdenum (Mo, Z=42) (Fig. 50)
- From Technetium (Tc, Z=43) to Cadmium (Cd, Z=48) (Fig. 51)
- From Indium (In, Z=49) to Xenon (Xe, Z=54) (Fig. 52)

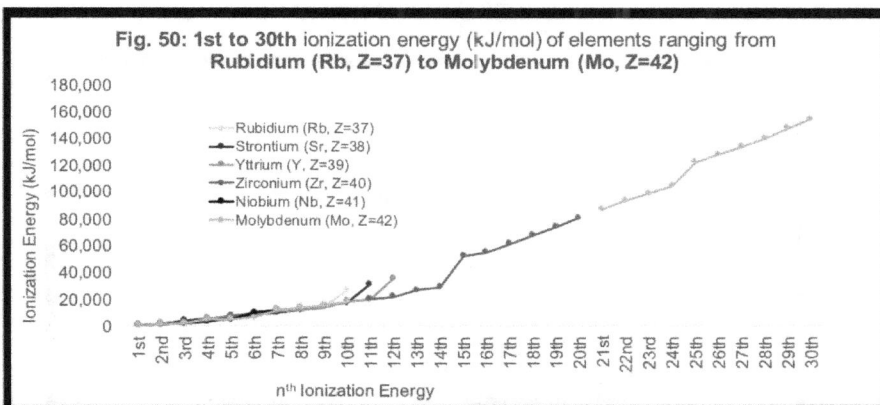

Fig. 50: 1st to 30th ionization energy (kJ/mol) of elements ranging from Rubidium (Rb, Z=37) to Molybdenum (Mo, Z=42)

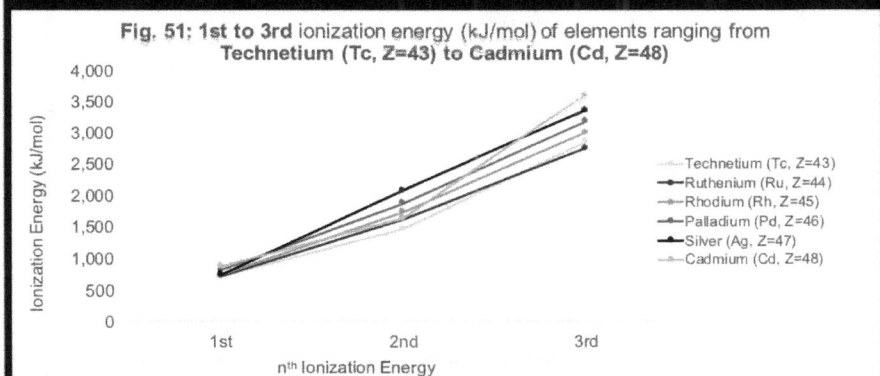

Fig. 51: 1st to 3rd ionization energy (kJ/mol) of elements ranging from Technetium (Tc, Z=43) to Cadmium (Cd, Z=48)

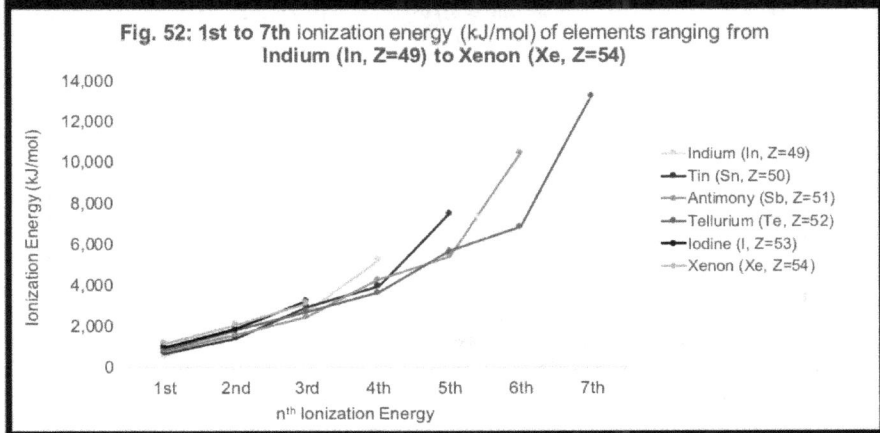

Fig. 52: 1st to 7th ionization energy (kJ/mol) of elements ranging from Indium (In, Z=49) to Xenon (Xe, Z=54)

Science180: Nonconformist, Rule-Breaker, and Accurate Demonstrator of Chemicals-Origin

Following are the trends:

- From Cesium (Cs, Z=55) to Neodymium (Nd, Z=60) (Fig. 53)
- From Promethium (Pm, Z=61) to Dysprosium (Dy, Z=66) (Fig. 54)
- From Holmium (Ho, Z=67) to Hafnium (Hf, Z=72) (Fig. 55)

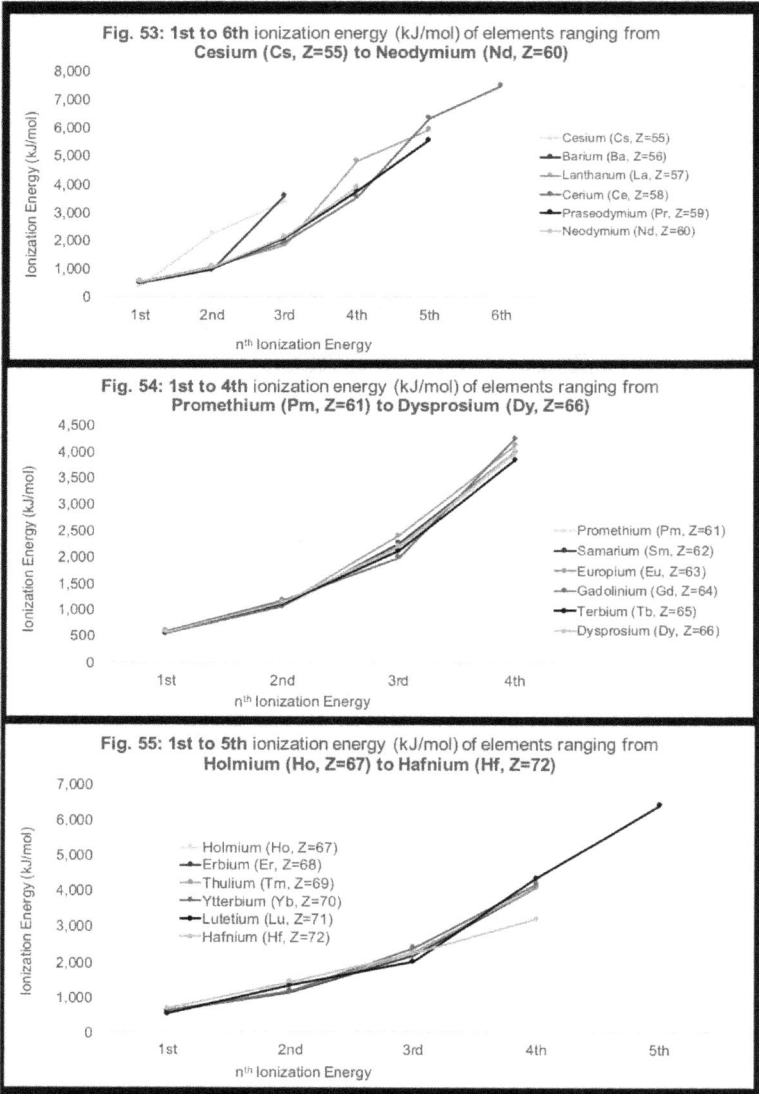

Fig. 53: 1st to 6th ionization energy (kJ/mol) of elements ranging from Cesium (Cs, Z=55) to Neodymium (Nd, Z=60)

Fig. 54: 1st to 4th ionization energy (kJ/mol) of elements ranging from Promethium (Pm, Z=61) to Dysprosium (Dy, Z=66)

Fig. 55: 1st to 5th ionization energy (kJ/mol) of elements ranging from Holmium (Ho, Z=67) to Hafnium (Hf, Z=72)

Next, I will present the trends:

- From Tantalum (Ta, Z=73) to Platinum (Pt, Z=78) (Fig. 56)
- From Gold (Au, Z=79) to Polonium (Po, Z=84) (Fig. 57)
- From Astatine (At, Z=85) to Thorium (Th, Z=90) (Fig. 58)

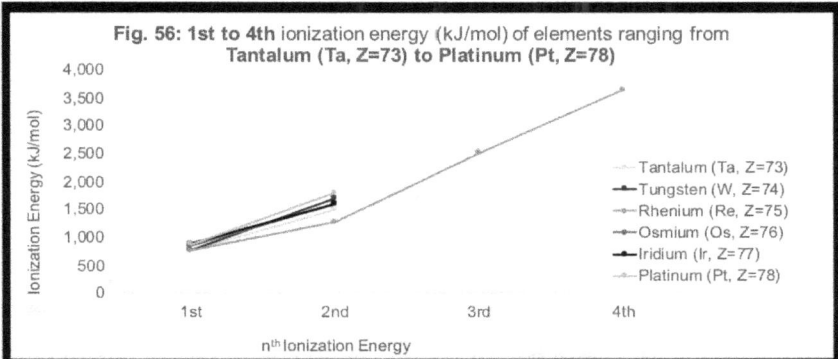

Fig. 56: 1st to 4th ionization energy (kJ/mol) of elements ranging from Tantalum (Ta, Z=73) to Platinum (Pt, Z=78)

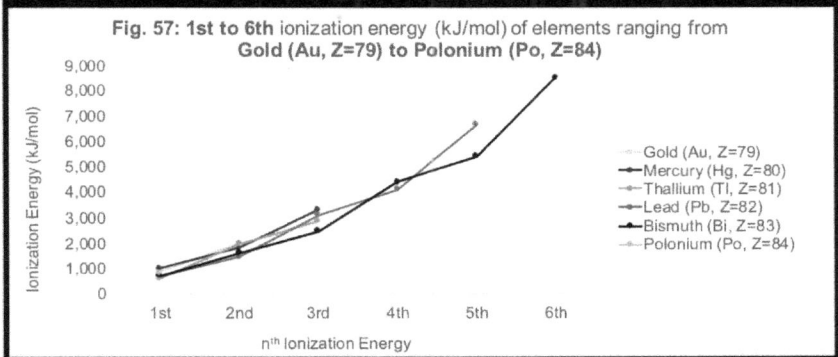

Fig. 57: 1st to 6th ionization energy (kJ/mol) of elements ranging from Gold (Au, Z=79) to Polonium (Po, Z=84)

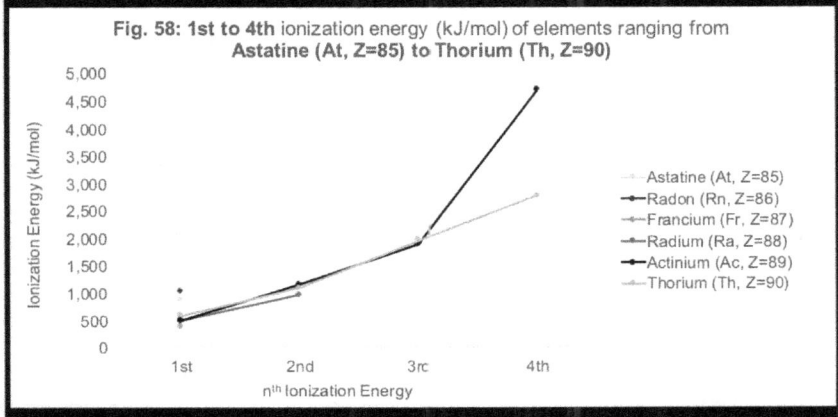

Fig. 58: 1st to 4th ionization energy (kJ/mol) of elements ranging from Astatine (At, Z=85) to Thorium (Th, Z=90)

The next trends are:

- From Protactinium (Pa, Z=91) to Curium (Cm, Z=96) (Fig. 59)

157

- From Berkelium (Bk, Z=97) to Nobelium (No, Z=102) (Fig. 60)
- From Lawrencium (Lr, Z=103) to Hassium (Hs, Z=108) (Fig. 61)

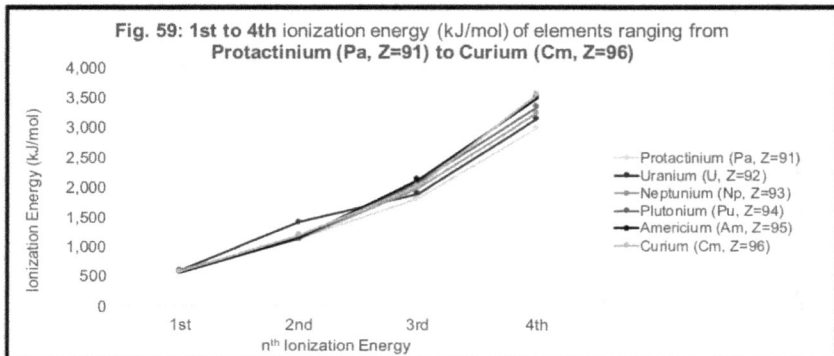

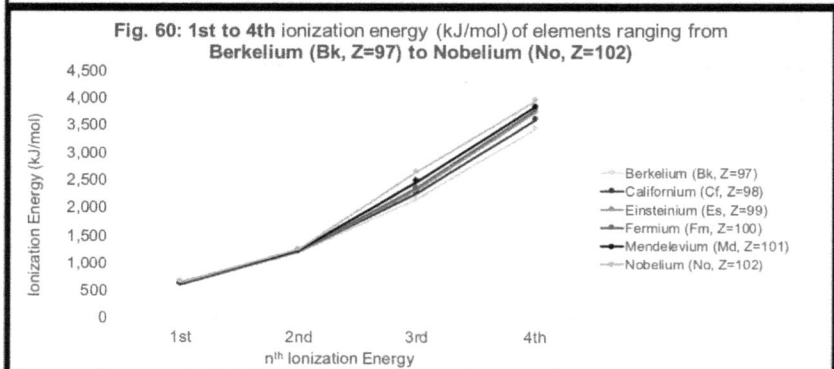

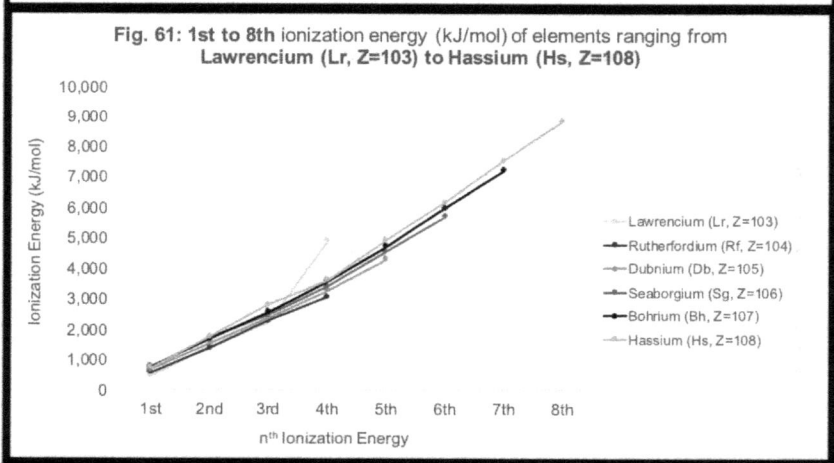

Fig. 59: 1st to 4th ionization energy (kJ/mol) of elements ranging from Protactinium (Pa, Z=91) to Curium (Cm, Z=96)

Fig. 60: 1st to 4th ionization energy (kJ/mol) of elements ranging from Berkelium (Bk, Z=97) to Nobelium (No, Z=102)

Fig. 61: 1st to 8th ionization energy (kJ/mol) of elements ranging from Lawrencium (Lr, Z=103) to Hassium (Hs, Z=108)

The last trend (Fig. 62) is From Meitnerium (Mt, Z=109) to Copernicium (Cn, Z=112):

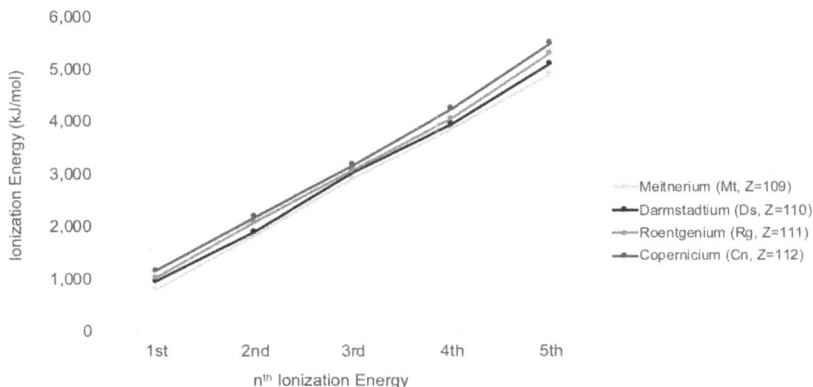

Fig. 62: 1st to 5th ionization energy (kJ/mol) of elements ranging from Meitnerium (Mt, Z=109) to Copernicium (Cn, Z=112)

18.3. First to 10th ionization energy

The 1st ionization energy of the chemical elements varies between 375.7 kJ/mol and 2,372.3 kJ/mol (Fig. 63). The smallest value was obtained with Cesium (Cs, Z=55), whereas the highest value was with Helium (He, Z=2), a noble gas. The second highest value (2080.7 kJ/mol) was obtained by Neon (Ne, Z=10) a noble gas. The 1st ionization energy of hydrogen is the 8th smallest (1312 kJ/mol). Most of the smaller 1st ionization energies are recorded with heavier elements, suggesting that, as atoms get bigger, the energy linking their outermost electrons to the nucleus generally decreases. In general, the 1st ionization energy of noble gases, nonmetals, and halogens are the highest and they decrease as their atomic number or mass increases. In contrast, the coefficient of variation of the 1st ionization energy of actinoids, lanthanoids, and metalloids is small, being less than 9%. The highest CV was observed with noble gases.

Fig. 63: 1st Ionisation Energy (kJ/mol) of the 118 chemical elements known as of 2020

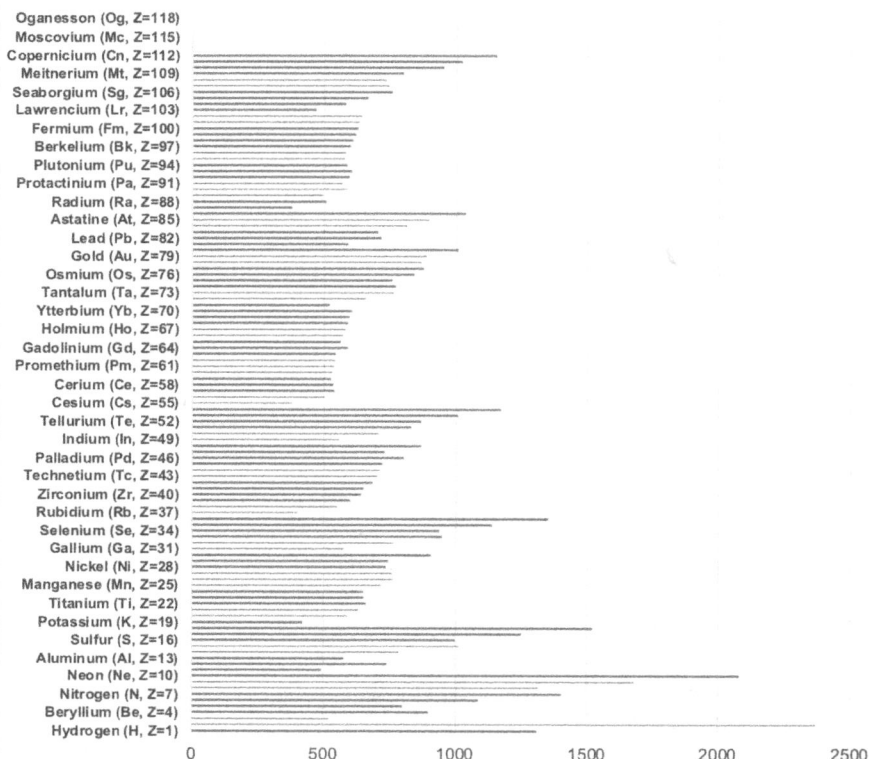

The 2nd ionization energy of the elements varies between 965.2 kJ/mol and 7,298.1 kJ/mol (Fig. 64). The smallest value was found with Barium (Ba, Z=56) while the highest value was seen with Lithium (Li, Z=3) followed by that (5250.5 kJ/mol) of Helium (He, Z=2). The CV of the 2nd ionization energy of actinoids and lanthanoids is less than 8%, meaning that the 2nd ionization energy of actinoids and lanthanoids varies less. In general, the highest 2nd ionization energies were recorded with the lightest chemical elements. For halogens and noble gases, a negative correlation exists between the mass of the elements and their 2nd ionization energy. Alkali metals and alkaline Earth metals also have similar negative tendencies.

Fig. 64: 2nd Ionisation Energy (kJ/mol) of the 118
chemical elements known as of 2020

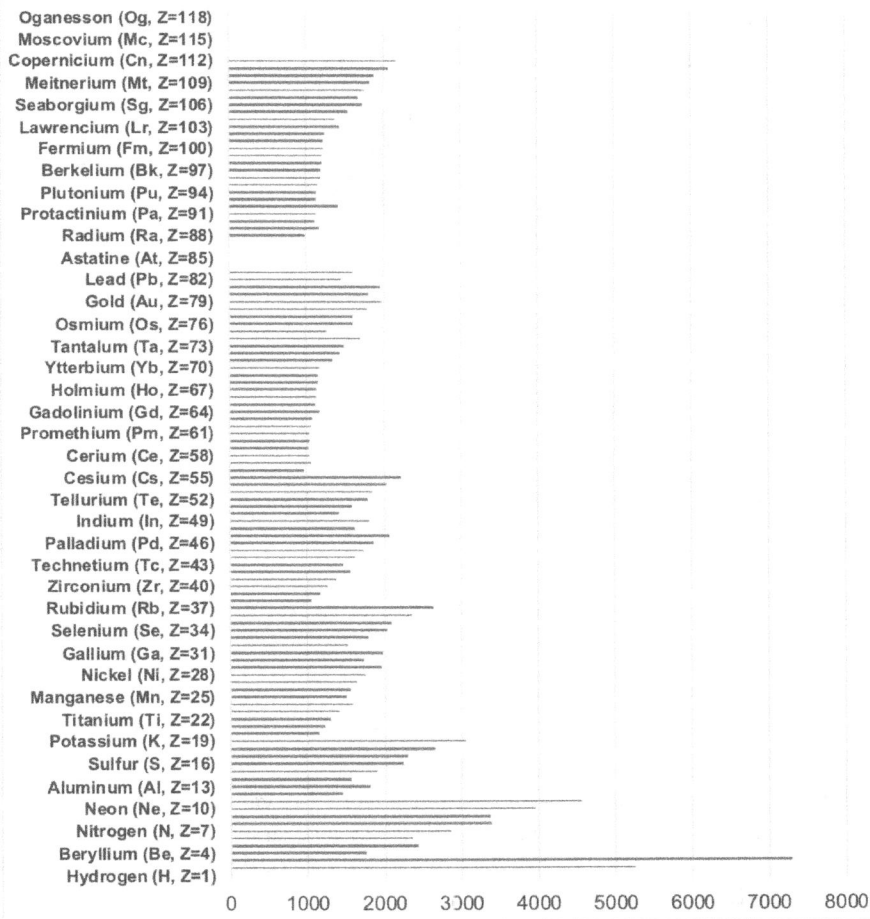

Protactinium (Pa, Z=91) recorded the smallest 3rd ionization energy (1814 kJ/mol) while Beryllium (Be, Z=4) recorded the highest value (14,848.7 kJ/mol) followed by that (11,815 kJ/mol) of Lithium (Li, Z=3) (Fig. 65). Just like the 1st and 2nd ionization energy, most of the highest 3rd ionization energies were generally recorded on lighter elements while the smallest values were found with heavier elements.

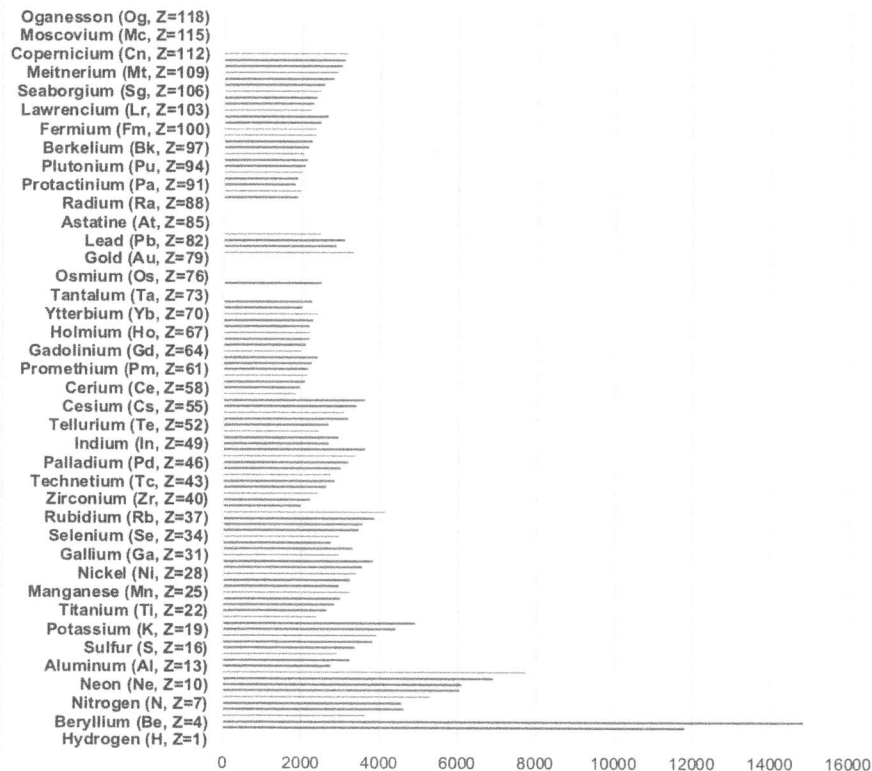

Fig. 65: 3rd Ionisation Energy (kJ/mol) of the 118 chemical elements known as of 2020

The 4th ionization energy of the chemical elements varies between 2780 kJ/mol and 25,025.8 kJ/mol (Fig. 66). Thorium (Th, Z=90) has the smallest value whereas Boron (B, Z=5) holds the highest value. The second highest value (21,006.6 kJ/mol) was found with Beryllium (Be, Z=4).

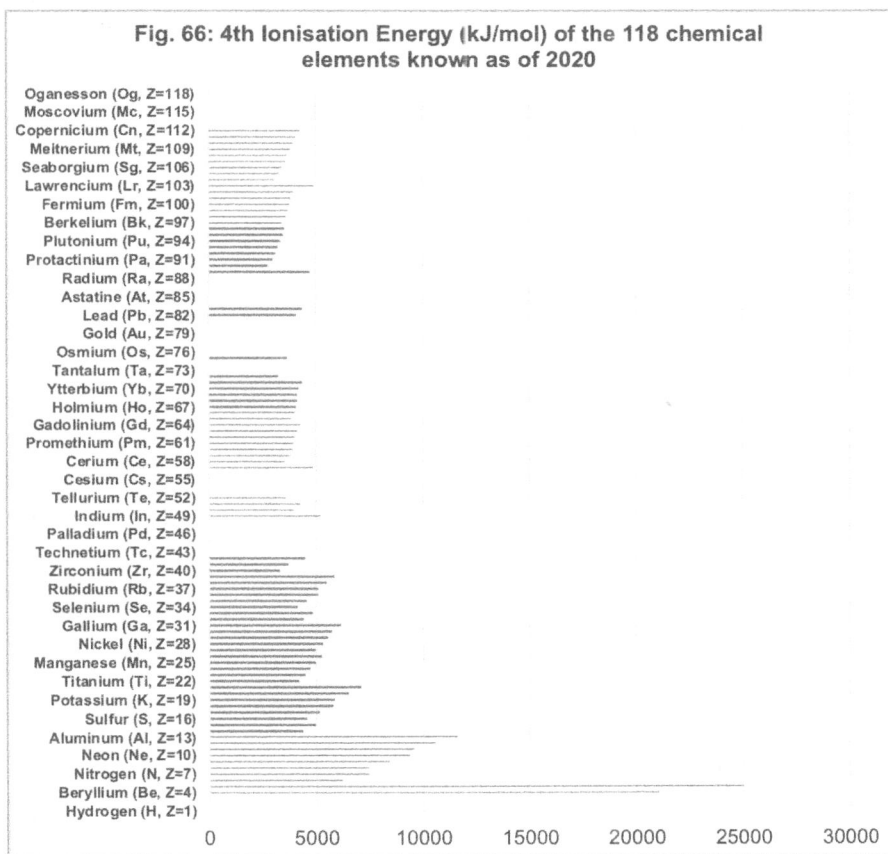

Fig. 66: 4th Ionisation Energy (kJ/mol) of the 118 chemical elements known as of 2020

Varying between 4305.2 kJ/mol and 37.831 kJ/mol, the 5th ionization energy (Fig. 67) is generally the highest with the lightest elements, while smaller for the heaviest ones. The smallest value was obtained with Dubnium (Db, Z=105), while the highest value was with Carbon (C, Z=6).

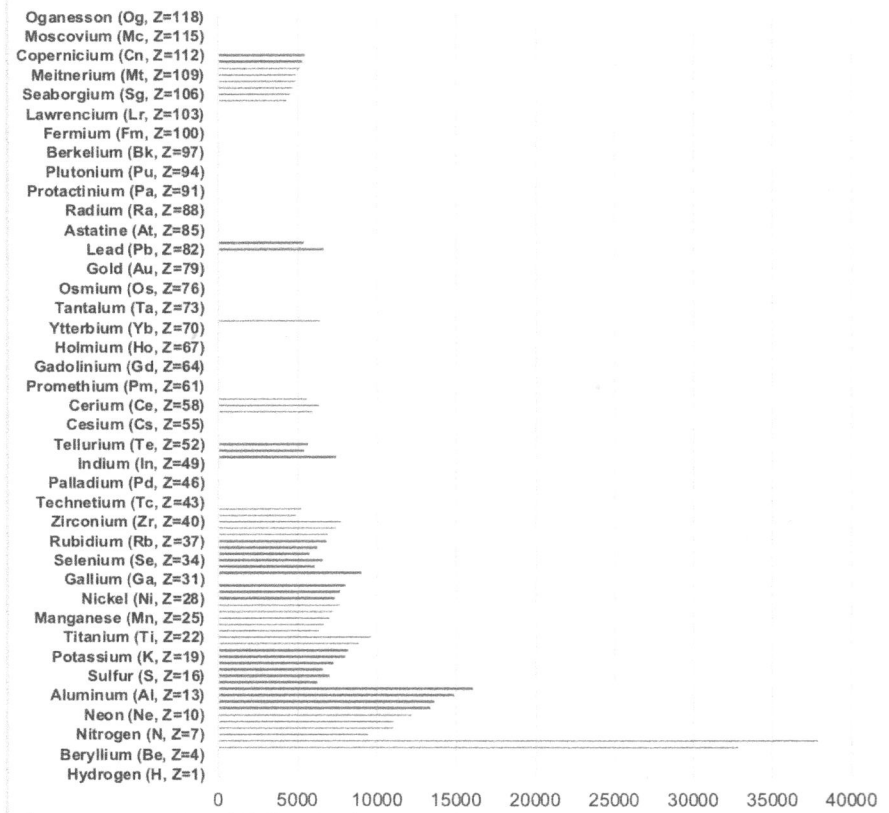

Fig. 67: 5th Ionisation Energy (kJ/mol) of the 118 chemical elements known as of 2020

The 6th ionization energy ranges between 5,715.8 kJ/mol and 53,266.6 kJ/mol (Fig. 68). Seaborgium (Sg, Z=106) has the smallest value, while Nitrogen (N, Z=7) has the highest value. Carbon (C, Z=6) recorded the second highest value (47,277 kJ/mol).

Nathanael-Israel Israel: Creator of Science180 Academy
(www.Science180Academy.com)

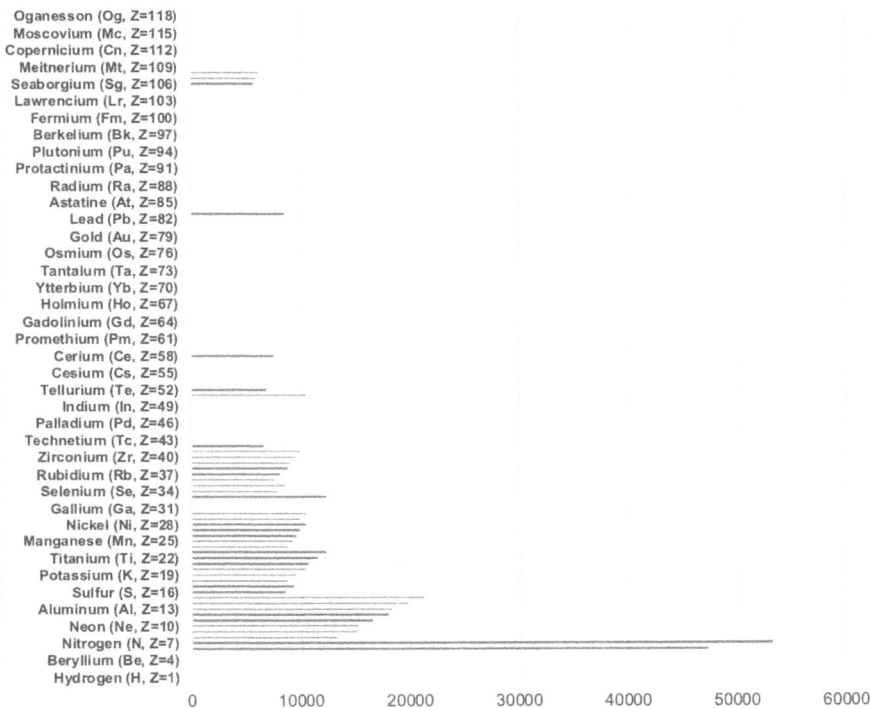

Fig. 68: 6th Ionisation Energy (kJ/mol) of the 118 chemical elements known as of 2020

The 7th ionization energy varies between 7,226.8 kJ/mol and 71,330 kJ/mol (Fig. 69). The smallest value was with Bohrium (Bh, Z=107), whereas the highest one was with Oxygen (O, Z=8), followed by that (64,360 kJ/mol) of Nitrogen (N, Z=7).

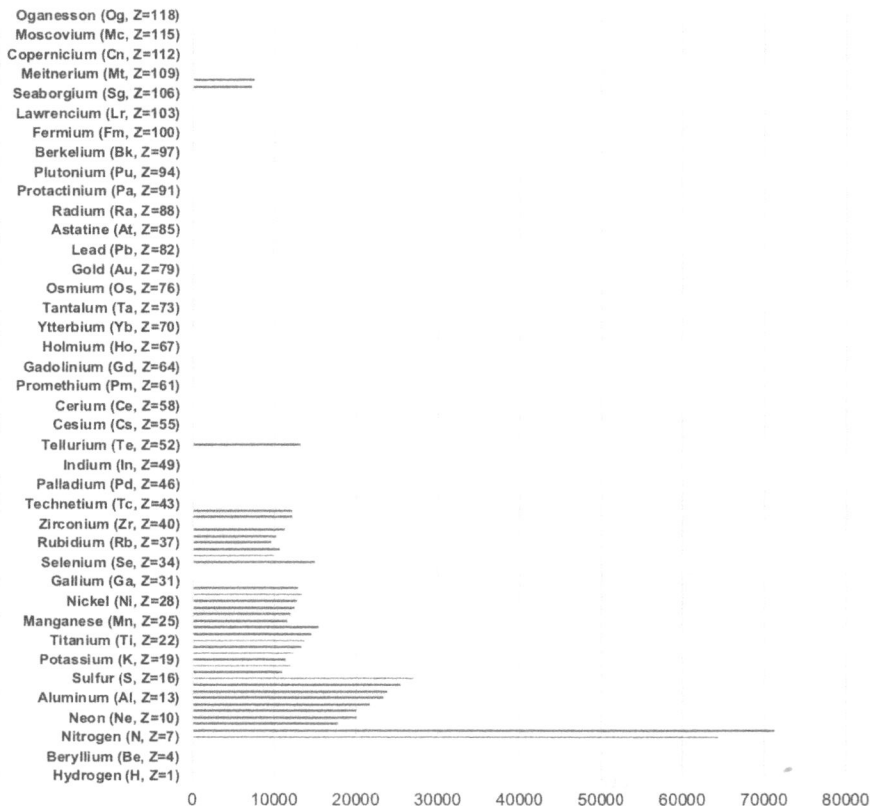

Fig. 69: 7th Ionisation Energy (kJ/mol) of the 118 chemical elements known as of 2020

The smallest 8th ionization energy (8,857.4 kJ/mol) was recorded with Hassium (Hs, Z=108) while the highest value (92,038.1 kJ/mol) was with Fluorine (F, Z=9) (Fig. 70). The second highest value (84,078 kJ/mol) was found with Oxygen (O, Z=8).

Fig. 70: 8th Ionisation Energy (kJ/mol) of the 118 chemical
elements known as of 2020

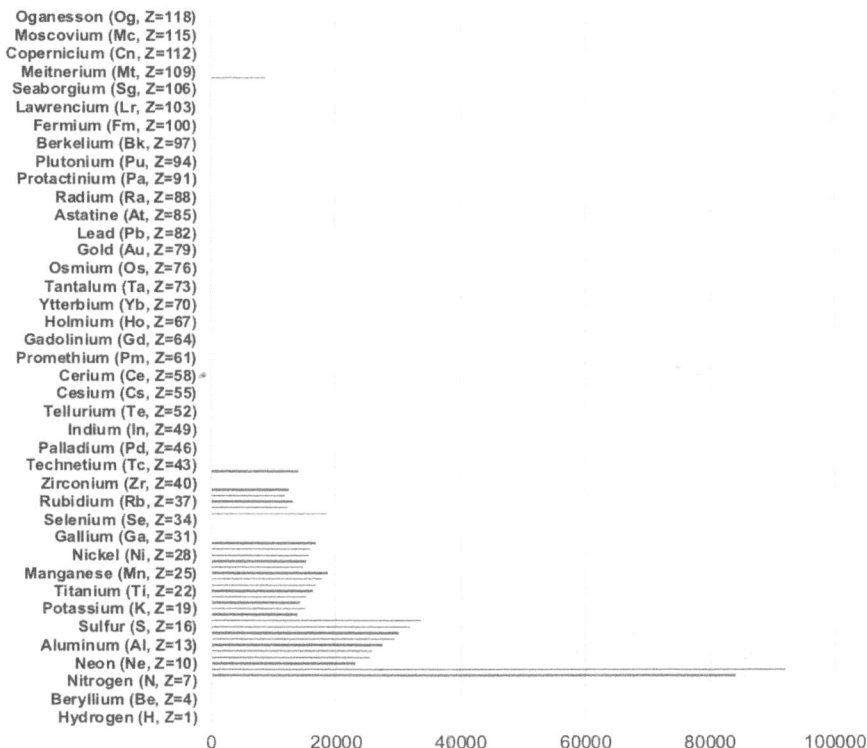

The smallest 9th ionization energy (14,110 kJ/mol) was recorded with Yttrium
(Y, Z=39) whereas the highest value (115,379.5 3 kJ/mol) was with Neon (Ne,
Z=10) followed by that (106,434.3 kJ/mol) of Fluorine (F, Z=9) (Fig. 71).

Fig. 71: 9th Ionisation Energy (kJ/mol) of the 118 chemical elements known as of 2020

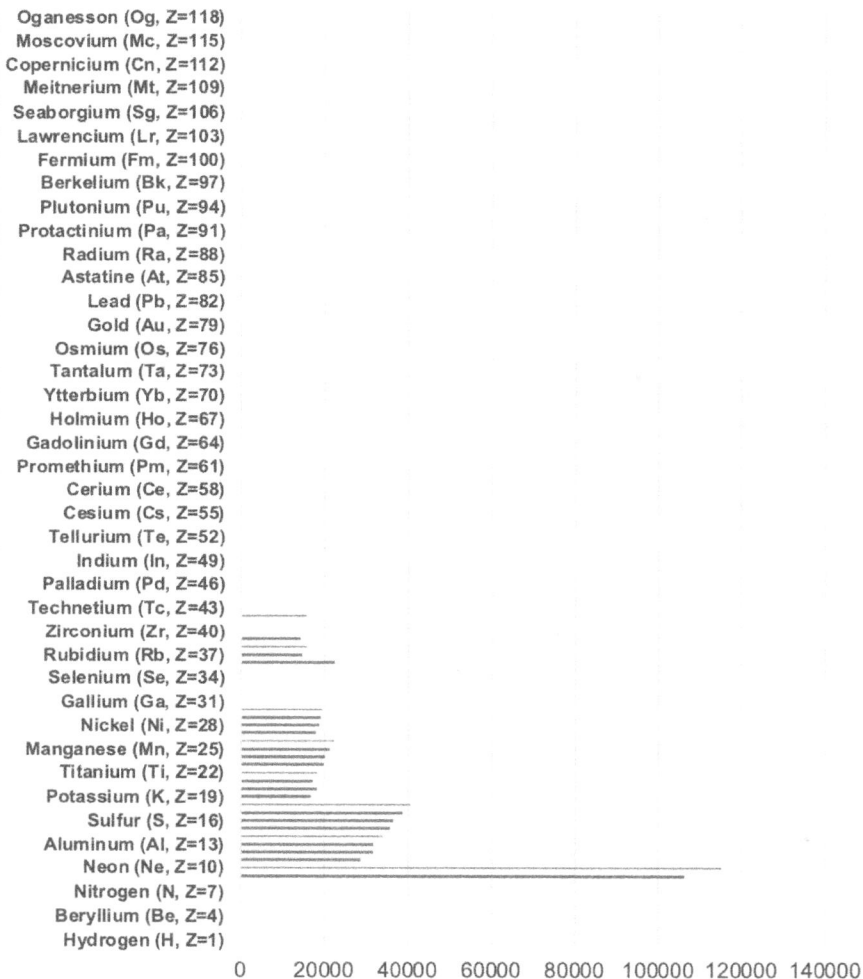

The 10th ionization energy varies between 17,100 kJ/mol and 141,362 kJ/mol. Strontium (Sr, Z=38) has the smallest value, while Sodium (Na, Z=11) has the highest one (Fig. 72). The second highest value (131,432 kJ/mol) was recorded with Neon (Ne, Z=10).

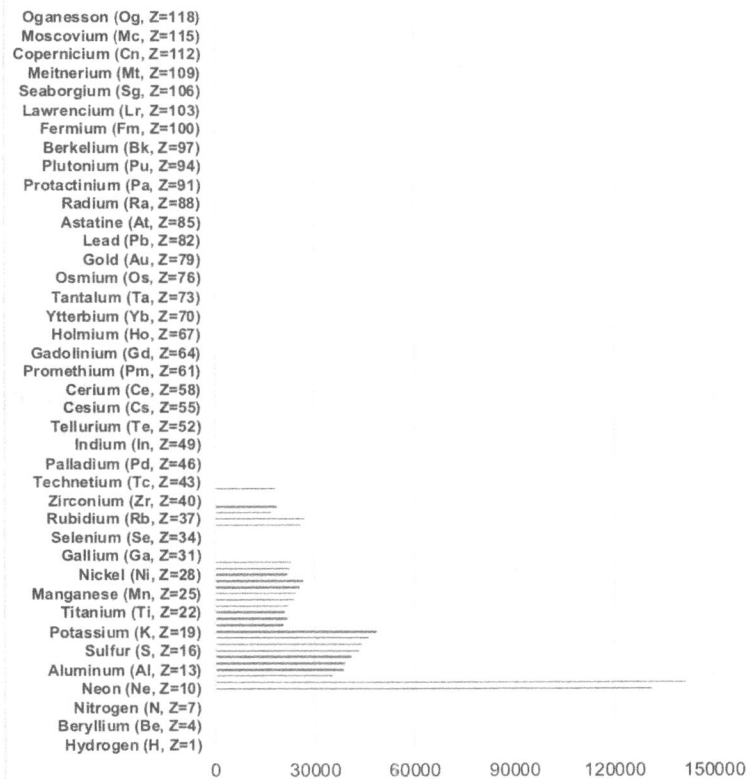

Fig. 72: 10th Ionisation Energy (kJ/mol) of the 118 chemical elements known as of 2020

18.4. Eleventh to 30th ionization energy (kJ/mol)

By the time I reached the 11th ionization, I noticed that in general, the 2 highest values of the nth ionization energies (e.g. 3rd, 4th) are usually recorded with the lightest chemical elements, while the smallest values are usually found with the heavier atoms. At the same time, many chemical elements do not have an ionization energy past the 5th ionization energy. Therefore, I quickly looked over the graphs I did for the 11th to the 30th ionization energy and I realized that their trends are similar to what I already mentioned above. To avoid spending time writing about the 11th to 30th ionization energies one by one, I decided not to study them as I did for the 1st to 10th ionization energy. Instead, I became interested in studying all of the ionization energies for each chemical element. My goal was to explore how the nth ionization energies vary for each element and then to see how the trends vary from one element to the other.

CHAPTER 19

CAN THE RADIUS OF CHEMICAL ELEMENTS BE A BIG PLAYER THAT BAILED SCIENCE OUT OF THE TROUBLING AND DANGEROUS DOUBTS ABOUT THE DECODING OF THE CHEMICALS-ORIGIN?

The atomic radius of a chemical element as the "*distance from the center of the nucleus to the outermost shell of the electron*" (Wikipedia, 2020h). Because the position of electrons is not well defined, the atomic radius is not precise, but can vary according to whether the atom is isolated or covalently bound with others. On the periodic table, the radius of the atoms is believed to generally decrease rightward along each period (row), but to increase down each group (column). Atomic radii are usually expressed in picometers (pm) which is equivalent to 10^{-12} m. Six kinds of atomic radii are usually mentioned in the literature:
- calculated radius
- covalent radius (single bond, double bond, triple bond radii, and other bonds found in super heavy elements)
- empirical radius
- ionic radius
- metallic radius
- van der Waals radius

In the following paragraphs, I focused on 4 types of radii:
- calculated radius
- covalent radius
- empirical radius
- van der Waals radius

To compare the 4 kinds of radius (calculated, covalent, empirical, and van der

Waals radius), I would like to say that, when it exists for a chemical, the van der Waals radius is usually higher than the other kinds of radius (Fig. 73). The empirical radius is usually smaller than the calculated radius. All regressions implicating the van der Waals radius are not strong ($r2<0.48$) (Fig. 74-76).

Fig. 73: Comparison of 4 kinds of radius (pm): calculated radius, covalent radius, empirical radius and van der Waals radius

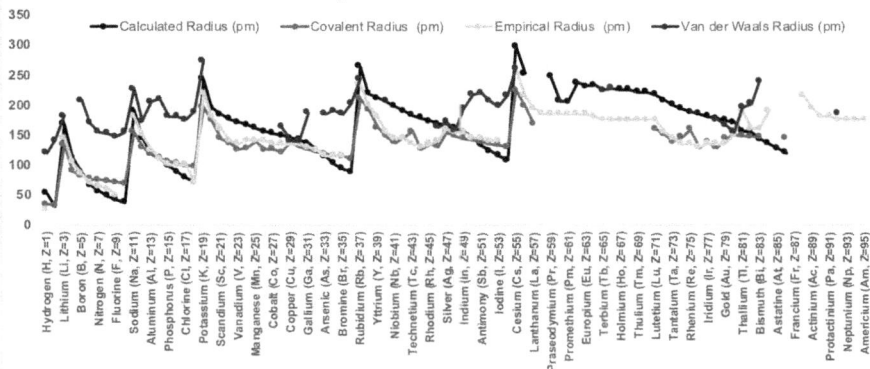

Fig. 74: Relationship between the Calculated Radius and Covalent Radius

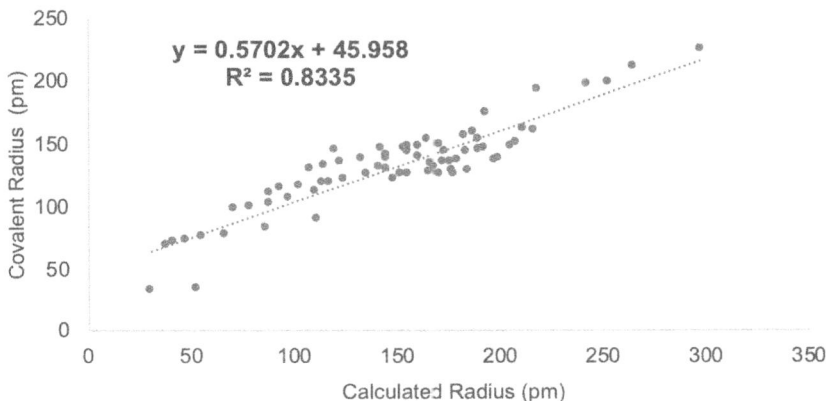

$$y = 0.5702x + 45.958$$
$$R^2 = 0.8335$$

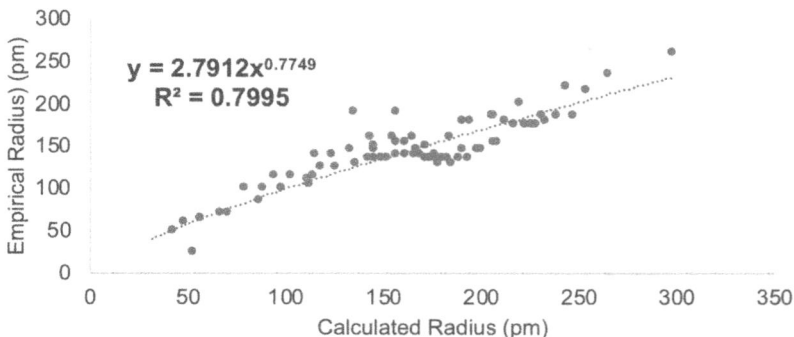

Fig. 75: Relationship between the Calculated Radius and Empirical Radius

$y = 2.7912x^{0.7749}$
$R^2 = 0.7995$

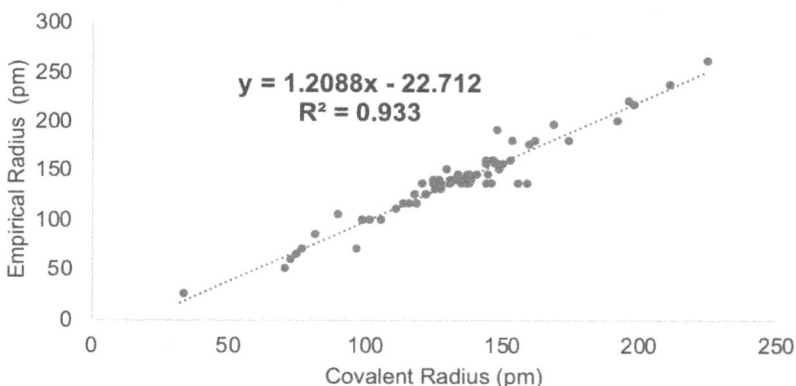

Fig. 76: Relationship between the Covalent Radius and Empirical Radius

$y = 1.2088x - 22.712$
$R^2 = 0.933$

19.1. Calculated radius

The references for the calculated atomic radius that I used is from (Clementi et al., 1967). I studied the calculated radius of 84 natural chemical elements, and it varies between 31 pm and 298 pm (Fig. 77). The calculated radius is not reported for any synthetic element. The highest calculated radius was found with Cesium (Cs, Z=55), an alkali metal, which recorded the smallest Mohs hardness, the smallest Brinell hardness, the smallest bulk modulus, and the smallest Young's modulus. This suggested that there may be a negative relationship between the size or radius and the hardness or modulus of some atoms. In contrast, the smallest radius was recorded on Helium (He, Z=2). The second smallest calculated radius (38 pm) was found with Neon (Ne, Z=10). Both Helium and Neon are noble gases. The next 5 smallest calculated radii were recorded with some of the key nonmetals on earth (i.e. C, H, O, N):

- Fluorine (F, Z=9), a halogen: 42 pm
- Oxygen (O, Z=8): 48 pm
- Hydrogen (H, Z=1): 53 pm
- Nitrogen (N, Z=7): 56 pm
- Carbon (C, Z=6): 67 pm

In general, the smallest radii were recorded with noble gases, halogens, and nonmetals. A positive relationship exists between the calculated radius and the atomic number of the alkali metals, alkaline earth metals, noble gases, and halogens. For lanthanoids, the calculated radius varies between 205 pm and 247 pm with a mean of 225.1 pm, and a CV of 5.1%.

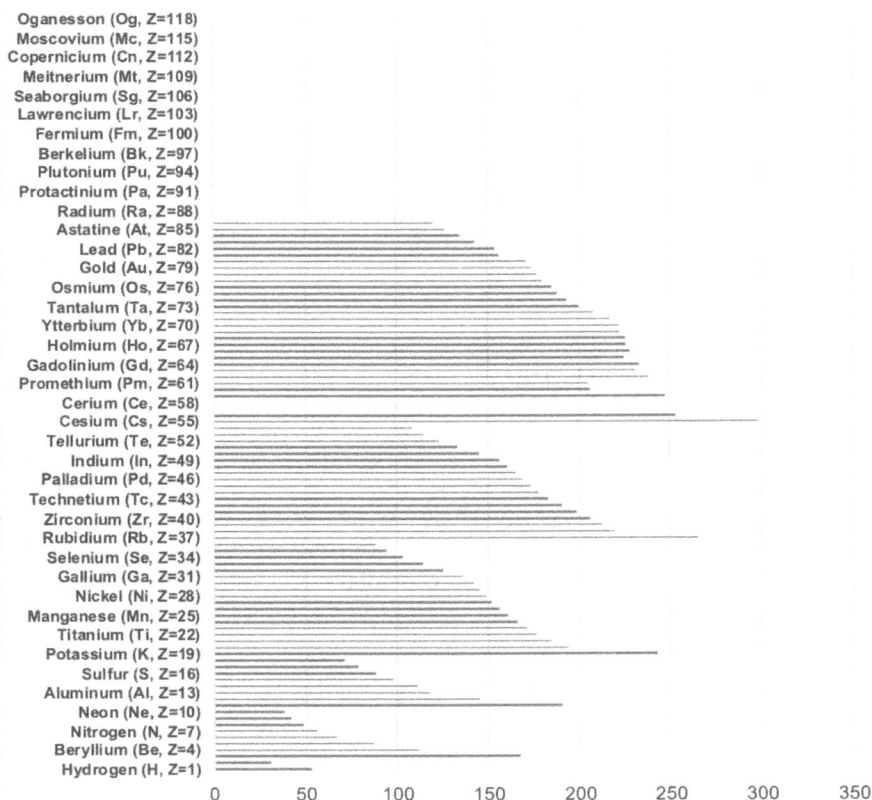

Fig. 77: Calculated Radius (pm) of the 118 chemical elements known as of 2020

19.2. Covalent radius

A covalent radius is measured on an atom that forms part of a covalent bond.

When two atoms are bound together, the length of the covalent bond between them is believed to be the sum of their covalent radius. The references that I used for the covalent radii are: Sanderson (1962), Sutton (1965), Huheey et al. (1993), Porterfield (1984), James and Lord (1992), and Riedel et al. (2005).

Data on covalent radius was available for 71 natural chemical elements but for no synthetic element. It varies between 32 pm and 225 pm (Fig. 78). Helium (He, Z=2) has the smallest covalent radius, followed by Hydrogen (H, Z=1): 34 pm. The radius reported for the other chemical elements is at least twice that of hydrogen. The other chemical elements which radius is less than 100 pm are:

- Neon (Ne, Z=10): 69 pm
- Fluorine (F, Z=9): 71 pm
- Oxygen (O, Z=8): 73 pm
- Nitrogen (N, Z=7): 75 pm
- Carbon (C, Z=6): 77 pm
- Boron (B, Z=5): 82 pm
- Beryllium (Be, Z=4): 90 pm
- Argon (Ar, Z=18): 97 pm
- Chlorine (Cl, Z=17): 99 pm

It is interesting that Boron, which is one of the hardest chemical elements, also has a smaller radius. The highest covalent radii were recorded by:

- Cesium (Cs, Z=55): 225 pm
- Rubidium (Rb, Z=37): 211 pm
- Barium (Ba, Z=56): 198 pm
- Potassium (K, Z=19): 196 pm
- Strontium (Sr, Z=38): 192 pm
- Calcium (Ca, Z=20): 174 pm
- Lanthanum (La, Z=57): 169 pm

In general, for each class of elements, the covalent radius and the atomic mass of the elements are positively correlated except for lanthanoids (for which only 2 data were collected), actinoids (for which no data exists) and the transition metals (which radius varies between 121 pm and 162 pm with a very small CV: 8.1%). In general, the smaller covalent radii were recorded with nonmetallic elements: noble gases, halogens, and nonmetals.

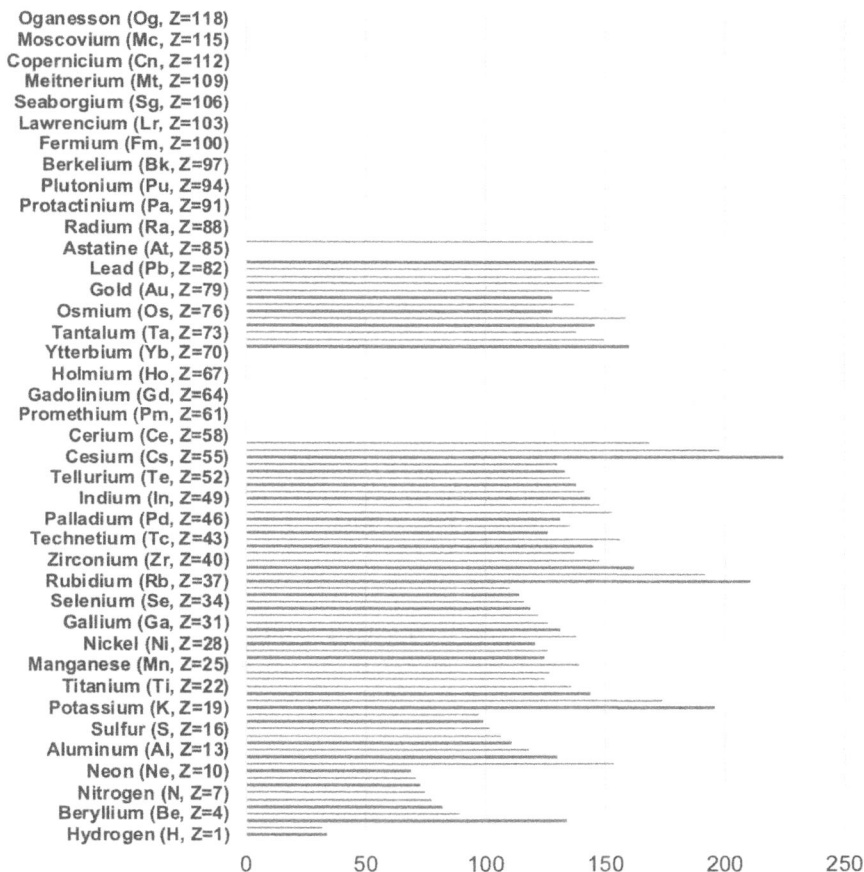

Fig. 78: Covalent Radius (pm) of the 118 chemical elements known as of 2020

19.3. Empirical radius

The raw data of the empirical radius I discussed here were from Slater (1964). Reported for 88 chemical elements. The empirical radius varies between 25 pm and 260 pm (Fig. 79). The only synthetic element which empirical radius was reported is Americium (Am, Z=95): 175 pm. Hydrogen (H, Z=1) recorded the smallest empirical radius. The next 7 smallest radii after hydrogen were recorded with:

- Fluorine (F, Z=9): 50 pm
- Oxygen (O, Z=8): 60 pm
- Nitrogen (N, Z=7): 65 pm
- Carbon (C, Z=6): 70 pm

- Argon (Ar, Z=18): 71 pm
- Boron (B, Z=5): 85 pm
- Phosphorus (P, Z=15): 100 pm

It is interesting to know that some of the key atoms in living organisms (carbon, hydrogen, oxygen, nitrogen, and phosphorous) are among the smallest. Their small size can also contribute to their ability to build compartments that other chemicals cannot traverse without using the proper gates. Similarly, the small size of hydrogen and oxygen can be one of the reasons water can diffuse through many materials and constitute a compound indispensable for life. In contrast, the highest radius was obtained with Cesium (Cs, Z=55). The 5 next highest empirical radii after that of Cesium were recorded with:

- Rubidium (Rb, Z=37): 235 pm
- Potassium (K, Z=19): 220 pm
- Barium (Ba, Z=56): 215 pm
- Radium (Ra, Z=88): 215 pm
- Strontium (Sr, Z=38): 200 pm

The high radius of potassium may explain why it is present at many "gates" in living organisms. Of the 6 noble gases, Argon (Ar, Z=18) is the only one which empirical radius was reported (71 pm). Besides the 5 noble gases, the other natural elements which empirical radius was not reported are:

- Astatine (At, Z=85), the heaviest halogen
- Francium (Fr, Z=87), the heaviest alkali metal

The coefficient of the transition metals, lanthanoids, and actinoids is smaller:

- Lanthanoids: 3.4%
- Actinoids: 4.1%
- Transition metals: 7.9%

On average, the mean radius of the following chemical groups is:

- Lanthanoids: 180.7 pm
- Actinoids: 179.3 pm
- Transition metals: 142.4 pm

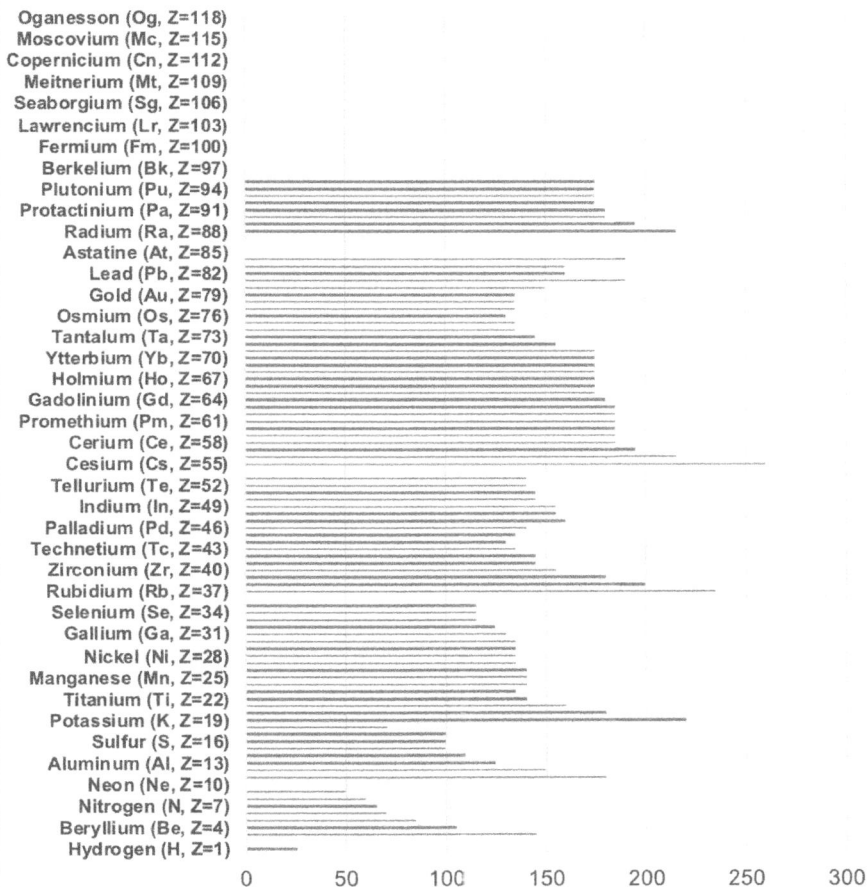

Fig. 79: Empirical Radius (pm) of the 118 chemical elements known as of 2020

19.4. Van der Waals radius

The references for the van der Waals radius that I used were Bondi (1964) and Mantina et al. (2009). Named after the physicist Johannes Diderik van der Waals, the van der Waals radius of an atom is defined as the *"radius of an imaginary hard sphere representing the distance of closest approach for another atom"*. Furthermore, the van der Waals interaction is believed to be the "attraction" between 2 atoms that are near one another, but which could not bind. Similarly, the van der Waals force is believed to be the force between 2 such atoms.

The van der Waals radius is reported for just 44 natural elements. It is available for all halogens but one (Astatine (At, Z=85)) and for all noble gases but one (Radon (Rn, Z=86)) and for all nonmetals. Most of the missing van der Waals radii were with metals. The van der Waals radius varies between 120 pm and 275 pm with the smallest value recorded on Hydrogen (H, $Z=1$) (Fig. 80). Indeed, the 7 smallest van der Waals radii were recorded with:

- Hydrogen (H, $Z=1$): 120 pm
- Zinc (Zn, $Z=30$): 139 pm
- Helium (He, $Z=2$): 140 pm
- Copper (Cu, $Z=29$): 140 pm
- Fluorine (F, $Z=9$): 147 pm
- Oxygen (O, $Z=8$): 152 pm
- Nitrogen (N, $Z=7$): 154 pm

The highest van der Waals radius was recorded with Potassium followed by Cesium. Indeed, the 10 highest van der Waals radii were recorded with:

- Potassium (K, $Z=19$): 275 pm
- Cesium (Cs, $Z=55$): 262 pm
- Rubidium (Rb, $Z=37$): 244 pm
- Bismuth (Bi, $Z=83$): 240 pm
- Sodium (Na, $Z=11$): 227 pm
- Antimony (Sb, $Z=51$): 220 pm
- Tin (Sn, $Z=50$): 217 pm
- Xenon (Xe, $Z=54$): 216 pm
- Silicon (Si, $Z=14$): 210 pm
- Boron (B, $Z=5$): 208 pm

Of the 4 kinds of radii that I studied, the smallest coefficient of variation (17.74%) was found with van der Waals, suggesting that as atoms are getting free or unbound from one another, their distance of closest approach for another atom tends toward a similar value. The fact that the van der Waals radius is higher than the other types of radii suggests that as atoms interact with others, their radius can decrease. To a larger extent, when the particles interact with others, their radius can change. Similarly, when nucleons are orbited by different kinds of leptons, their size can change. This also partially explains the proton radius puzzle.

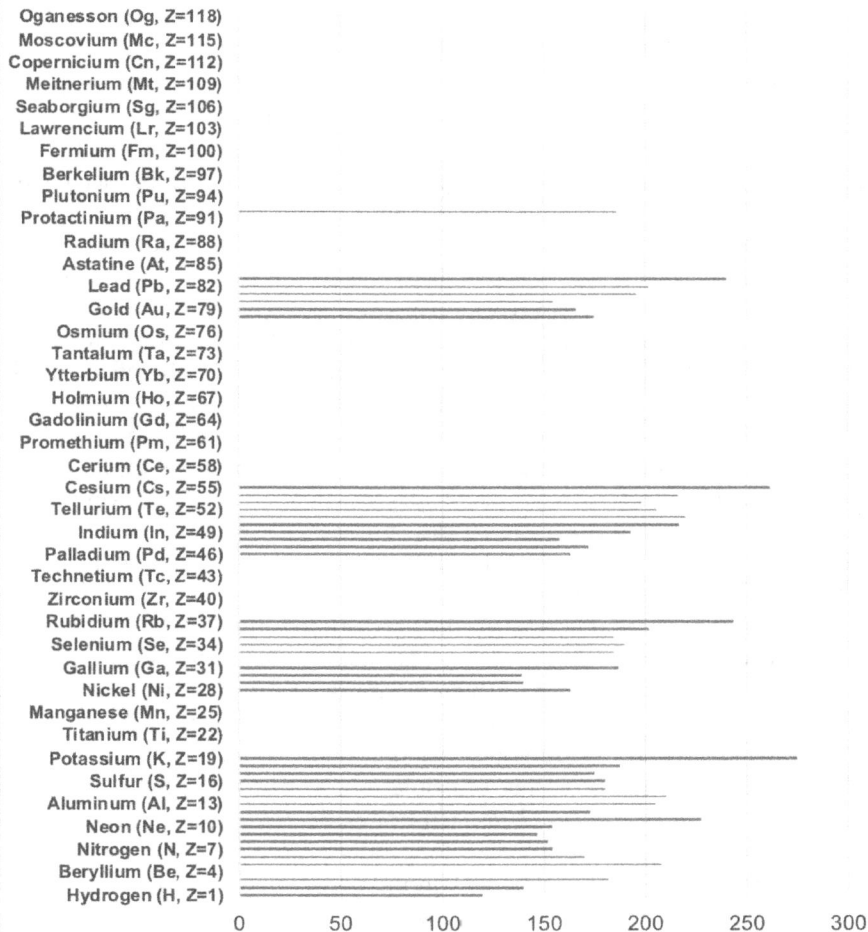

Fig. 80: Van der Waals Radius (pm) of the 118 chemical elements known as of 2020

19.5. Ratio "radius of secondary bodies / radius of their primary body"

In *"Turbulent Origin of the Universe"*, I was interested in checking whether the ratio of the radius of the celestial bodies to that of their primary body follows any law. I found that in the Solar System, the ratio of the radius of the celestial bodies to that of their primary bodies varies between 2.87E-09 and 0.511. The highest ratio was recorded between Charon (the biggest Plutonian satellite) and Pluto. The ratio of the radius of the planets to that of the Sun ranges between 0.00171 and 0.10272. No satellite in the Solar System is bigger than its primary planet. In fact, about 99% of the 210 satellites known in the Solar System as of 2020 have a radius inferior to 6% of the radius of their primary planet.

In my book that I mentioned in the previous paragraph, I showed that, a balance exists between the size of a secondary body and its primary body. When the characteristics of the primary body and its secondary bodies are not balanced, the constitution of these bodies can be affected and their nature can change if these bodies are forced to stay in their system without respecting the natural rules that sustain their formation. Another way of expressing this is that the bodies in each system, small or large, were formed in "harmony" with the constraints that the precursor of such a system had to go through. Efforts to introduce changes in the systems of bodies could have affected the nature of these bodies regardless of their characteristics (mass, energy, speed, etc.). Therefore, forcing a bigger body to orbit another one that is not as big as that it was supposed to naturally orbit is like forcing a rich man to depend on a poor man. Forcing a small body to be orbited by a bigger body is like forcing a poor person who has 3 children to adopt 100 additional children. That parent cannot afford that and therefore, both the children and the parent will be in trouble and may collapse. A similar problem could have happened and is still happening to atoms or particles that are forced to orbit bodies which size is different than the one they were made to naturally orbit. When such artificial orbital movements (e.g. revolution) are forced on particles or clusters of particles, the characteristics of the bodies involved can change, including their size, speed, radiation, and other things that cannot be seen physically. That is one of the problems some scientists are facing with the changes of the radius of some particles when they are orbited by other particles smaller or bigger than the ones they were made to naturally orbit. In the sections related to EMC and proton radius puzzle, I better elaborated on these trends. For, bodies were not made to orbit others by chance, but according to certain laws in harmony with the formation and functioning of the universe.

The misunderstanding of this principle is also causing a lot of problems in medical and health sciences and in chemical engineering. For chemical engineers are mixing incompatible substances to produce drugs, which medical doctors prescribe for their patients to take and, in the end, people are consuming products which, once they arrive in the digestive system, create various kinds of dangerous turbulences and artificial interactions with the denatured "natural" processes. Consequently, some drugs or medications that some people take cause them more harm and destroy life more than the pharmaceutical industries that are prospering under the sales of these products can ever know and/or acknowledge. It should not be surprising that, despite the scientific advances in many fields, human health has not really improved, but more incurable and complicated diseases are increasing across the globe and, worse, in nations that are so-called "developed", where some dangerous and mortal diseases are found, while some are neglected by the international audience until the day that some of them will become global pandemics! Yet, people will not comprehend that what they eat, drink, and breath every day is causing more damage to them than the medicine they try to take to cure their diseases. To put it another way, instead of controlling their diet and consuming things that agree with living processes, human beings consume what they want, and then, expect to live longer or healthier even if they can watch and pick some clues from their weight that something is not right. Far from offending anyone, because what I am saying is a pure natural truth that cannot be changed, I elaborated on these things in another book. For the forces and power which control natural processes are stronger and less unchangeable by human beings, no matter the efforts they may apply to replace or make artificial things to do better than natural things.

'Science180 Academy' Success Strategy
SCIENCE180 PUBLISHING: AUTHORS WANTED

Science180 Publishing, the American publishing company that published the groundbreaking discovery about the origin of the universe, of life, and of chemicals spearheaded by Dr. Nathanael-Israel Israel, really wants to publish your book(s) regardless of your field of expertise. This is a unique opportunity for:

- established authors
- people aspiring to become authors
- people who have written a book or are wanting to write one and need help with anything regarding publishing
- people who are not well known, inexperienced
- people whose books are viewed as nonconformist, controversial, or unconventional
- people who do not have enough resources or knowledge to navigate the publishing process
- people who are struggling to find an affordable, experienced, and high-quality publisher

Although Science180 Publishing is based in the USA, it can publish your books within your budget regardless of your geographical location. Science180 Publishing is highly interested in your document and possibly helping you publish it. Please visit Science180Publishing.com to explore how we may assist you. No matter the content of your book, as far as it is original, not promoting anything illegal, not duplicating anyone else's idea, Science180 Publishing can help you publish it in the USA. Please contact us asap and see how we can help.

To start your journey of publishing your book with Science180 Publishing, please visit Science180Publishing.com today.

CHAPTER 20

ARE THE STATE AND OCCURRENCE OF CHEMICAL ELEMENTS POWERFUL ENOUGH TO HELP US ON OUR JOURNEY TO CRACK THE CHEMICALS-ORIGIN?

20.1. State at STP (Standard Temperature and Pressure: 273 K)

State of matter is the form in which matter can exist. Matter is usually classified according to 4 states characterized as (Wikipedia, 2020i):

- Solids have a fixed volume and shape, with particles close together and fixed into place.
- Liquids have a fixed volume, but a variable shape that adapts to fit their container and their particles are close together but move freely.
- Gases have a volume and shape adapting both to fit their container, while their particles are neither close together nor fixed in place.
- Plasmas have variable volumes and shapes and contain particles that can move around freely such as: neutral atoms, ions, and freely-moving disassociated electrons believed to be ripped away from their nuclei.

However, while referring to chemical elements, plasma is not mentioned. Usually, by changing the pressure and temperature, the states of chemical elements can be changed, suggesting that one of the players in the formation of various chemical elements could have been a process that created different small pockets of space occupied by subatomic and atomic particles and then allowed the precursors of all matter to be spiraled, rolled, shaped, squeezed, and compressed differently. Solids can transition to liquid when they are subdued to a temperature equal or higher than their melting point. When liquid is subdued to a temperature equal or higher than their boiling point, they can become a gas. When gases are heated high enough, they can enter a plasma state in which electrons are so energized that they are believed to leave their parent atoms.

182

CHAPTER 20: STATE AND OCCURRENCE OF CHEMICAL ELEMENTS

The environmental conditions that contributed to shaping, squeezing, and compressing the types of matter, could have not allowed the gases to be as compressed as the liquids nor the liquids as much as the solids. The differences in these states of matter could have been affected by the position and turbulence of their precursors. Plasmas are like a mixture of particles, some of which would have not been fully aggregated into normal atoms. Hence, the constitutive subatomic particles of plasma are mixed and spread all over its volume. For instance, if the conditions were met, the free electrons in plasma that some people believe were ripped away from their nuclei are electrons that would not have initially orbited any nuclei in the beginning. They are like secondary bodies formed without being able to properly orbit a primary nucleus to form a normal atom. The turbulence intensity was smaller than that which completed the formation of some rocky planets. The small intensity of the turbulence in the stars could not have allowed the aggregation of all their constitutive particles into a cluster of primary subatomic particles (e.g. nucleons) orbited by secondary subatomic particles (e.g. electrons or other leptons). That is why stars like the Sun are believed to be made of plasma. Regions where the turbulence could have been higher in those stars could have allowed the formation of heavier elements but due to the overall low intensity turbulence of the precursors of stars, their main constituents are particles dispersed into a plasma. I provided more details in *"Turbulent Origin of the Universe"*.

The high intensity of the turbulence that led to the formation of the Earth and the Earth's particles did not fit conditions that favored the presence of pure plasma, hence a plasma state is said to not freely exist under normal conditions on Earth, but it can be commonly generated by processes like lightning, electrical sparks, fluorescent lights, and neon lights. For under the environmental conditions on Earth, particles in a plasma could be transformed into normal particles and chemical elements like those found on Earth. It should not be surprising that the particles that the Sun "projected" into space could be transformed into other particles once they reached environments having a turbulence level different than that of the Sun. Other types of subatomic particles and atoms not found on Earth could be present in the plasma of stars for the level of turbulence could have allowed the aggregation of subatomic particles into other kinds of atoms not found in an environment dominated by higher turbulence. Similarly, in environments where the turbulence could have been stronger and precursors of particles were heavily squeezed, denser subatomic elements and atoms could be found, even denser than those currently known. In other words, it is a mistake for scientists to believe that chemical elements are the same from one stellar system to another and from one galaxy to another.

The atoms and the subatomic particles of the precursors of each state of matter were gathered together differently and brought close to one another at different speeds according to the conditions of their formation. The atoms and molecules of solids were brought closer together than those of a fluid. Consequently, atoms of solids are held together more tightly than those in liquids.

Science180: Source of Unconventional Wisdom and Knowledge on the Origin of Chemicals

The intermolecular cohesive forces in fluids are not sufficient to hold their constitutive atoms together. Hence, fluids usually flow under the influence of some stress and instability.

Molecules or atoms in gases are much farther apart than those of liquids. This makes gases more compressible and at the same time, they can also expand indefinitely when the external pressure applied to them is removed. Although liquids are considered to be relatively incompressible, during the formation of the universe, their precursors were greatly compressed until a certain "limit" or equilibrium was reached. However, if the proper amount of pressure is applied to any liquid, I believe that a liquid can be compressed and the nature of its atoms or molecules can change.

Of the 99 chemical elements which states of matter were clearly defined at the time I studied them, 11 are gas and 2 are liquid at the standard temperature and pressure of 273 K. The 11 gases are 3 nonmetals, 2 halogens, and all 6 noble gases:

- Hydrogen (H, $Z=1$), a nonmetal
- Helium (He, $Z=2$), a noble gas
- Nitrogen (N, $Z=7$), a nonmetal
- Oxygen (O, $Z=8$), a nonmetal
- Fluorine (F, $Z=9$), a halogen
- Neon (Ne, $Z=10$), a noble gas
- Chlorine (Cl, $Z=17$), a halogen
- Argon (Ar, $Z=18$), a noble gas
- Krypton (Kr, $Z=36$), a noble gas
- Xenon (Xe, $Z=54$), a noble gas
- Radon (Rn, $Z=86$), a noble gas

The liquid chemical elements are:

- Bromine (Br, $Z=35$), a halogen
- Mercury (Hg, $Z=80$), a transition metal

In other words, of the 5 halogens, 2 are gas, 1 is liquid, and 2 are solid:

- Fluorine (F, $Z=9$): gas
- Chlorine (Cl, $Z=17$): gas
- Bromine (Br, $Z=35$): liquid
- Iodine (I, $Z=53$): solid
- Astatine (At, $Z=85$): solid

Therefore, of the 18 chemical elements usually classified as nonmetals (the 6 noble gases, 5 halogens, and 7 nonmetal) 11 are found as gas, 1 as liquid, and 6 as solid. Besides the 12 nonmetals mentioned above (11 gases and 1 liquid), the remaining 6 nonmetals found in a solid state at the standard temperature and pressure of 273 K are:

- Carbon (C, $Z=6$)

- Phosphorus (P, Z=15)
- Sulfur (S, Z=16)
- Selenium (Se, Z=34)
- Iodine (I, Z=53)
- Astatine (At, Z=85)

Besides, Mercury (Hg, Z=80), a transition metal, all of the metals and metalloids which data were available are solid at the standard temperature and pressure of 273 K. However, although 5 synthetic elements were defined as solid (i.e. Americium (Am, Z=95), Curium (Cm, Z=96), Berkelium (Bk, Z=97), Californium (Cf, Z=98), and Einsteinium (Es, Z=99)), the state of the chemical elements which atomic mass (number of protons) is higher than 99 was unknown at the time I collected the data:

- Fermium (Fm, Z=100)
- Mendelevium (Md, Z=101)
- Nobelium (No, Z=102)
- Lawrencium (Lr, Z=103)
- Rutherfordium (Rf, Z=104)
- Dubnium (Db, Z=105)
- Seaborgium (Sg, Z=106)
- Bohrium (Bh, Z=107)
- Hassium (Hs, Z=108)
- Meitnerium (Mt, Z=109)
- Darmstadtium (Ds, Z=110)
- Roentgenium (Rg, Z=111)
- Copernicium (Cn, Z=112)
- Nihonium (Nh, Z=113)
- Flerovium (Fl, Z=114)
- Moscovium (Mc, Z=115)
- Livermorium (Lv, Z=116)
- Tennessine (Ts, Z=117)
- Oganesson (Og, Z=118)

The fact that Mercury is a metal but liquid at room temperature holds a secret about its formation.

20.2. Occurrence of chemical elements

Chemical elements are sometimes classified into 3 categories according to their occurrence:

- primordial naturally occurring elements
- transient elements which can occur from the decay of others
- synthetic elements produced technologically and which are not known to

occur naturally

Among the natural elements, the following are said to be transient, meaning they can also occur through decay of other elements:

- Technetium (Tc, Z=43)
- Promethium (Pm, Z=61)
- Polonium (Po, Z=84)
- Astatine (At, Z=85)
- Radon (Rn, Z=86)
- Francium (Fr, Z=87)
- Radium (Ra, Z=88)
- Actinium (Ac, Z=89)
- Protactinium (Pa, Z=91)
- Neptunium (Np, Z=93)

Of the 24 artificial or synthetic elements, the following are said to be transient meaning they can occur from the decay of other elements:

- Americium (Am, Z=95)
- Curium (Cm, Z=96)
- Berkelium (Bk, Z=97)
- Californium (Cf, Z=98)

However, because I do not believe in the secular theories of the genesis of chemical elements, I did not spend too much time on the occurrence of the elements.

CHAPTER 21

BOILING POINT AND MELTING POINT

The boiling point of an element is the temperature at which its liquid form is at equilibrium with the gaseous form at an air pressure of 1 atmosphere.

To download the content of this chapter, visit www.Science180.com/BoilingPointMeltingPoint

CHAPTER 22

CAN A SINGLE VARIABLE MAKE THE OTHERS BOW TO ITS POWER BY LIGHTING IN THE HEART OF CHEMISTRY THE LAMP OF UNDERSTANDING THAT OPENED THE GATE TO UNRAVEL THE ORIGIN OF CHEMICAL PARTICLES? DISCOVER WHAT THE ABUNDANCE OF CHEMICALS IS SAYING ... BUT SCIENTISTS ARE NOT LISTENING

The abundance of a chemical element measures its occurrence relative to all other elements in a given environment. I extensively studied the abundance of chemical elements in the following bodies or systems of bodies:

- universe or in our galaxy
- Sun
- human beings
- oceans
- Earth's crust
- lava
- atmosphere of the planets in the Solar System
- atmosphere and crust of the Moon

However, for the sake of time, I did not present all of the data here, but just some excerpts that can directly help understand the formation of the chemical elements. Although by studying the celestial bodies I had a good impression of how chemical elements could have been formed, it was only after carefully studying their chemical abundance in various environments that the processes of their origin and formation became clear to me. Instead of first enumerating the steps of the formation of the chemical elements, I decided to present the journey of my findings, hoping that it will help the readers to better know the methodology I used to unearth the formation of some key chemical elements and compounds before I generalized the process to other subatomic particles, composite particles including: atoms, molecules, minerals, rocks, and the celestial bodies and space that contain them.

22.1. Abundance in the universe or in our galaxy

Based on the current scientific theories used to estimate the abundance of chemical elements in the universe or in our galaxy (Periodictable.com, 2014a), 3 chemical elements are believed to account for 99% of the mass of the chemicals in the universe:

- Hydrogen (H, Z=1), a nonmetal: 75%
- Helium (He, Z=2), a noble gas: 23%
- Oxygen (O, Z=8), a nonmetal: 1%

The other chemical elements which abundance (with respect to mass) in the universe or in our galaxy is significant include:

- Carbon (C, Z=6), a nonmetal: 0.5%
- Neon (Ne, Z=10), a noble gas: 0.13%
- Iron (Fe, Z=26), a transition metal: 0.11%
- Nitrogen (N, Z=7) a nonmetal: 0.1%
- Silicon (Si, Z=14), a metalloid: 0.07%
- Magnesium (Mg, Z=12), an alkaline earth metal: 0.06%
- Sulfur (S, Z=16) a nonmetal: 0.05%
- Argon (Ar, Z=18), a noble gas: 0.02%
- Calcium (Ca, Z=20), an alkaline earth metal: 0.007%
- Nickel (Ni, Z=28), a transition metal: 0.006%
- Aluminum (Al, Z=13), a metal: 0.005%
- Sodium (Na, Z=11), an alkali metal: 0.002%

The abundance of any of the other chemical elements is less than 0.002%. More than 99.7% of the chemical elements that are abundant in the universe are believed to be nonmetallic (i.e. noble gases, halogens, or nonmetals). The universe could be dominated by nonmetals because the original matter, which was the precursor of all kinds of matter, was unable to convert all of itself into metallic elements. It would have required more energy and turbulence to spiral and gather together all of the precursors of particles into heavier and denser metallic elements. The way the turbulent events that birthed the elements acted could have allowed a certain proportion of the precursors of matter to be converted into denser elements, while most of them could have stayed in less dense forms. Hence bigger and denser elements are not usually the most abundant but rare. Later, I will better elaborate on the underlying cause of this trend.

22.2. Abundance in the Sun

The abundance of chemicals in the Sun is also called solar abundance. I noticed that a strong relationship exists between the abundance of chemical elements in the Sun and that in the universe or in our galaxy ($r2=1$). This should not be surprising because what is known about the universe away from the Solar System is mostly stars. Therefore, most of the estimations of chemical presence or abundance in the universe or in our galaxy would have been measured or estimated on stars. Consequently, as the Sun is a star, the estimation of the elements for the other stars are said to be similar. However, I do not personally think that the constitution of all stars in the universe would be the same. For, according to the data that I analyzed on the celestial bodies in the Solar System, the composition of stars and other bodies in the universe would depend on the nature of the turbulence that took place during their formation. And because the data I dealt with showed that turbulence would vary depending on the position, location, size, and movement of the precursors of bodies, it would be fair to state that the original turbulence across the bodies in the universe would vary. Consequently, the constitution and abundance of the elements born from these turbulences would not be expected to be the same everywhere.

The raw data I used for the abundance of chemical elements in the Sun are from Periodictable.com (2014b). Due to time constraints and the strong correlation between the abundance of chemical elements in the universe or our galaxy and that in the Sun, I did not detail the abundance of the chemical elements in the Sun. Together, hydrogen and helium are said to account for 98% of the mass of the Sun. By decreasing order of their abundance, the 20 most abundant chemicals in the Sun are postulated to be:

- Hydrogen (H, $Z=1$): 75%
- Helium (He, $Z=2$): 23%
- Oxygen (O, $Z=8$): 0.9%
- Carbon (C, $Z=6$): 0.3%
- Nitrogen (N, $Z=7$): 0.1%
- Neon (Ne, $Z=10$): 0.1%
- Iron (Fe, $Z=26$): 0.1%
- Silicon (Si, $Z=14$): 0.09%
- Magnesium (Mg, $Z=12$): 0.07%
- Sulfur (S, $Z=16$): 0.04%
- Nickel (Ni, $Z=28$): 0.008%
- Argon (Ar, $Z=18$): 0.007%
- Calcium (Ca, $Z=20$): 0.007%
- Aluminum (Al, $Z=13$): 0.006%
- Sodium (Na, $Z=11$): 0.004%
- Chromium (Cr, $Z=24$): 0.002%

- Manganese (Mn, Z=25): 0.001%
- Chlorine (Cl, Z=17): 8.E-04%
- Phosphorus (P, Z=15): 7.E-04%
- Potassium (K, Z=19): 4.E-04%

Noble gases constitute about 23.1% of the mass of the Sun, while the elements typically called nonmetals constitute 76.3% of the Sun's mass. Together all nonmetallic elements (noble gases and nonmetals) constitute about 99.4% of the Sun's mass.

22.3. Abundance in human beings

The abundance of the chemical elements in human beings (Periodictable.com, 2014c) varies between 0 and 61%, with Oxygen (O, Z=8) being the most abundant (61%). More than 97.9% of the mass of human beings is made of nonmetals (Fig. 83), and 3 of these nonmetals account for about 94% of the mass of human beings: Oxygen (O, Z=8), Carbon (C, Z=6), and Hydrogen (H, Z=1). No synthetic element, lanthanoid, or noble gas was found in human beings.

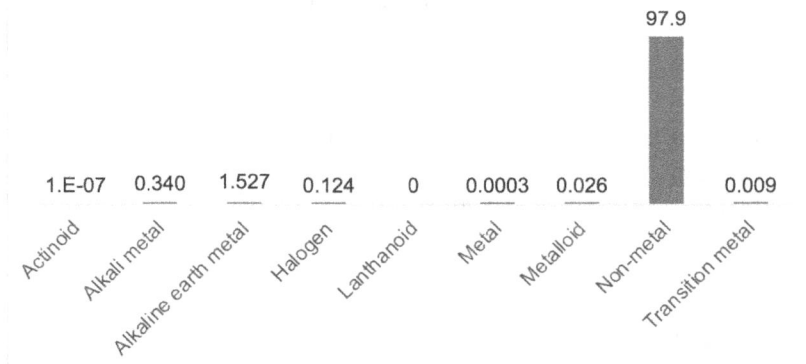

Fig. 83: Sum of the Abundance of chemical elements in human beings (%)

Of the 118 chemical elements known as of 2025, only 41 (meaning 34.75% of them) can be found at a measurable amount equal to or higher than 1.E-13%, which is the abundance of Radium (Ra, Z=88). The abundance of 28 chemical elements is reported to be 0, while that of 49 elements is not reported at all, not even in a trace. By decreasing abundance order, the 15 most abundant chemical elements in human beings are:

1. Oxygen (O, Z=8), a nonmetal: 61%
2. Carbon (C, Z=6), a nonmetal: 23%
3. Hydrogen (H, Z=1), a nonmetal: 10%
4. Nitrogen (N, Z=7), a nonmetal: 2.6%

5. Calcium (Ca, Z=20): 1.5%
6. Phosphorus (P, Z=15), a nonmetal: 1.1%
7. Sulfur (S, Z=16), a nonmetal: 0.2%
8. Potassium (K, Z=19): 0.2%
9. Sodium (Na, Z=11): 0.14%
10. Chlorine (Cl, Z=17): 0.12%
11. Magnesium (Mg, Z=12): 0.027%
12. Silicon (Si, Z=14): 0.026%
13. Iron (Fe, Z=26): 0.006%
14. Fluorine (F, Z=9): 0.0037%
15. Zinc (Zn, Z=30): 0.0033%

The abundance of any other chemical element found in human beings is not higher than 4.6E-04%. The most abundant actinoid in human beings is Uranium (U, Z=92), contributing to 1E-07% of the human mass. I was a little surprised that our bodies contain Uranium, which is known as very radioactive. I tried to find the organ where Uranium is most concentrate but could not.

As far as alkali metals are concerned, the least abundant is Francium (Fr, Z=87), with an abundance reported as 0%, while the most abundant are:
- Potassium (K, Z=19): 0.2%
- Sodium (Na, Z=11): 0.14%

The alkaline earth metals in human beings are dominated by:
- Calcium (Ca, Z=20): 1.5% followed by
- Magnesium (Mg, Z=12): 0.027%

The most abundant halogens in human beings are:
- Chlorine (Cl, Z=17): 0.12% and
- Fluorine (F, Z=9): 0.0037%

Concerning the other 3 halogens, while Bromine (Br, Z=35) and Iodine (I, Z=53) were recorded at a concentration smaller than 0.0003%, Astatine (At, Z=85) was not found at all.

Of the typical metals, the most abundant were:
- Lead (Pb, Z=82): 1.7E-04%
- Aluminum (Al, Z=13): 9E-05%
- Tin (Sn, Z=50): 2E-05%

The most abundant metalloids are:
- Silicon (Si, Z=14): 0.026%
- Boron (B, Z=5): 7E-05%, and a trace of
- Arsenic (As, Z=33): 5E-06%

The most abundant transition metals in human beings are:

- Iron (Fe, Z=26): 0.006% and
- Zinc (Zn, Z=30): 0.0033%

Among the heaviest chemical elements (atomic numbers > 38) found in human beings, Lead (Pb, Z=82) is the most abundant: 0.00017%. This may also be why lead intoxication is a big deal. Human beings do not incorporate into their bodies the most abundant chemical elements in nature, but elements that align with their physiological needs. In my book on the origin of life, I elaborated on the chemical elements found in most living organisms.

Before I close this segment on the abundance of chemical elements in human beings, I would like to quickly address their location in the human body. Indeed, chemical elements are found in different concentrations in different parts of the human body. While some chemical elements like Carbon (C, Z=6) are found everywhere in human bodies, others are found in specific locations. For instance, some of the locations where the chemical elements are found include:

- bones and teeth (e.g. Fluorine (F, Z=9))
- enzymes (e.g. Potassium (K, Z=19), Manganese (Mn, Z=25) and Iron (Fe, Z=26))
- body liquids (e.g. Chlorine (Cl, Z=17))
- proteins (e.g. Sulfur (S, Z=16)), and
- urine and bones (e.g. Phosphorus (P, Z=15)).

Sodium (Na, Z=11) is found in all liquids and tissues. Hydrogen (H, Z=1) and Oxygen (O, Z=8) are found in all liquids, tissues, bones, and proteins. Nitrogen (N, Z=7) is found in all liquids, tissues, and proteins. Magnesium (Mg, Z=12) is fundamentally found in lungs, kidney, liver, thyroid, brain, muscles, and heart. In addition to all of the places where Magnesium is found, Calcium (Ca, Z=20) is also found in bones. The types of chemicals present in specific locations and the characterization of these elements could help to understand some of the functionalities of the organs present at these locations.

22.4. Abundance of chemical elements in the Earth's crust

Before I start discussing the composition of the Earth's crust, I would like to comment a little bit on the composition of the crust of the planets. Unlike the atmosphere, the crust of most planets in the Solar System have not been analyzed. Efforts have been made to analyze the crust of the Earth and Mars. Moreover, since 1958, several missions have been made to the Moon (Earth's satellite) and some samples have been returned to Earth for analysis. To my knowledge, as of 2025, no mission has been able to collect yet a crust sample from any planet aside from Earth and Mars. In other words, apart from the Earth, the other planet which surface has been extensively studied is Mars. Some missions to the celestial bodies looked into their atmosphere and to estimate their composition based on analytical tools such as spectroscopy. Some of those data showed that Mars surface is composed of silicates. What is known about the chemical composition of the other planets is mostly based on what was found in their atmosphere.

Although theories have been made to try to explain the composition of the interior of the planets, it is important to mention that they are based on pure assumptions. For instance, on Earth, many efforts have been made to explore the bottom of the Earth by digging the soil. As of 2025, the deepest mine on Earth is in South Africa and is about 3.9 km deep. Besides mining, other techniques have been used to drill the Earth's crust. As of 2025, the deepest drill into the Earth's crust is the Kola Superdeep Borehole (in Northwest Russia) which is about 12 km deep and 23 cm in diameter (Elders, 1989). The previous author reported that the drilling project started in 1970 with an aim of drilling as deep as possible into the Earth's crust. About twenty-four years after the drilling started, the drilling team reached a point where the earth's temperature was 180°C, which was too hot for them to continue the drilling. Therefore, they stopped the project in 1994. Knowing the radius of the Earth (6378.14 km at the equator and 6356.75 km at the pole), the 12 km hole that the Russians scientists have drilled into Earth is nothing compared to what is needed to reach the center of the Earth. It represents just 0.19% of the Earth's radius. This information means that scientists don't know much about the interior of the Earth. Yet, they behave as if they know everything. Moreover, the radius of Jupiter for instance is 11.2 times that of the Earth, suggesting that it will surely be harder for scientists to empirically know the interior of the giant planets like Jupiter. If we are unable to understand the interior of the Earth, which is closer to us, how much more difficult would it be to understand the interior of other celestial bodies which are further from us? By the time the drilling in Russia ended, they found some surprising results including the discovery of water many kilometers beneath some impermeable layers. If the drilling was not stopped, the scientists would have found many more surprises. These data suggest that the scientific models regarding the interior of the planets surely contain errors.

Because scientists are still unable to know and reach the surface of the 4 giant planets in the Solar System, the chemical composition of the atmosphere of these planets are not fully understood. Those who have ever traveled in an airplane in temperate zones during winter can testify that, when the sky is filled with snow, it is impossible to know what is on the ground. Similarly, when the sky is very cloudy, it can be impossible to even see the Sun despite its brightness. Likewise, the fact that the sky of giant planets is filled with gases that prevent the view of the surface of these planets means that many things are still unknown about the chemical composition of these planets. Consequently, the data available on the chemical composition of the atmosphere of the planets may contain mistakes.

Many authors have estimated the abundance of chemical elements in the Earth's crust. Indeed, studied for 108 chemical elements, the abundance of chemical elements in the Earth's crust Dayah (1997) varies between 0 and 46%.

CHAPTER 22: ABUNDANCE OF CHEMICAL ELEMENTS, A MASTER KEY TO CRACK CHEMICALS-ORIGIN

Oxygen is the most abundant element. By decreasing order of abundance, the most abundant chemical elements in the Earth's crust are:

1. Oxygen (O, Z=8): 46%
2. Silicon (Si, Z=14): 27%
3. Aluminum (Al, Z=13): 8.1%
4. Iron (Fe, Z=26): 6.3%
5. Calcium (Ca, Z=20): 5%
6. Magnesium (Mg, Z=12): 2.9%
7. Sodium (Na, Z=11): 2.3%
8. Potassium (K, Z=19): 1.5%
9. Titanium (Ti, Z=22): 0.66%
10. Carbon (C, Z=6): 0.18%
11. Hydrogen (H, Z=1): 0.15%
12. Manganese (Mn, Z=25): 0.11%
13. Phosphorus (P, Z=15): 0.099%
14. Fluorine (F, Z=9): 0.054%
15. Sulfur (S, Z=16): 0.042%

Unlike the elements in the universe or our galaxy, or those in the Sun or the oceans where in general 2 elements account for more than 97% of the mass, in the Earth's crust, the 2 most dominant elements (Oxygen and Silicon) account for just 73% of the mass. Three other elements account for at least 5% each, meaning that more elements are abundant in the Earth's crust than in the universe, or in the Sun or in the oceans.

In general, nonmetals (dominated by oxygen) account for more than 46.5% of the mass of the Earth's crust. Dominated by Silicon, the metalloids account for more than 27%. Argon is the most dominant noble gas in the Earth's crust.

22.5. Abundance in lava

I was interested in studying the abundance of chemical elements in lava because I thought it could help me to have a glimpse at the composition of the interior of the Earth. Although no one can confidently say the depth from which most lava erupt, at least lava hint at the kind of chemicals beneath the crust, yet it may be impossible to know the exact composition of the Earth's interior.

Wikipedia defined lava as a "molten rock (magma) that has been expelled from the interior of some planets (including Earth) and satellites". Upon their eruption, volcanoes release magma, which after cooling down, become solid rocks (e.g. igneous rocks) such as basalt and granite. Lava has been found to be up to 100,000 times more viscous than water (Pinkerton and Bagdassarov, 2004; INIST, 2008).

Although lava on Earth is dominated by silicate minerals, other types of lava called unusual lavas have been observed and they include:

- Carbonatite lavas recorded for instance in a carbonatite volcano in Tanzania
- Iron oxide lavas recorded in Sweden (Harlov et al., 2002).

- Sulfur lava recorded in Chile (Guijón et al., 2011).

A volcano is defined by Wikipedia as a "rupture in the crust of a planetary-mass object, such as Earth, that allows hot lava, volcanic ash, and gases to escape from a magma chamber below the surface." The main chemical elements found in silicate melts are: silicon, oxygen, aluminum, alkalis (sodium, potassium, calcium), magnesium, and iron (Watson et al., 2006). The dominance of silicates in both the Earth's crust and in magma suggested to me that the crust and the interior of the Earth could have been made from a similar material but which could have gone through different changes due to some environmental factors and the intensity of the turbulence of their precursors. The precursor of the crust and that of the interior of the Earth could have belonged to different layers.

Although most lava are dominated by silicates, lava in the deep interior of the Earth may be dominated by chemical compounds different from ordinary silicates, which dominate the Earth's crust. Considering the data available on lava and the constitution of some volcanic rocks, some rocks in the Earth's crust and in the crust of some planets could have been made by the cooling or hardening of lava or its precursor.

22.6. Abundance in oceans

Studied for 104 chemical elements, the abundance of chemical elements in oceans (Dayah, 1997) varies between 0 and 86%. No synthetic element was found in the oceans, but as pollution increases, they may be everywhere. The most abundant elements in oceans is Oxygen (O, Z=8) followed by hydrogen, which together make up the water molecules which dominate all waters. Together, oxygen and hydrogen account for about 97% of the mass of the oceans. Indeed, the 15 most abundant elements in oceans are:

1. Oxygen (O, Z=8), a nonmetal: 86%
2. Hydrogen (H, Z=1), a nonmetal: 11%
3. Chlorine (Cl, Z=17), a halogen: 2%
4. Sodium (Na, Z=11), an alkali metal: 1.1%
5. Magnesium (Mg, Z=12), an alkaline earth metal: 0.13%
6. Sulfur (S, Z=16), a nonmetal: 0.093%
7. Potassium (K, Z=19), an alkali metal: 0.042%
8. Bromine (Br, Z=35), a halogen: 6.7E-03%
9. Carbon (C, Z=6), a nonmetal: 2.8E-03%
10. Strontium (Sr, Z=38), an alkaline earth metal: 8.1E-04%
11. Boron (B, Z=5), a metalloid: 4.4E-04%
12. Calcium (Ca, Z=20), an alkaline earth metal: 4.2E-04%
13. Fluorine (F, Z=9), a halogen: 1.3E-04%
14. Silicon (Si, Z=14), a metalloid: 1E-04%
15. Nitrogen (N, Z=7), a nonmetal: 5.0E-05%

Argon is the dominant noble gas in the oceans. It is interesting to note that nitrogen (which is very dominant in the Earth's atmosphere) and silicon (which is dominant in the Earth's crust) are just a trace in the oceans. In general, more than 97.1% of the chemical elements in the oceans are nonmetals (Fig. 84).

Fig. 84: Sum of the Abundance of chemical elements in oceans (%)

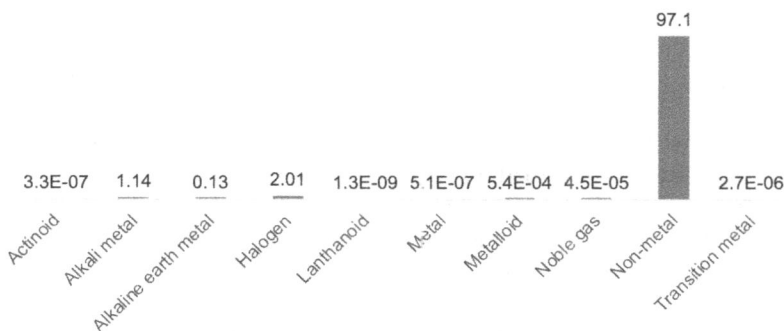

Actinoid	Alkali metal	Alkaline earth metal	Halogen	Lanthanoid	Metal	Metalloid	Noble gas	Non-metal	Transition metal
3.3E-07	1.14	0.13	2.01	1.3E-09	5.1E-07	5.4E-04	4.5E-05	97.1	2.7E-06

22.7. Abundance of chemical elements in the crust and atmosphere of the Moon

Rocks brought back to Earth from the Moon showed that the composition of the Moon's crust by weight is:

- Oxygen: 60%
- Silicon: 16-17%
- Aluminum: 6-10%
- Calcium: 4-6%
- Magnesium: 3-6%
- Iron: 2-5%
- Titanium: 1-2%
- Other elements present in amounts much smaller than 1%

The abundance of these chemical elements is said to be similar to that of the Earth's crust, therefore pointing to a common origin of the Moon and Earth that all theories have failed to explain so far. In *"Turbulent Origin of the Universe"*, I explained how the Moon and the Earth descended from the same precursor: the precursor of the Earth-Moon System. Because a piece of rock cannot reflect the composition of the entire Moon's crust, conclusions drawn from the similarities between the abundance of the chemical elements in the Earth's crust and the Moon's crust must be done carefully. Even as for Earth, a same rock can have different compositions according to its location. For instance, all granites are not the same. Therefore, the Moon's rocks brought back to Earth, which are not even taken from many locations of the Moon surface, should not be sufficient to conclude the composition of the Moon. The rest of this segment of the Moon, focused on its atmosphere.

Also called the lunar atmosphere, but not to be confused with the Moon's crust, the Moon's atmosphere contains many atoms and ions. The chemical composition of the Moon has been one of the data used by some scientists to try to explain the formation of that satellite. Therefore, I was also interested in comparing the atmosphere of the Moon and that of the Earth to see if they contain any information that can help me know what might have happened to their common precursor.

Unlike the planets which atmospheric composition is expressed in percent (%) or part(s) per million (ppm), the atmospheric composition of the Moon is in number of particles per cm3 (Fig. 85). These units alone made it a little harder to compare the composition of the atmosphere of the Moon to that of the Earth. It also made me to think that something significant is different in their atmosphere or the way it was studied. To manage this challenge, I ranked the constituents of the atmosphere from the most abundant to the least abundant. To the most abundant element, I gave the 1st rank and to the second most abundant, I gave the 2nd rank and so on and so forth. Then, I compared the rank of the constituents of the Moon and the Earth.

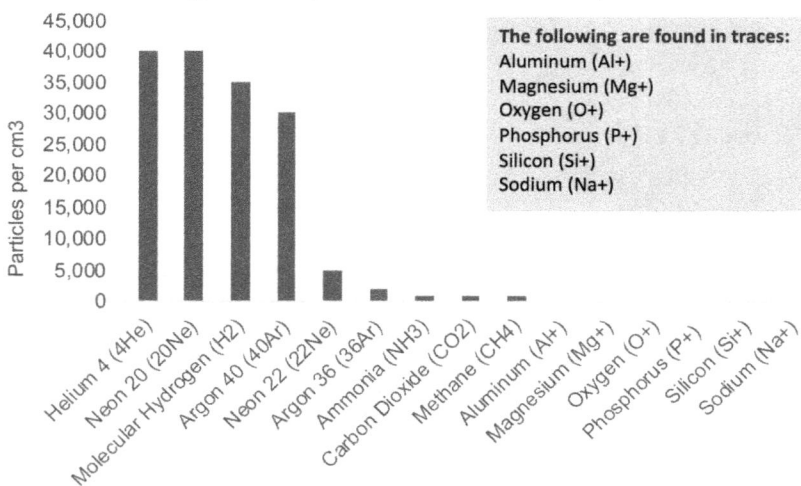

Fig. 85: Composition of the lunar atmosphere

The following are found in traces:
Aluminum (Al+)
Magnesium (Mg+)
Oxygen (O+)
Phosphorus (P+)
Silicon (Si+)
Sodium (Na+)

I found that, unlike the atmosphere of the Earth, the atmosphere of the Moon is dominated by helium and neon, two of the naturally occurring noble gases lighter than nitrogen and oxygen (which dominate the Earth's atmosphere). The atmosphere of the Moon is richer in noble gases than that of the Earth maybe because of the smaller rotational speed of the Moon which also implies that the precursor of the Moon's atmosphere could not have rotated fast, but slowly.

Water is the 3rd most abundant constituent of the Earth's atmosphere, whereas in the atmosphere of the Moon, molecular hydrogen is the 3rd most abundant constituent. However, on Earth, molecular hydrogen is the 10th most abundant gas. Table 7 compares the composition of the lunar atmosphere and the Earth's atmosphere.

Table 7: Comparison of the composition of the lunar atmosphere and the Earth's atmosphere

Name	Moon's atmosphere		Earth's atmosphere	
	Particles per cm^3	Rank	(ppm, by volume)	Rank
Helium 4 (^{4}He)	40000	1st	5.24	8th
Neon 20 (^{20}Ne)	40000	1st ex	18.18	6th
Molecular Hydrogen (H_2)	35000	3rd	0.55	11th
Argon 40 (^{40}Ar)	30000	4th	9340	4th
Neon 22 (^{22}Ne)	5000	5th	18.18	6th ex
Argon 36 (^{36}Ar)	2000	6th		
Ammonia (NH_3)	1000	7th		
Carbon Dioxide (CO_2)	1000	7th ex	400	5th
Methane (CH_4)	1000	7th ex	1.7	9th
Aluminum (Al+)	Trace	8th		
Magnesium (Mg+)	Trace	8th ex		
Oxygen (O+)	Trace	8th ex	209500	2nd
Phosphorus (P+)	Trace	8th ex		
Silicon (Si+)	Trace	8th ex		
Sodium (Na+)	Trace	8th ex		
Nitrogen (N_2) (ppm)			780800	1st
Water (H_2O) (ppm)			10000	3rd
Krypton (Kr) (ppm)			1.14	10th

Source of the raw data: NASA (2018).

Argon is the 4th most abundant gas in the atmosphere of both the Moon and the Earth. A trace of many heavier chemical elements is found in the atmosphere of the Moon, but not in that of the Earth. Some of the elements are: aluminum, magnesium, oxygen, phosphorus, silicon, and sodium. All of these elements are heavier and denser than nitrogen, the most abundant gas in the Earth's atmosphere. To put it another way, the Moon's atmosphere contains heavier elements that are not even found in the Earth's atmosphere. The distance that the precursor of the Moon traveled after escaping the precursor of the Earth may have favored the formation of denser particles as the turbulence in the precursor of the Moon was developing. The small size of the precursor of the Moon (as compared to that of the Earth) may have favored the squeezing of the precursors of its constituents, which yielded heavier atoms just like the precursors of Mars also did. The formation of those heavier elements in the atmosphere of the Moon could have worked against the formation of lighter nonmetals like nitrogen and oxygen, which are abundant in the Earth's atmosphere, but are just in traces in the atmosphere of the Moon. Probably, after the precursor of the Moon split from the precursor of the Earth-Moon system, the precursors of atoms in the precursor of the Moon's atmosphere could have kept them from being shaken and molded a little longer than the precursor in the Earth's atmosphere before reaching the orbit of the Moon. The extra time they took before the formation of the Moon was finalized could have contributed to the formation of denser atoms in the Moon's atmosphere. Probably, the Moon's crust would contain heavier elements than the Earth's.

22.8. Abundance of chemical particles in the atmosphere of the planets in the Solar System

Before I present the abundance of the chemicals in the atmosphere of the planets, I will first talk a little bit about the mass of the atmosphere of the planets in the Solar System. Indeed, all of the planets in the Solar System have an atmosphere. Some atmospheres are very tiny, while others are very large. The surface of some planets (e.g. giant planets) has not been well defined or ever seen and consequently, the innermost limit of their atmosphere is not well known. For instance, the giant planets (e.g. Jupiter and Saturn) are sometimes called giant gases because scientists have not empirically seen or accessed their crust so far, therefore they are believed to be mostly made of gas. Nevertheless, it may be difficult to rule out that they have no crust. The atmospheric composition is the relative composition by volume of gases and of any other constituents in the atmosphere.

Scientists are not able to measure the mass of the atmosphere of all the planets yet. For instance, as of 2025, the surface of the giant gas planets (e.g. Jupiter and Saturn) is unknown but believed to be filled with gas. The mass of the atmosphere of the 4 innermost planets (Mercury, Venus, Earth, and Mars) to vary between 10,000 kg and 4.8E+20 kg (NASA, 2018). The lightest atmosphere is that of Mercury, whereas the heaviest atmosphere is that of Venus. While the atmosphere of Venus is 94.1 times heavier than that of the Earth, that of Mars is 0.49% that of the Earth. However, the atmosphere of Mercury is very tiny, being 1.96078E-15 times that of the Earth. Table 8 illustrates the mass of the atmosphere of the planets in the Solar System (NASA, 2018).

Table 8: Mass of the atmosphere of the planets in the Solar System

Name	Mass of Atmosphere (kg)
Mercury	10000
Venus	4.8E+20
Earth	5.1E+18
Mars	2.5E+16
Jupiter	
Saturn	
Uranus	
Neptune	
Pluto	

The mass of Venus' atmosphere may have something to do with the circumstance of its formation. As a little comparison, the mass of the Moon is 734.6 x10^20 kg, meaning that the Moon is 153.05 times heavier than Venus' atmosphere. In other words, the atmosphere of Venus is 0.65% the mass of the Moon (Earth's satellite). More than 99% of the mass of the Earth's atmosphere is found below 100 km of latitude above the sea level, which is defined as the Karman line, which marks the beginning of space. However, even beyond that line, several chemical compounds are still found in space, but as the altitude increases, the distance between the chemical compounds increases as well.

At this point, I will now present the abundance of chemical particles in the atmosphere of the planets in the Solar System. The name, number, and content or abundance of the chemical elements and compounds in the atmosphere of the planets in the Solar System are presented in Table 9.

Table 9: Atmospheric composition (% by volume) of the planets in the Solar System

Chemicals	Mercury	Venus	Earth	Mars	Jupiter	Saturn	Uranus	Neptune	Pluto
Aluminum (Al)	Trace								
Ammonia (NH_3)					0.026	0.0125			
Ammonia hydrosulfide (NH_4)HS (Aerosols)					Yes	Yes	Yes	Yes	
Ammonia ice (Aerosols)					Yes	Yes	Yes	Yes	
Argon (Ar)	Trace	0.007	0.934	1.6					
C_2H_x Hydrocarbons									Yes
Calcium (Ca)	Trace								
Carbon Dioxide (CO_2)	Trace	96.5	0.041	95.32					
Carbon Monoxide (CO)		1.7E-03		0.08					0.05
Ethane (C_2H_6)					5.8E-04	7.E-04		1.5E-04	
Helium (He)	6	1.2E-03	5.24E-04		10.2	3.25	15.2	19	
Hydrogen (Molecular)(H_2)	22		5.5E-05		89.8	96.3	82.5	80	
Hydrogen Deuteride (HD)					2.8E-03	1.1E-02	1.48E-02	1.92E-02	
Hydrogen-Deuterium-Oxygen (HDO) also called heavy water				8.5E-05					
Iron (Fe)	Trace								
Krypton (Kr)	Trace		1.14E-04	3.E-05					
Magnesium (Mg)	Trace								
Methane (CH_4)			1.7E-04		0.3	0.45	2.3	1.5	0.5
Methane ice (?) (Aerosols)							Yes	Yes	
Neon (Ne)	Trace	7.E-04	1.82E-03	2.5E-04					
Nitrogen (N_2)	Trace	3.5	78.08	2.7					99
Nitrogen Oxide (NO)				0.01					
Oxygen (O_2)	42		20.95	0.13					
Potassium (K)	0.5								
Sodium (Na)	29								
Sulfur Dioxide (SO_2)		0.015							
Water (H_2O)	Trace	0.002	1	0.021	4.E-04				
Water ice (Aerosols)					Yes	Yes	Yes	Yes	
Xenon (Xe)	Trace			8.E-06					

Source of the raw data: NASA (2018).

Figure 86 illustrates the abundance of the chemicals in the planets as mentioned in the above table. The chemicals that are in trace amounts are not mentioned in the graph.

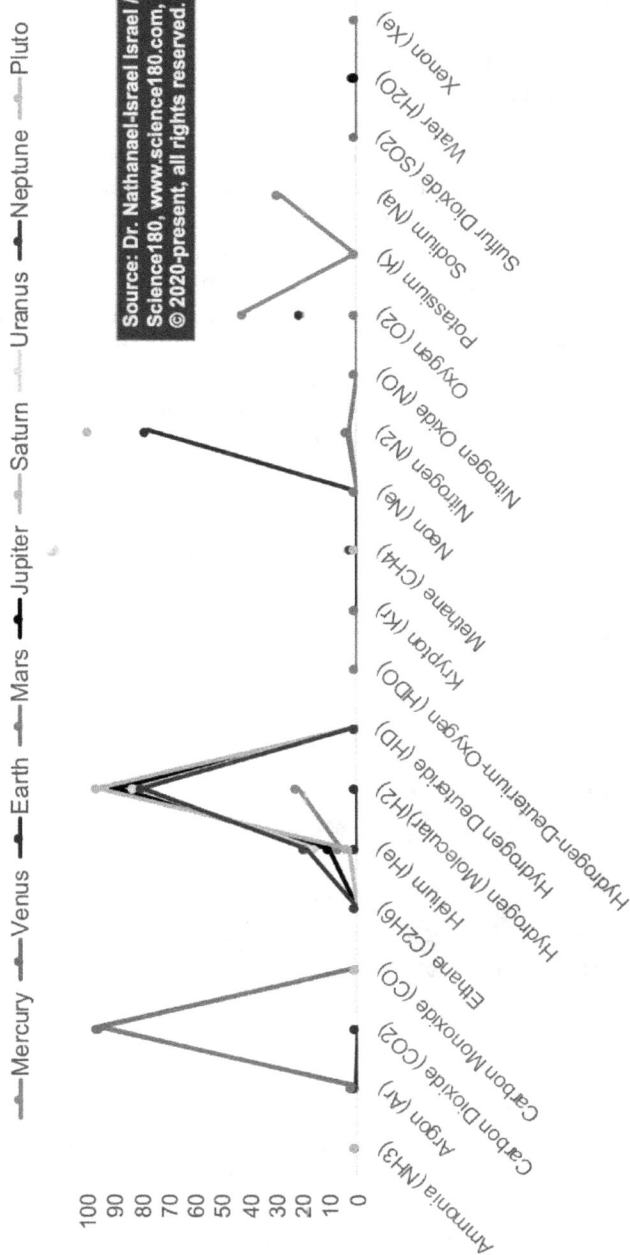

Fig. 86: Abundance (%) of the most abundant chemical elements and compounds in the atmosphere of planets in the Solar System

The number of chemical elements in the atmosphere of the planets depends on their location and varies between 4 and 15 (Fig. 87). The highest number was found in the atmosphere of Mercury, whereas the lowest number was with the atmosphere of Pluto. The number of chemical elements found in the atmosphere of the giant planets is the same: 7. The atmosphere of the 4 innermost planets contains more elements than that of the 4 giant planets, suggesting that the position of the precursors of the planetary systems could have affected the number of atoms that the precursors of their atoms yielded. One thing that the innermost planets have in common is their high orbital speed, which suggested that the speed of the precursors of the planets could have affected how their constitutive particles could have been split or aggregated into microscopic entities that produced the various atoms they contain today. The fact that many more chemical elements were formed in the atmosphere of Mercury than in the atmosphere of any other planet in the Solar System could explain why the abundance of the most abundant compound in its atmosphere is the least among the most abundant chemicals found in the atmosphere. In other words, it is not by chance that most elements present in the atmosphere of Mercury are in traces. In the next paragraphs, I better elaborated on this trend.

Fig. 87: Number of chemical elements in the atmosphere of the planets in the Solar System

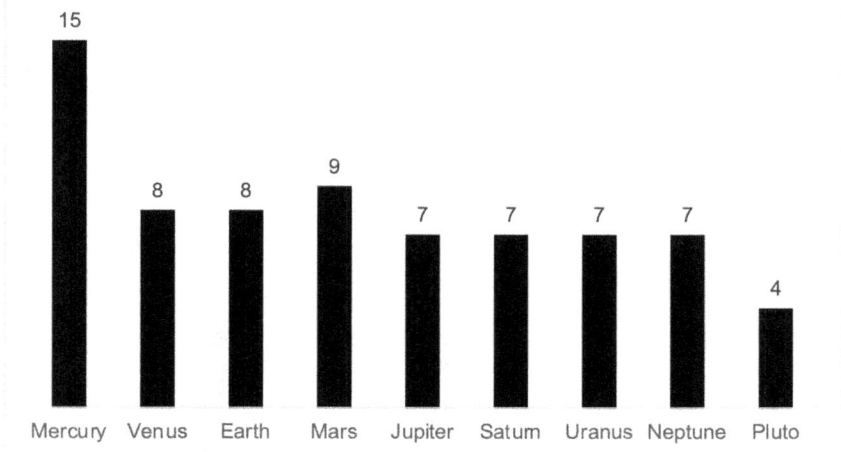

Before addressing the origin of the chemical elements, I thought about quickly recapitulating here their trends in the atmosphere of the planets. The atmosphere of the 5 terrestrial planets (Mercury, Venus, Earth, Mars, and Pluto) abounds with denser atoms than the atmosphere of the giant planets. Indeed, oxygen followed by sodium and hydrogen are the main elements in the atmosphere of Mercury. Not only is Sodium recorded only in the atmosphere of Mercury, but its abundance is very high (29%). The atmosphere of Venus is dominated by carbon dioxide followed by nitrogen. The Earth's atmosphere is dominated by nitrogen followed by oxygen. The atmosphere of Mars is dominated by CO_2 (95.32%)

followed by nitrogen. The atmosphere of Pluto, the outermost terrestrial planet in the Solar System, is dominated by nitrogen (99%).

Just like the atmosphere of the other giant planets, the atmosphere of Jupiter is dominated by molecular hydrogen (89.8%), and to a smaller extent by helium as well (10.2%). Probably, during the genesis of the atmosphere of Jupiter, about 10.2% of the volume of the precursors of matter in the precursor of the atmosphere of Jupiter could have been converted into a few atoms of helium but about 89.8% of those precursors became molecular hydrogen. The atmosphere of Saturn is dominated by hydrogen gas (96.3%). During the genesis of Saturn's atmosphere, fewer proportions of the precursor of matter could have been converted into helium (about 3.25%) and consequently, a higher proportion of it was molded into molecular hydrogen (98.3%). That is why the hydrogen content on Saturn is higher than that on Jupiter. Additionally, more hydrogen deuteride was formed on Saturn than on Jupiter's atmosphere. Of the 4 giant planets, Saturn is the least dense. The environmental or spatial conditions which affected the density of Saturn also affected the way the precursors of the matter which became Saturn's atmosphere were shaped and molded. In other words, just as the factors which shaped the precursor of Saturn were not able to make that planet very dense, they also were not able to make most of the atoms in the atmosphere of Saturn very dense. As of 2020, hydrogen is the least dense atom. In other words, the most abundant chemical on the atmosphere of the least dense planet is the least dense atoms. To put it into other words, the environmental factors that affected the formation of the celestial bodies also affected the particles, atoms, molecules, and chemical compounds in them.

The helium content of the atmosphere of Uranus is higher than that of Jupiter and Saturn. In contrast, its hydrogen content is smaller. During the formation of the atmosphere of Uranus, higher proportions of the precursor of matter could have been converted into heavier atoms than hydrogen. For instance, the abundance of helium and methane (subsequently the abundance of carbon) on Uranus is higher than that on Saturn or Jupiter. Consequently, a less proportion of that precursor of atoms was converted into hydrogen. Hence, abundance of hydrogen on Uranus (82.5%) is lower than that on Saturn (96.3%) and on Jupiter (89.8%).

Of the atmosphere of the 4 giant planets, that of Neptune contains less molecular hydrogen (80%), but more helium (19%). It contains less methane than Uranus' atmosphere. I think that during the formation of Neptune's atmosphere, fewer atoms were converted into carbon and consequently were gathered together with hydrogen to form more methane on Neptune than on Uranus. In contrast, a higher proportion of the precursor of the matter of Neptune was converted into helium than for any other planet in the Solar System. That is why the abundance of helium in the atmosphere of Neptune is higher than that in the atmosphere of any other planet. Because on Neptune a higher proportion of the precursor of matter was converted into helium, a lesser proportion was converted into hydrogen. Consequently, the hydrogen content of the atmosphere in Neptune is smaller than that in the atmosphere of any of the other giant planets. The fact that Neptune is denser than any of the other giant planets (see details in *"Turbulent Origin of the Universe"*) could also explain why its atmosphere has more helium (which is denser than hydrogen) than the atmosphere of the other giant planets.

Based on what I was sensing when I deeply looked at the data on chemical elements, I knew that with a strategical thinking and review of the data, I would unearth some critical information that can explain the pathways of the formation of chemical elements with the perspective of turbulence almost just as I did it for the celestial bodies. Therefore, although time was passing, I spent more time analyzing the chemical elements in the atmosphere from other angles as explained below.

Now, I will compare the most abundant constituents in the atmosphere of the planets. The data I presented so far showed that, the most abundant chemicals in the atmosphere depend on the planets. For instance, while oxygen (O_2) is the most abundant constituent of the atmosphere of Mercury (42%), carbon dioxide (CO_2) is the most abundant compound in the atmosphere of Venus (96.5%) and Mars (95.32%). In contrast, the most abundant constituent in the atmosphere of the Earth is nitrogen (N_2) (78.08%). Pluto, the outermost terrestrial planet, is dominated by nitrogen (N_2), which constitutes 99% of its atmosphere. Finally, molecular hydrogen (H_2) is the most abundant constituent in the atmosphere of the giant planets: Jupiter (89.8%), Saturn (96.3%), Uranus (82.5%), and Neptune (80%). In other words, the atmosphere of the terrestrial planets (Mercury, Venus, Earth, and Pluto) is dominated by heavier chemical elements (e.g. Carbon (C, Z=6), Nitrogen (N, Z=7), Oxygen (O, Z=8), and Sodium (Na, Z=11)). In contrast, the atmospheres of the 4 giant planets are dominated by the lightest chemical elements: Hydrogen (H, Z=1) followed by Helium (He, Z=2).

I was interested in studying the abundance of the chemical elements, but I was slowed down by the fact that the data are available for compounds and some gases instead of their constitutive element. Therefore, to try to assess the abundance of the chemical compounds in the atmosphere, I ordered them by decreasing abundance into the:

- 1st most abundant
- 2nd most abundant
- 3rd most abundant

Because it was very difficult to properly estimate the abundance of the chemical elements using the abundance of the compounds in which they are found, I used the atomic mass and radius of the chemical elements to order their abundance. While doing so, I also considered the abundance of the compounds. Then, I did Table 10:

Table 10: Most abundant chemicals in the atmosphere of the planets in the Solar System

Planets	#1 Most abundant compound	2nd most abundant compound	3rd most abundant compound	Order and abundance of the 3 most abundant chemical elements (%)	Chemical elements present in atmosphere
Mercury	Oxygen (O_2) 42%	Sodium (Na): 29%	Molecular Hydrogen (H_2): 22%	O (Z=8), Na (Z=11), H (Z=1): 93%	Hydrogen (H, Z=1), Helium (He, Z=2), Carbon (C, Z=6), Nitrogen (N, Z=7), Oxygen (O, Z=8), Neon (Ne, Z=10), Sodium (Na, Z=11), Magnesium (Mg, Z=12), Aluminum (Al, Z=13), Argon (Ar, Z=18), Potassium (K, Z=19), Calcium (Ca, Z=20), Iron (Fe, Z=26), Krypton (Kr, Z=36), Xenon (Xe, Z=54)
Venus	Carbon Dioxide (CO_2): 96.5%	Nitrogen (N_2) 3.5%	Sulfur Dioxide (SO_2): 0.015%	O (Z=8), C (Z=6), N (Z=7): 100%	Hydrogen (H, Z=1), Helium (He, Z=2), Carbon (C, Z=6), Nitrogen (N, Z=7), Oxygen (O, Z=8), Neon (Ne, Z=10), Sulfur (S, Z=16), Argon (Ar, Z=18)
Earth	Nitrogen (N_2) 78.08%	Oxygen (O_2): 20.95%	Water (H_2O): 1%	N (Z=7), O (Z=8), H (Z=1): 99.03%	Hydrogen (H, Z=1), Helium (He, Z=2), Carbon (C, Z=6), Nitrogen (N, Z=7), Oxygen (O, Z=8), Neon (Ne, Z=10), Argon (Ar, Z=18), Krypton (Kr, Z=36)
Mars	Carbon Dioxide (CO_2): 95.32%	Nitrogen (N_2) 2.7%	Argon (Ar): 1.6%	O (Z=8), C (Z=6), N (Z=7): 98.02%	Hydrogen (H, Z=1), Deuterium (D, Z=1), Carbon (C, Z=6), Nitrogen (N, Z=7), Oxygen (O, Z=8), Neon (Ne, Z=10), Argon (Ar, Z=18), Krypton (Kr, Z=36), Xenon (Xe, Z=54)
Jupiter	Molecular Hydrogen (H_2): 89.8%	Helium (He): 10.2%	Methane (CH_4): 0.3%	H (Z=1), He (Z=2), C (Z=6): 100%	Hydrogen (H, Z=1), Deuterium (D, Z=1), Helium (He, Z=2), Carbon (C, Z=6), Nitrogen (N, Z=7), Oxygen (O, Z=8), Sulfur (S, Z=16)
Saturn	Molecular Hydrogen (H_2): 96.3%	Helium (He): 3.25%	Methane (CH_4): 0.45%	H (Z=1), He (Z=2), C (Z=6): 99.55%	Hydrogen (H, Z=1), Deuterium (D, Z=1), Helium (He, Z=2), Carbon (C, Z=6), Nitrogen (N, Z=7), Oxygen (O, Z=8), Sulfur (S, Z=16)
Uranus	Molecular Hydrogen (H_2): 82.5%	Helium (He): 15.2%	Methane (CH_4): 2.3%	H (Z=1), He (Z=2), C (Z=6): 97.7%	Hydrogen (H, Z=1), Deuterium (D, Z=1), Helium (He, Z=2), Carbon (C, Z=6), Nitrogen (N, Z=7), Oxygen (O, Z=8), Sulfur (S, Z=16)
Neptune	Molecular Hydrogen (H_2): 80%	Helium (He): 19%	Methane (CH_4): 1.5%	H (Z=1), He (Z=2), C (Z=6): 99%	Hydrogen (H, Z=1), Deuterium (D, Z=1), Helium (He, Z=2), Carbon (C, Z=6), Nitrogen (N, Z=7), Oxygen (O, Z=8), Sulfur (S, Z=16)
Pluto	Nitrogen (N_2): 99%	Methane (CH_4): 0.5%	Carbon monoxide (CO): 0.05%	N (Z=7), C (Z=6), H (Z=1): 99.55	Hydrogen (H, Z=1), Carbon (C, Z=6), Nitrogen (N, Z=7), Oxygen (O, Z=8)

Z= number of protons = number of electron of chemical elements. Molar mass (g/mol): Hydrogen (H, Z=1): 1; Helium (He, Z=2): 4; Carbon (C, Z=6): 12; Nitrogen (N, Z=7): 14; Oxygen (O, Z=8): 16; Sodium (Na, Z=11): 23.

At this point, I will present the relationship between the 1st, the 2nd, and the 3rd most abundant chemical compounds and elements in the atmosphere of the planets. Indeed, as I was carefully visualizing the graphs of the trends of the chemical elements, I noticed a relationship between the abundance of the elements within the atmosphere of each planet and from one planet to the other. Because the abundance data was not available for all chemical compounds, I had to focus on what was reported. Because the abundance of the 3 most abundant chemical compounds varies between 93% and 100%, and many other compounds shared a very small percentage, I decided to focus the regression on the 3 most abundant compounds in the atmosphere of the planets, knowing that I will care for the other elements later. Fig. 88 and Fig. 89 show the regression between the top 3 most abundant compounds in the atmosphere of each planet.

The abundance data showed that the abundance of the 1st most abundant compound in the atmosphere of each planet negatively affected that of the 2nd most abundant compound (r2>0.99) (Fig. 88) and that of the 3rd most abundant compound (r2=0.99) (Fig. 89). The relationship between the abundance of the 2nd and 3rd most abundant compounds is not strong (r2=0.6).

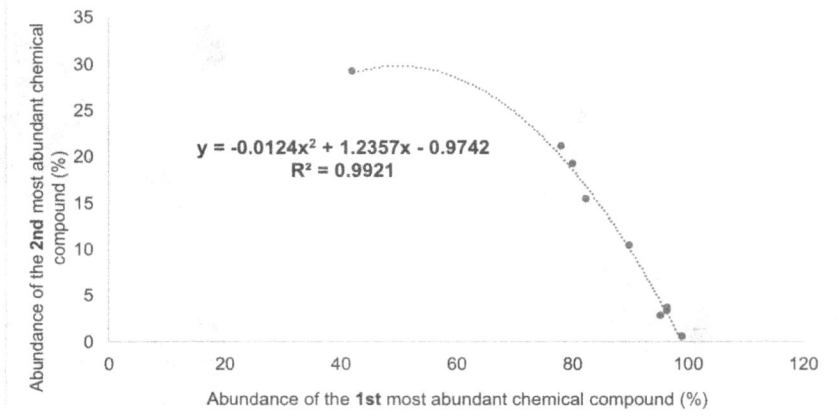

Fig. 88: Relationship between the abundance of the **first** most abundant and the **2nd** most abundant chemical compounds in the atmosphere of the planets

$y = -0.0124x^2 + 1.2357x - 0.9742$
$R^2 = 0.9921$

Abundance of the **2nd** most abundant chemical compound (%)

Abundance of the **1st** most abundant chemical compound (%)

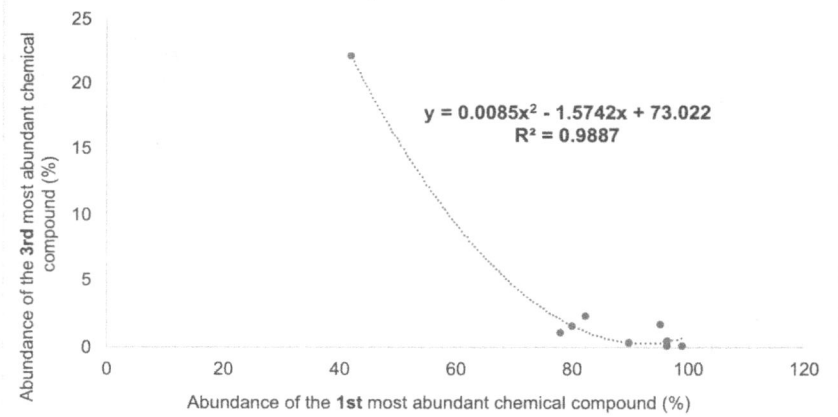

Fig. 89: Relationship between the abundance of the **first** most abundant and the **3rd** most abundant chemical compounds in the atmosphere of the planets

$y = 0.0085x^2 - 1.5742x + 73.022$
$R^2 = 0.9887$

Abundance of the **3rd** most abundant chemical compound (%)

Abundance of the **1st** most abundant chemical compound (%)

Based on the global trends of all the data I have on the chemical elements and compounds, I felt like the precursors of the atoms could have competed for resources. In other words, as the pool of precursors of atoms was being transformed into various atoms, some of them were differentiated into particular atoms, while the remaining were converted into others in such a way that the more precursors were converted into specific atoms, the amount of the precursors of atoms left to be converted into other types of atoms decreased. In the end, when the abundance of certain chemical elements increased, that of the others likely decreased. For it is from the same bulk of precursors of atoms that all diverse forms of atoms were molded. What defined the type of atoms that are dominant in each atmosphere were the environmental conditions. Therefore, when these environmental conditions favored the formation and dominance of certain atoms and consequently the compounds in which they are present, these conditions are not always favorable to the abundance of atoms of other elements. In the end, the abundance of the chemical elements differs from one planet to the other. In environments where the conditions were more "similar", the most abundant chemical elements and compounds were formed with an abundance belonging to the same order of magnitude. That is why, because they were formed in the most turbulent regions of the precursor of the bodies orbiting the Sun, the precursors of the giant planets yielded atmospheres dominated by the same chemical elements: hydrogen followed by helium. However, because the precursors of these elements "competed" for the same resources (precursors of atoms) to form or shape themselves, their abundance varied according to the other elements that were formed in their environment.

Because the #1 most abundant compound on Mercury did not have a very huge share in the total abundance as the #1 most abundant compound in the atmosphere of the other planets, the abundance of the 2nd and 3rd most abundant compounds on Mercury are higher than that in the atmosphere of the other planets. The message I am conveying here is that, because the most abundant compound on Mercury took a small share of the pool of the precursors of atoms on Mercury, the 2nd and 3rd most abundant compound in the atmosphere of that planet had a bigger share. This trend is one of the key elements that suggested to me that there may have been a kind of "competition" of the precursors of the atoms for how many precursors of atoms can be differentiated into certain types of atoms. To put it in another way, from the same pool of precursors of atoms, some were differentiated into certain atoms, while others were differentiated into other kinds of atoms. Before closing this section, I need to emphasize that, because a chemical element is said to be absent in the atmosphere does not mean that it does not exists even in traces. For, although some chemicals exist in the atmosphere of the planets, they were not detected because their level of abundance is beneath the threshold that the measurement equipment could detect.

Another Book by Nathanael-Israel Israel:
SCIENCE180 ACCURATE SCIENTIFIC PROOF OF GOD

THE FIRST AND THE ONLY SCIENTIFIC BOOK THAT TALKS TO ANTI-CREATIONISTS, EVOLUTIONISTS, BIG BANG PROPONENTS, ATHEISTS, AND ALL OTHER FREETHINKERS AND RATIONALISTS ABOUT THE UNIVERSE FORMATION AND THEY BEG TO KNOW MORE ABOUT GOD, THE CREATOR, THAT THEY DENY.

As you read this historic book, you will:

- Scientifically know what is the one clear sign you should always pay attention to in your efforts to decipher the primary cause and the key drivers of the fundamental processes responsible for the universe-formation

- Discover the only way to scientifically know if God exist, and if so, which of the thousands of beings worshipped across the globe is the true God

- Accurately answer the most critical universe-origin and life-origin questions so you can stop standing in tension with consequential question marks including those related to religion and reason or the so-called war between science and the Bible

- Definitively answer all your doubts about the source or author of the universe and life … (learn more at Science180.com/godproof)

- Understand that religion or faith, reason or science can coexist and can be properly reconciled to accurately lead you to the correct source of everything in the universe

- Satisfy your burning desire for freedom from beliefs and scientific theories about the universe-origin and life-origin that suffocate you and bind your mind, faith, unbelief, heart, and education

- Scientifically set on fire all false theories or dogmas about the existence of God, the Creator, that are enslaving humankind

- Whether you are a believer, unbeliever, freethinker, administrator, politician, curriculum designer, curriculum specialist, education policymaker, teacher, librarian, school board member, researcher, parent, student, clergy, or a layperson, as long as you are really seeking to scientifically understand the rational proof of the existence of God, *"Science180 Accurate Scientific Proof of God"* is the much-admired book written for great people just like you! Grab your copy today and start reading it! Don't wait any longer!

Dr. Nathanael-Israel Israel is a Beninese-born American scientist, entrepreneur, and international consultant

CHAPTER 23

IF YOU ARE A SCIENTIST, PAY ATTENTION TO THIS CRUCIAL LESSON FROM THE COMPARISON OF THE ABUNDANCE OF CHEMICAL ELEMENTS– AND AVOID THE DANGEROUS MISTAKE ALL CHEMISTS AND PHYSICISTS HAVE MADE– UNKNOWINGLY?

Here, I will compare the abundance of the chemical elements in the universe or our galaxy, the Sun, human beings, oceans, the Earth's crust, and in the atmosphere of the planets, then, I will deduce the machinery of the formation of the chemicals in the next chapter. By the time I reached this point of the analysis of the chemical elements, I felt like I needed to compare their abundance in the universe, the Sun, the Earth's crust, the oceans, human beings, and in the atmosphere of the planets. For tying together all of the abundances of the chemical elements according to their location can give a global perspective of how they were formed as components of the major shakeup that the universe went through during its formation. Using key recapitulative graphs, I will point out the main findings. As you read the following segments, you will understand how the diversity of the chemical elements in the universe is a footprint of the various scales of turbulence that occurred in different "pockets" or portions of space of the precursor of matter, which was shaped or molded differently by the major turbulence that occurred in the early universe, and which is the root of the split-gathering of matters into bodies of different characteristics, ranging from subatomic particles to galaxy clusters passing by stars, planets, asteroids, and various systems of bodies (e.g. stellar systems, planetary systems, satellite systems, asteroids systems, etc.) that altogether form the complex and well-organized world that human beings have been trying to decode and understand since the beginning of time.

23.1. General trends

The 15 most abundant chemical elements in the universe or in our galaxy, the Sun, human beings, oceans, the Earth's crust, and atmosphere of the planets have an atomic number smaller than 39 (Fig. 90). In other words, the most abundant chemical elements in nature are not the heaviest elements. Of the 38 chemical elements having the smallest atomic number, implying the 38 lightest chemical elements, 13 are not among the 15 most abundant in the universe, our galaxy, the Sun, the oceans, the Earth's crust, and in human beings (Fig. 90). By increasing order of their atomic number, these 13 chemical elements, which, except Krypton (Kr, Z=36), happen to be more abundant in the Earth's crust, are:

- Lithium (Li, Z=3), mostly abundant in the Earth's crust: 0.0017%
- Beryllium (Be, Z=4), mostly abundant in the Earth's crust: 1.9E-04%
- Scandium (Sc, Z=21), highest abundance was in the Earth's crust: 2.6E-03%
- Vanadium (V, Z=23), highest abundance was in the Earth's crust: 0.019%
- Chromium (Cr, Z=24), highest abundance was in the Earth's crust: 0.014%
- Cobalt (Co, Z=27), highest abundance was in the Earth's crust: 0.003%
- Copper (Cu, Z=29), highest abundance was in the Earth's crust: 0.007%
- Gallium (Ga, Z=31), highest abundance was in the Earth's crust: 1.9E-03%
- Germanium (Ge, Z=32), highest abundance was in the Earth's crust: 1.4E-04%
- Arsenic (As, Z=33), highest abundance was found in the Earth's crust: 2.1E-04%
- Selenium (Se, Z=34), highest abundance was found in the Earth's crust and in human beings: 5E-06%
- Krypton (Kr, Z=36), highest abundance was found in the Universe or in our Galaxy: 4E-06%
- Rubidium (Rb, Z=37), highest abundance was found in the Earth's crust: 0.006%

These data suggested that having a small atomic number is not sufficient for a chemical element to be abundant in nature. However, the biggest or heaviest chemical elements are usually the least abundant.

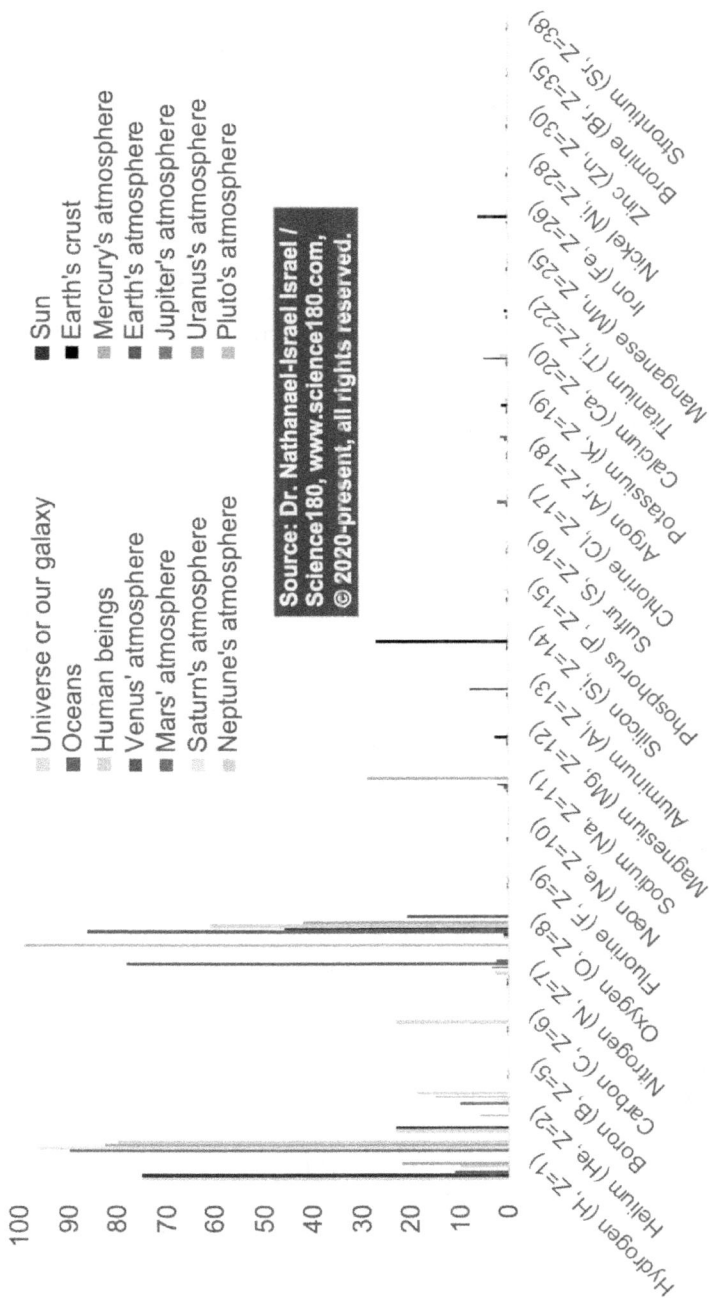

Fig. 90: Most dominant chemical elements in the universe or our galaxy, the Sun, oceans, Earth's crust, and human beings

The chemical elements which abundance is higher than 20% in the environment where they were measured are: hydrogen, helium carbon, oxygen, and silicon. Although Fig. 90 does not clearly show the data about chemicals in the atmosphere of the planets, to avoid any confusion, I would like to recall here that, as I explained in the following sections, carbon dioxide (CO_2) is more than 95% in the atmosphere of some planets (i.e. Venus and Mars), while molecular hydrogen (H_2) is more than 80% in the atmosphere of the 4 giant planets (Jupiter, Saturn, Uranus, and Neptune). In contrast, nitrogen (N_2) dominates the atmosphere of the Earth (78.08%) and Pluto (99%). Oxygen is most abundant in the atmosphere of Mercury (42%) and of the Earth (20.95%). Finally, sodium (Na), the lightest metal found in the atmosphere is abundant in the atmosphere of Mercury (29%), while no metal is recorded in the atmosphere of any other planet.

In the universe and in the Sun, the abundance of hydrogen is higher than 75% just as it is the case in the atmosphere of the 4 giant planets. Yet, the Earth's crust contains just 0.15% hydrogen and some trace in its atmosphere (5.5E-05%). Similarly, while hydrogen is not dominant in oceans, the Earth's crust, and human beings, as it accounts for less than 12%, oxygen is very abundant. For instance, oxygen makes up 86% of the oceans' mass, 46% of the Earth's crust, and 61% in humans. However, oxygen is said to make up just 1% of the mass of the universe or 0.9% that of the Sun. After hydrogen, helium is mostly found in the universe and the Sun where it accounts for 23% of their weight. The highest abundance of carbon was found in the atmosphere of some planets and in human beings where it accounts for 23%, coming just after oxygen, the most dominant chemical elements in human beings. Silicon is most abundant in the Earth's crust where it comes as the second most abundant chemical element (27%), just after oxygen. Yet, silicon is not found in the atmosphere of any planet, not even that of the Earth or of the terrestrial planets, which crust are postulated to be rich in silicates. In the Sun, the abundance of silicon is said to be 0.09%, while in the universe it is said to be 0.07%. In human beings, the abundance of silicon is 0.026%, while in oceans, it is 1E-04%.

To better see the trends, I narrowed down the number of chemical elements or compounds in the previous comparative graph (Fig. 90) to the 7 most abundant in each system to obtain Fig. 91. On that graph, the tendency of the abundance in the Sun and that in the universe or our galaxy are similar. In fact, when I tried to separate the abundance of the Sun from that in the universe or our galaxy, the trends of the abundance in the universe were overshadowed by that in the Sun, which can confuse some people as they may not even know that the data on the elements in the universe were also shown in the graph. The abundance of chemicals in the universe and in the Sun is similar because what is known about the chemical composition of the universe is basically that of the stars. For the sky is dominated by stars and most of the exoplanets and exo asteroids may not be clearly seen from the Earth. Therefore, because the chemical composition of the stars have some similarities, the composition of the universe itself, which can be seen as the mean composition of the stars in the universe, is likely similar to that

Nathanael-Israel Israel: Has had the honor to be Acknowledged the First Human Being that Scientifically Reconciled Science and Biblical Creation

of the Sun, the closest star to the Earth. However, considering what I know about the planets and satellites in the Solar System, I personally think that the composition of the stars may differ according to their position, size, and speed. Smaller and faster stars may be denser and richer in dense and heavy elements than bigger and slower stars.

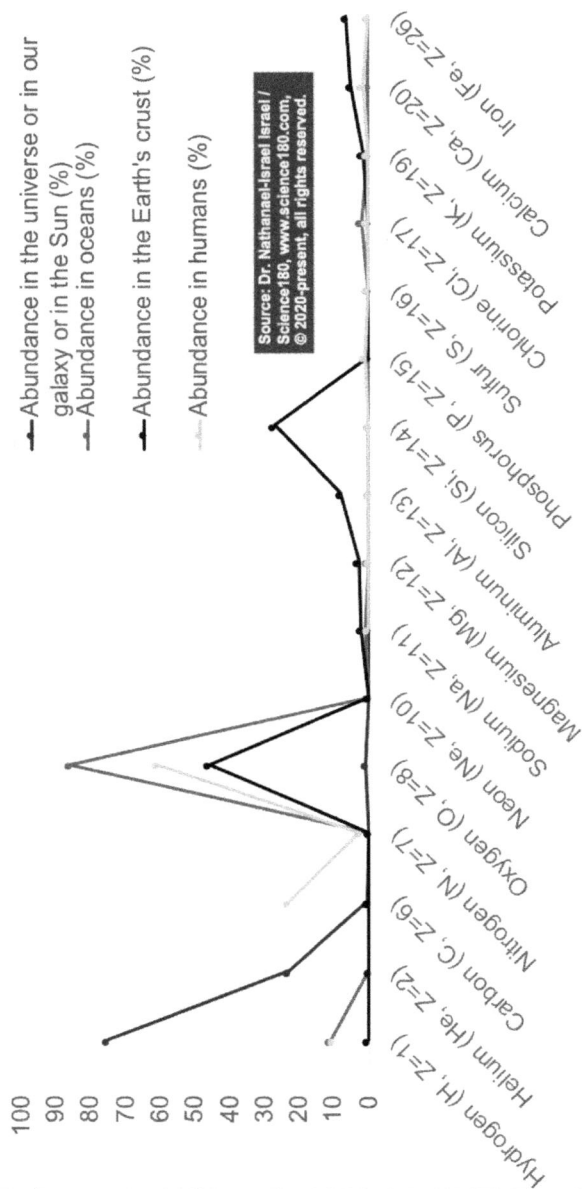

Fig. 91: Comparison of the abundance (%) of the 7 most abundant chemical elements in the universe or in our galaxy or the Sun, human beings, oceans, and in the Earth's crust

In Fig. 91, I wished to add the data of the atmospheres, but I felt like it would make the graph more complex and complicate the understanding. Therefore, I addressed the atmospheric data as I dealt with each element separately. For more clarity, the following segments detailed the abundance of the 15 most abundant

Nathanael-Israel Israel: Has had the honor to be Acknowledged the First Human Being that Scientifically Reconciled Science and Biblical Creation

chemical elements in the universe or our galaxy, the Sun, the Earth's crust, oceans, human beings, and in the atmosphere of the planets. My goal by doing those graphs (Fig. 92-Fig. 118) is to present the data in a more visible and comparative way so that the readers better enjoy the presentation of the data and perceive some of their trends.

23.2. Hydrogen (H, Z=1)

Hydrogen in nature is found in association with many other elements. For instance, besides being found in the form of hydrogen gas (H_2), hydrogen in the atmosphere of the planets is present in various compounds:

- Ammonia (NH_3)
- Ammonia hydrosulfide (NH_4)HS
- Ammonia ice (aerosol)
- C_2H_x Hydrocarbons
- Ethane (C_2H_6)
- Hydrogen Deuteride (HD)
- Hydrogen Deuterium Oxygen (HDO)
- Methane (CH_4)
- Methane ice (?) (aerosol)
- Water (H_2O)
- Water ice (aerosol)

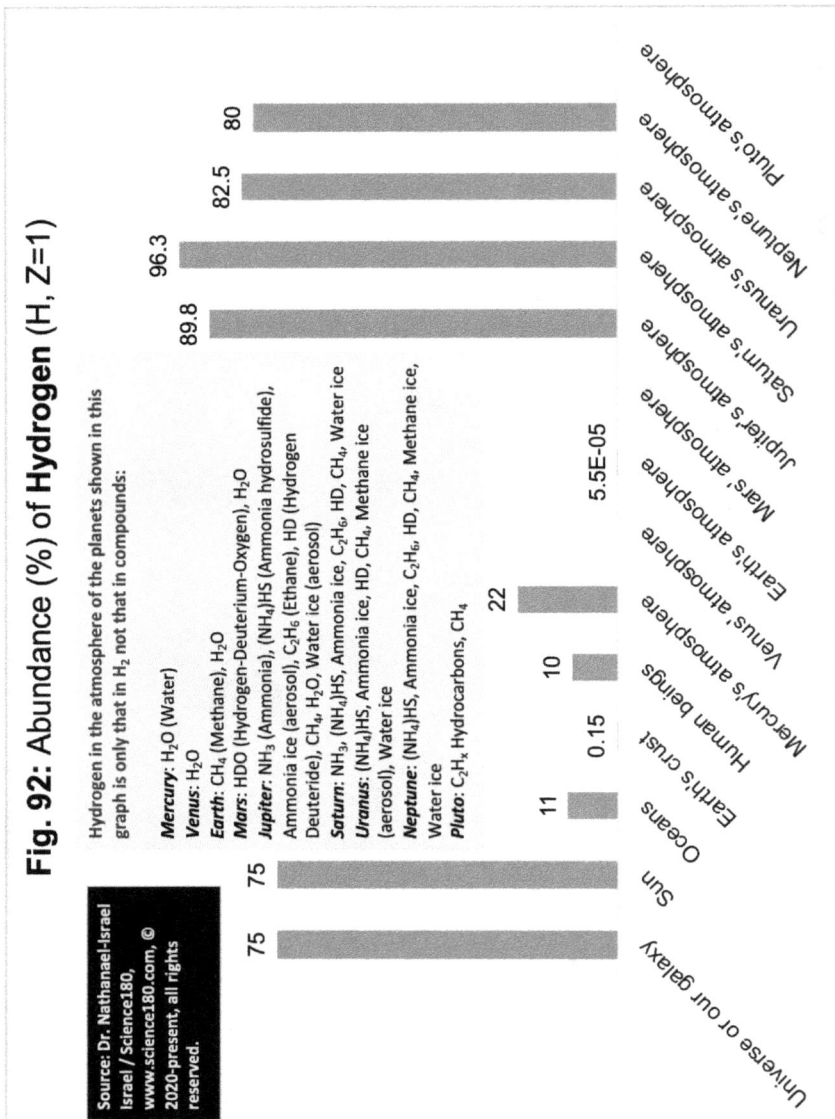

Fig. 92: Abundance (%) of Hydrogen (H, Z=1)

Hydrogen in the atmosphere of the planets shown in this graph is only that in H_2 not that in compounds:

Mercury: H_2O (Water)
Venus: H_2O
Earth: CH_4 (Methane), H_2O
Mars: HDO (Hydrogen-Deuterium-Oxygen), H_2O
Jupiter: NH_3 (Ammonia), $(NH_4)HS$ (Ammonia hydrosulfide), Ammonia ice (aerosol), C_2H_6 (Ethane), HD (Hydrogen Deuteride), CH_4, H_2O, Water ice (aerosol)
Saturn: NH_3, $(NH_4)HS$, Ammonia ice, C_2H_6, HD, CH_4, Water ice
Uranus: $(NH_4)HS$, Ammonia ice, HD, CH_4, Methane ice (aerosol), Water ice
Neptune: $(NH_4)HS$, Ammonia ice, C_2H_6, HD, CH_4, Methane ice, Water ice
Pluto: C_2H_x Hydrocarbons, CH_4

The presence and abundance of each of these compounds depend on the planets. All of these compounds are not present in the atmosphere of all the planets. In general, hydrogen is more abundant in the atmosphere of the giant planets than in the Sun, the Earth's atmosphere, or Earth's crust. On Earth, it is more abundant in the ocean than in the crust. In general, the minimum abundance of hydrogen is encountered in the atmosphere of most terrestrial planets except that of Mercury (Fig. 92). Considering everything I have said so far, hydrogen is

Nathanael-Israel Israel: Has had the honor to be Acknowledged the First Human Being that Scientifically Reconciled Science and Biblical Creation

more abundant in the environment of bigger bodies (e.g. giant planets and Sun), but it can also be found in environments where heavy metals are found (e.g. Mercury's atmosphere). I felt like its abundance is defined by the level of turbulence and the other chemicals that are formed in its environment. When the energy or movement in some precursors of atoms are not strong enough to form heavier and denser atoms, hydrogen is abundantly made. After heavier elements were made, hydrogen could also be produced from the left over of the precursors of atoms that could not be molded into heavier or denser atoms. Hence, the high abundance of hydrogen in Mercury's atmosphere, the only atmosphere where metals were found. Of course, metals are abundant in the crust, but Mercury's atmosphere is the only atmosphere which contains metals. The abundance of hydrogen in the environment and the presence of oxygen can define the formation of water. Hence, some atmospheres which are abundant with hydrogen lack a significant amount of water because oxygen is limiting. Similarly, although the atmosphere of the 4 giant planets is rich in hydrogen, they are poor and some are even deficient in water because they are not rich in oxygen because of factors I already abundantly addressed in previous sections.

Another chemical compound usually formed with hydrogen is methane (CH_4). Methane exists in the atmosphere of all the planets from Earth to Pluto except Mars. The lack of methane in the atmosphere of Mars may be due to the fact that the majority of the carbon atoms formed on Mars were preferably associated with oxygen to form carbon dioxide (CO_2), which occupies 95.32% of the volume of Mars' atmosphere and other carbon atoms associated with other oxygen atoms to form carbon monoxide (CO), which abundance on Mars' atmosphere is just 0.08%. In other words, the conditions of the precursor of Mars and/or that of the Martian atmosphere could have favored the dominant formation of carbon and oxygen, which are bound together to mostly form CO_2 instead of carbon binding with hydrogen to form methane. The few hydrogen atoms which were formed in the atmosphere of Mars bound with some oxygen atoms to form water (H_2O), which is just 0.021% of the Martian atmosphere, while the remaining hydrogen atoms on Mars were associated with deuterium (a hydrogen isotope) and oxygen to form the heavy water (Hydrogen-Deuterium-Oxygen) which as of 2025, is significantly found in Mars' atmosphere only.

Of the 4 giant planets, although hydrogen was abundant (Fig. 92), carbon was not, hence the abundance of hydrogen in their atmosphere was not enough to abundantly form methane (CH_4), which formation was limited by the lack of carbon atoms. Hence the abundance of methane in the atmosphere of the giant planets is not higher than 2.3%. Finally, methane is not dominant on Pluto's atmosphere because the conditions of that planet abundantly favored the formation of nitrogen. The few carbon and hydrogen atoms which were formed on Pluto's atmosphere bound together to yield the small amount of methane present in the Plutonian atmosphere.

Similarly, hydrogen deuteride is mostly abundant in the atmosphere of the 4 giant planets by chance. I think that, as the precursors of atoms in the giant planets were being abundantly molded into protium (the ordinary hydrogen atom), which later combined with one another to abundantly form the hydrogen gas (H2) in the atmosphere of the giant planets, some of those precursors of hydrogen atoms were shaped into deuteride. All that could be needed for deuterium to form was a regular proton to associate with a neutron before being orbited by an electron of the precursor of deuterium atoms being split in such a way that its nucleus contains a neutron. As a reminder, the difference between protium (the ordinary hydrogen) and deuterium (a hydrogen isotope) is that the latter has one proton and one neutron in its nucleus, while the former has only one proton in its nucleus. In other words, as protons and neutrons mixed to form atoms, the precursor of deuterium got a neutron on top of the regular proton found in protium. Then, the hydrogen isotope deuterium associated with protium. In the atmosphere of Mars, deuterium atoms were formed, but instead of being stable and associated with protium only to form hydrogen deuteride, they instead were combined with both protium and oxygen to form the heavy water noticed on Mars. But because the conditions in the precursors of the atmosphere of the giant planets did not favor the formation of oxygen, protium atoms bound with the deuterium atoms to form the hydrogen deuteride diatomic molecules.

23.3. Deuterium (D or ^{2}H)

Also called hydrogen-2, heavy hydrogen, deuterium (represented by D or ^{2}H) is found not only in the atmosphere of many planets. Deuterium accounts for about 0.02% (0.03% by mass) of all the naturally occurring hydrogen in the oceans, while protium accounts for more than 99.98% (Wikipedia, 2020j). Deuterium's mean abundance in ocean water is reported as 155.76 ppm (from Vienna Standard Mean Ocean Water). The molar mass of deuterium is 2.014 g/mol. Its density at STP (0°C, 101.325 kPa) is 0.180 kg/m^3. In liquid form, its density is 162.4 kg/m^3, while in the form of gas, its density is 0.452 kg/m^3.

Deuterium is found in hydrogen deuteride in the atmosphere of the 4 giant planets: Jupiter, Saturn, Uranus, and Neptune. In the atmosphere of the planets, deuterium is found in hydrogen deuteride (HD) and in hydrogen-deuterium-oxygen (HDO). Indeed, because its molecule consists of one hydrogen isotope ^{1}H (protium) and one isotope ^{2}H (deuterium), hydrogen deuteride is a diatomic molecule, a substance which proper molecular formula is ^{1}H^{2}H, but which is usually simplified as HD. Unlike the ordinary hydrogen, which molar mass is about 1 g/mol, the molar mass of hydrogen deuteride is 3.02204 g/mol. Hydrogen deuteride, which is a gas, has been significantly found in the atmosphere of the 4 giant planets (Fig. 93):

- Jupiter (28 ppm),
- Saturn (110 ppm),
- Uranus (148 ppm), and
- Neptune (192 ppm).

Fig. 93: Abundance (% by volume) of **Hydrogen Deuteride** (HD) in the atmosphere of the planets in the Solar System

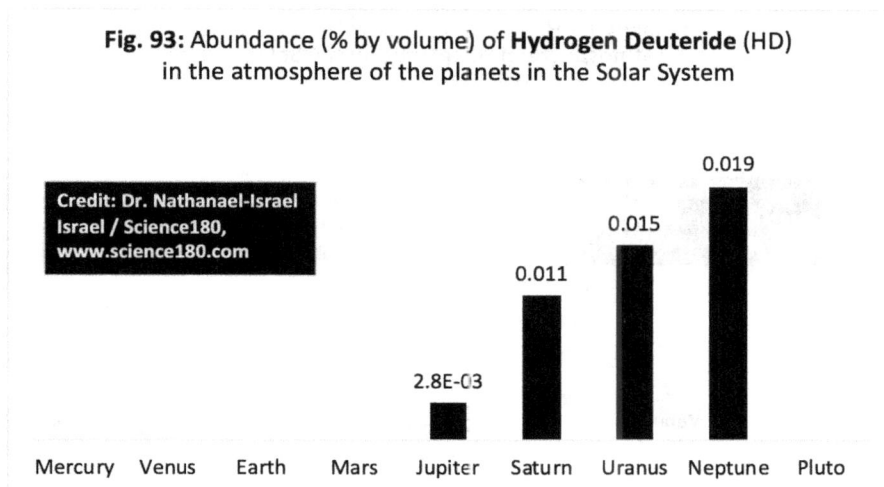

Credit: Dr. Nathanael-Israel Israel / Science180, www.science180.com

Mercury	Venus	Earth	Mars	Jupiter	Saturn	Uranus	Neptune	Pluto
				2.8E-03	0.011	0.015	0.019	

Annotated ^2H_2O, D_2O and also called heavy water or deuterium oxide, hydrogen deuterium oxygen is a type of water composed of oxygen and deuterium (hydrogen isotope having one proton and one neutron). It is called heavy water because, instead of containing the ordinary hydrogen-1 isotope (^1H or H, also called protium) that constitutes most of the hydrogen in normal water, the heavy water contains the hydrogen isotope deuterium (^2H or D) also known as heavy hydrogen. Consequently, the molecular weight of hydrogen deuterium oxygen (20 g/mol) is higher than that of ordinary water (18 g/mol).

As of 2020, the planet which atmosphere has the highest level of hydrogen deuterium oxygen (HDO) is Mars (0.85 ppm) (Fig. 94). A significant level of HDO has not been measured on any other planet. As shown in the previous section, hydrogen deuteride was not found on Mars. This suggested that on Mars, the atoms of hydrogen that might have been combined to form hydrogen deuteride were preferably combined with oxygen to form the hydrogen deuterium oxygen (HDO).

Fig. 94: Abundance (% by volume) of **Hydrogen Deuterium Oxygen** (HDO or D_2O) also called **Heavy Water**, in the atmosphere of the planets in the Solar System

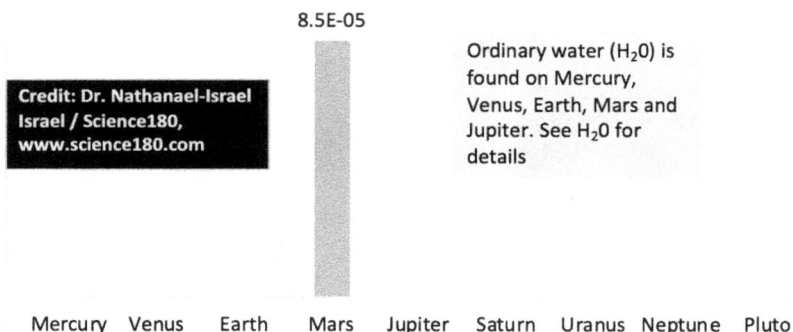

8.5E-05

Credit: Dr. Nathanael-Israel Israel / Science180, www.science180.com

Ordinary water (H_2O) is found on Mercury, Venus, Earth, Mars and Jupiter. See H_2O for details

Mercury Venus Earth Mars Jupiter Saturn Uranus Neptune Pluto

23.4. Helium (He, Z=2)

Helium is said to be more abundant in the universe and in the Sun than in the atmosphere of the giant planets (Fig. 95). While the helium concentration in the atmosphere of the Earth is very low (5.24 ppm), it is the second most abundant gas in the atmosphere of the giant planets. For instance, on Saturn, it is 3.5% of the atmosphere, whereas on Neptune, it is 19%. Finally, it is important to mention that the helium concentration in the atmosphere of Venus (12 ppm) and Mercury (6%) is higher than that in the atmosphere of Earth (Fig. 95). In other words, Helium is more abundant in the atmosphere of the giant planets than in that of most of the terrestrial planets. The only terrestrial planet which atmosphere abounds with more helium than the atmosphere of a giant planet is Mercury. Indeed, the helium abundance in Mercury's atmosphere is higher than that in Saturn's atmosphere. The conditions that favored the formation of hydrogen could have also favored that of helium.

Finally, although scientists think that helium can be formed by a nuclear fusion of hydrogen for instance, the majority of the natural helium could not have been formed by just fusing hydrogen. Although fusion reactions have failed to explain the genesis of all chemical elements, scientists still unfortunately cling to them as a major pathway for chemical genesis. Very soon, I will show you how helium was formed.

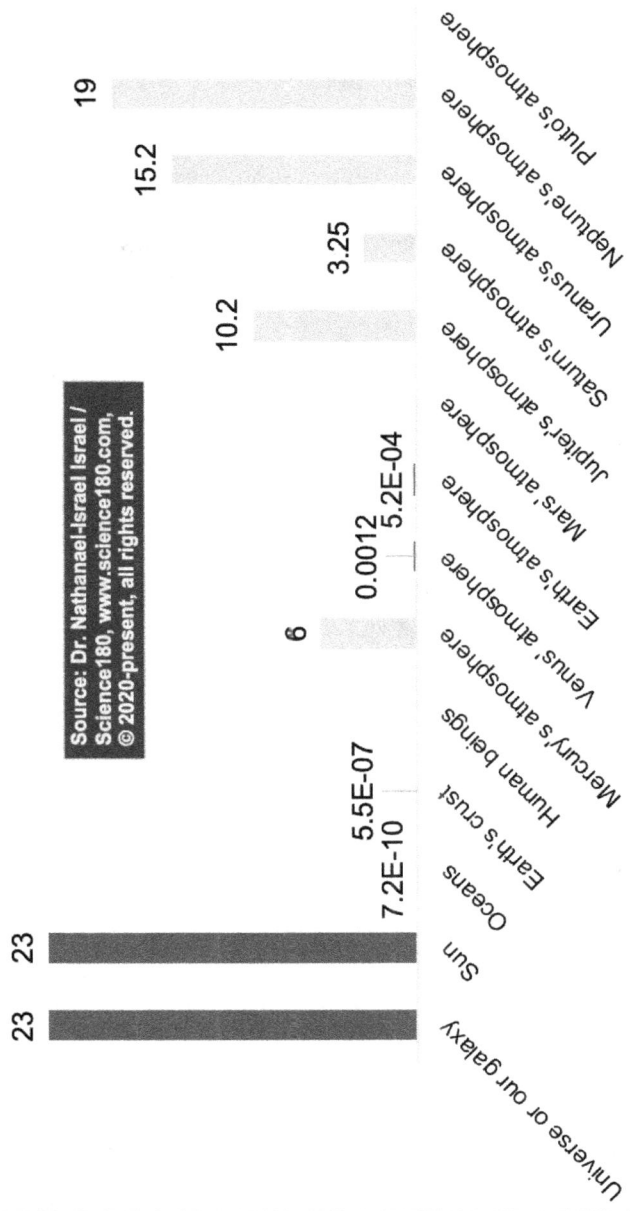

Fig. 95: Abundance (%) of Helium (He, Z=2)

23.5. Boron (B, Z=5)

Boron is most abundant in the Earth's crust followed by the oceans (Fig. 96). It is rare in the Sun and the universe and is not even found in the atmosphere of the planets because more squeezing may be required for its precursors than the places it is not found could have provided. Although it is not found in the Earth's atmosphere, it is found in significant amount in human beings, suggesting that human beings are able to incorporate it into their bodies through their diet and not through breathing. The oceans are poorer in boron than the Earth's crust because less turbulence could have occurred in the precursor of the oceanic water than in the precursor of the Earth's crust, and that the precursor of boron required a stronger turbulence before being tightly squeezed into boron.

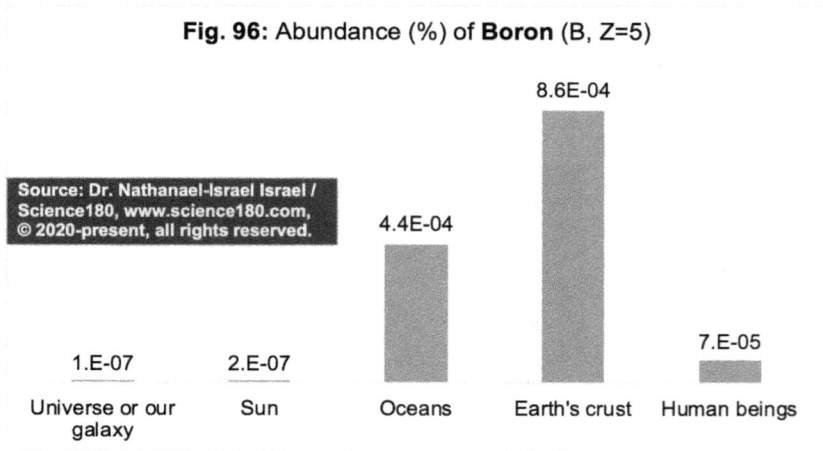

Fig. 96: Abundance (%) of **Boron** (B, Z=5)

8.6E-04

4.4E-04

7.E-05

1.E-07

2.E-07

| Universe or our galaxy | Sun | Oceans | Earth's crust | Human beings |

23.6. Carbon (C, Z=6)

Carbon is found in many forms including:

- CO_2 (Carbon Dioxide)
- CO (Carbon Monoxide)
- C_2H_6 (Ethane)
- CH_4 (Methane)
- Methane ice
- C_2H_x Hydrocarbons

Abundantly present in the form of CO_2 in the atmosphere of Venus (96.5%) and in that of Mars (95.32%), and to a smaller extent in human beings (23%) (Fig. 97), carbon is not found much in the universe (implying in the stars including the Sun), the oceans, or the Earth's crust. In the atmosphere of the Earth, CO_2 constitutes about 0.04% of the gases. If the atmosphere of the Earth was not fortunately poor in CO_2, some animals might have not been able to live on Earth. Although plants use CO_2 to build their carbohydrates and other carbon-based compounds, CO_2 can be toxic to some animals including human beings. Surely, the Earth's atmosphere contains enough CO_2 to satisfy the needs of plants without jeopardizing the life of the animals. It is a great thing that plants uptake CO_2 and

CHAPTER 23: CRUCIAL LESSONS FROM THE COMPARISON OF THE ABUNDANCE OF THE CHEMICAL ELEMENTS

release O_2. If the Earth's atmosphere was too rich in CO_2 like the atmosphere of the two planets that are closer to it (Venus and Mars), life as known today would surely be impossible on Earth. This suggests that the composition of the atmosphere of the planets is not just by chance.

Why is CO_2 abundant in Venus' atmosphere? Venus is the hottest planet in the Solar System and its average temperature is 737 Kelvin. Of the 118 chemical elements known by 2025, the highest melting point (3773.2 Kelvin) was recorded on carbon, meaning that carbon does not melt easily. Some people may think that, to be abundant in Venus' atmosphere, a chemical element needs to be able to sustain a very high temperature. A high melting point of carbon, one of the constituents of CO_2, could not alone explain the high level of CO_2 in Venus' atmosphere. Because CO_2 is also abundant on Mars, a planet colder than Earth, heat alone cannot be sufficient to explain the presence and abundance of CO_2. Other factors must be involved in the formation of CO_2.

Significant amounts of carbon monoxide are found in the atmosphere of 3 planets: Venus (17 ppm), Mars (800 ppm), and Pluto (500 ppm). Carbon monoxide is very toxic to certain animals including human beings. This suggests that astronauts who may be visiting Mars soon, as some countries like the USA are planning, may face a intoxication or toxicity problems if they were to be exposed to the high level of CO in the Martian atmosphere.

Probably, some of the carbon on Venus and Mars which were unable to associate with two oxygen atoms to form CO_2, have been associated with just one oxygen atom to form CO. Because of the higher content of CO_2 in the atmosphere of Venus and Mars, I deducted that the conditions on Venus and Mars favored the formation of CO_2 than CO. As explained later, on Pluto, carbon atoms associated with hydrogen atoms to form several kinds of hydrocarbons, whereas in the atmosphere of other planets, it is associated with oxygen to form various carbon oxides.

Considering what I said above, the high density of carbon and the energy that was needed to form it could have been one of the factors that did not favor its abundance in certain places, such as in the Sun, the universe, the 4 giant planets of the Solar System, etc. Human beings have the highest concentration of carbon because of their ability to incorporate it through food. The abundance of compounds containing carbon may have been limited by the scarcity of the other elements found in those compounds.

Fig. 97: Abundance (%) of **Carbon** (C, Z=6)

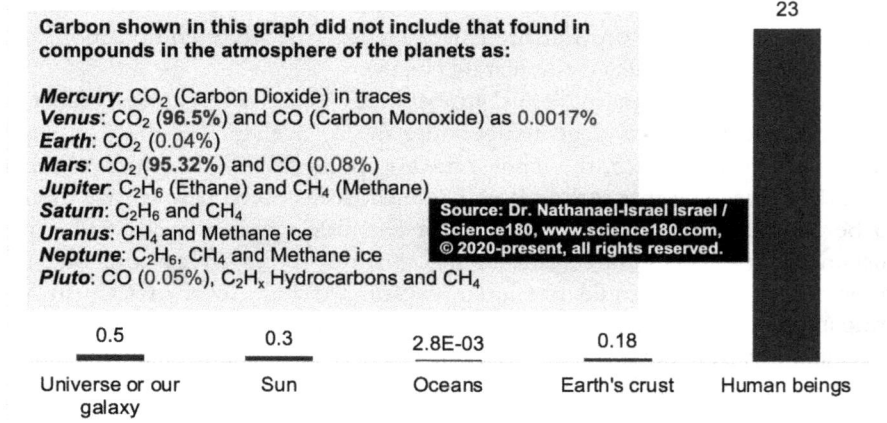

Carbon shown in this graph did not include that found in compounds in the atmosphere of the planets as:

Mercury: CO_2 (Carbon Dioxide) in traces
Venus: CO_2 (96.5%) and CO (Carbon Monoxide) as 0.0017%
Earth: CO_2 (0.04%)
Mars: CO_2 (95.32%) and CO (0.08%)
Jupiter: C_2H_6 (Ethane) and CH_4 (Methane)
Saturn: C_2H_6 and CH_4
Uranus: CH_4 and Methane ice
Neptune: C_2H_6, CH_4 and Methane ice
Pluto: CO (0.05%), C_2H_x Hydrocarbons and CH_4

Universe or our galaxy	Sun	Oceans	Earth's crust	Human beings
0.5	0.3	2.8E-03	0.18	23

23.7. Nitrogen (N, Z=7)

Of the celestial bodies and atmospheres that I studied, nitrogen is mostly found in the atmosphere of the Earth (78.08%) and Pluto (99%) (Fig. 98). While a trace of nitrogen is found on Mercury, it accounts for 3.5% and 2.7% in the atmosphere of Venus and Mars respectively (Fig. 98). The presence of Nitrogen on Pluto greatly surprised many scientists after the data collected by New Horizons spacecraft in 2015 showed that. Many scientists including some at NASA believed that the N_2 on Pluto is the product of solids at the surface warmed up and evaporated into the atmosphere during Pluto's closest approach to the Sun. They did not think that nitrogen could be found in the atmosphere of such a remote planet. Nitrogen is not found in the atmosphere of any giant planets. The factors that positively affected the density of Pluto and that of the other terrestrial planets could have also affected the precursors of atoms during the formation of Pluto's atmosphere. In the end, nitrogen, a dense atom, was more abundantly formed on Pluto than hydrogen, helium, and other smaller atoms. And because most of the precursors of atoms in the precursors of the atmosphere of Pluto were molded or converted into nitrogen, few other chemical elements were formed. Hence nitrogen occupies 99% of Pluto's atmosphere.

Although nitrogen is 78.08% of the Earth's atmosphere (Fig. 98), its content in human beings is 2.6%, meaning that nitrogen is more concentrated in the Earth's atmosphere than in human beings. Similarly, Silicon (Si, Z=14) is 27% of the Earth's crust, yet it is just 0.026% in human beings, meaning that Silicon is more concentrated in the Earth's crust than in human beings. In contrast, although carbon dioxide in the Earth's atmosphere or in the air that human beings breathe is just 0.04%, carbon is 23% of human beings, meaning that carbon is more concentrated in human beings than in the air. These trends mean that the abundance of a chemical element in the atmosphere of the Earth does not imply its abundance in human beings. In other words, human beings do not passively consume chemical elements in the air or in the Earth's crust, but their systems selectively uptake what is needed for the biological processes that sustain life.

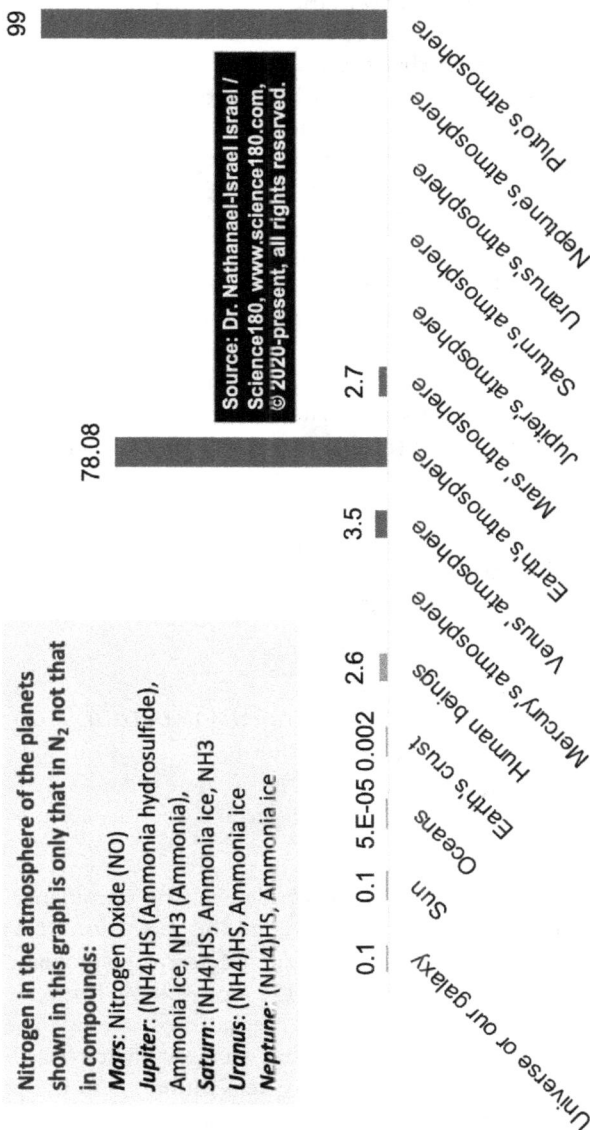

Fig. 98: Abundance (%) of Nitrogen (N, Z=7)

Nitrogen in the atmosphere of the planets shown in this graph is only that in N_2 not that in compounds:

Mars: Nitrogen Oxide (NO)
Jupiter: (NH4)HS (Ammonia hydrosulfide), Ammonia ice, NH3 (Ammonia),
Saturn: (NH4)HS, Ammonia ice, NH3
Uranus: (NH4)HS, Ammonia ice
Neptune: (NH4)HS, Ammonia ice

23.8. Oxygen (O, Z=8)

Based on the data I studied, the highest abundance of oxygen was found on Earth where it dominates the oceans by 86%, the Earth's crust by 46% and the Earth's atmosphere by 20.95% (Fig. 99). Human beings are 61% made of oxygen. The atmosphere of Mercury is the only other place where a higher abundance of oxygen (42%) was also found in the Solar System. Considering the crucial role that oxygen plays in the physiology of human beings and other living organisms, it is important to underline that the Earth is a privileged place to host human beings. Some scientists are trying to find living organisms on other planets and even on other stellar systems, but, to my understanding of the universe, life on Earth is not by chance. Therefore, we also need to be grateful for the Earth's environmental conditions, which match our needs. Else, if human beings were made in any of the other environments I studied in the Solar System, life (as known on Earth) could have already been extinguished, for those environments do not contain the proper amount of the chemical elements required to power and sustain the biological reactions and chains that sustain life as known on Earth.

According to the environment it is found in, oxygen associates with other chemical elements to form diverse compounds. For instance, in the atmosphere of the Earth, oxygen atoms associated with carbon atoms to form CO_2, or with hydrogen to form water (H_2O). Oxygen atoms can also bind to one another to form the oxygen gas (O_2), that living organisms breathe, or to form the ozone (O_3), which is said to shield the Earth from some toxic radiations coming from the Sun or from other bodies in space. In the atmosphere of the planets, oxygen is present in different compounds (Fig. 99):

- Carbon dioxide (CO2),
- Carbon monoxide (CO),
- Hydrogen deuteride (HD),
- Hydrogen deuterium oxygen (HDO or D2O),
- Nitrogen oxide (NO),
- Sulfur dioxide (SO2),
- Water (H2O), and
- Water ice (aerosol).

On the atmosphere of the giant planets, the dominance of hydrogen and the lack of oxygen atoms could have limited the formation of water. Hence, the small amount of oxygen in the atmosphere of the 4 giant planets is sequestered like aerosol in those giant bodies where the atmospheric pressure is also very high.

The fact that oxygen that is the second most abundant element in the Earth's atmosphere is the #1 most abundant element in the Earth's crust suggests that there was a higher turbulence or harsher conditions in the precursor of the Earth, which yielded other denser atoms that could have favored its abundance in the crust.

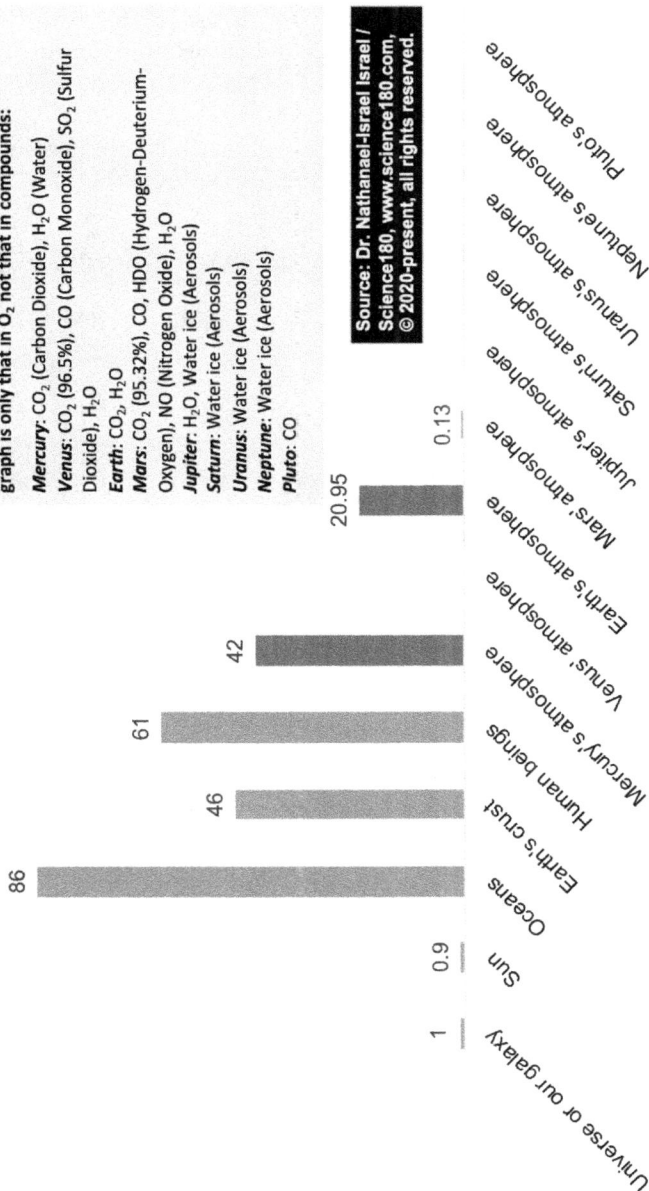

Fig. 99: Abundance (%) of **Oxygen (O, Z=8)**

Oxygen in the atmosphere of the planets shown in this graph is only that in O_2, not that in compounds:

Mercury: CO_2 (Carbon Dioxide), H_2O (Water)

Venus: CO_2 (96.5%), CO (Carbon Monoxide), SO_2 (Sulfur Dioxide), H_2O

Earth: CO_2, H_2O

Mars: CO_2 (95.32%), CO, HDO (Hydrogen-Deuterium-Oxygen), NO (Nitrogen Oxide), H_2O

Jupiter: H_2O, Water ice (Aerosols)

Saturn: Water ice (Aerosols)

Uranus: Water ice (Aerosols)

Neptune: Water ice (Aerosols)

Pluto: CO

Because many precursors of atoms were differentiated into oxygen in the precursor of the Earth's crust, and others into heavier atoms, a small portion was left to be converted into smaller atoms. Hence, the Earth's crust is rich in heavy atoms, but lacks lighter ones. Consequently, nitrogen, which was abundantly formed in the Earth's atmosphere, could not be abundantly formed in the Earth's crust for the conditions in the precursor of the Earth's crust were suitable for the abundant formation of atoms heavier and denser than nitrogen.

23.9. Fluorine (F, Z=9)

Fluorine is one of the smaller elements but which are rare in nature. Its highest concentration has been observed in the Earth's crust (Fig. 100).

Fig. 100: Abundance (%) of **Fluorine** (F, Z=9)

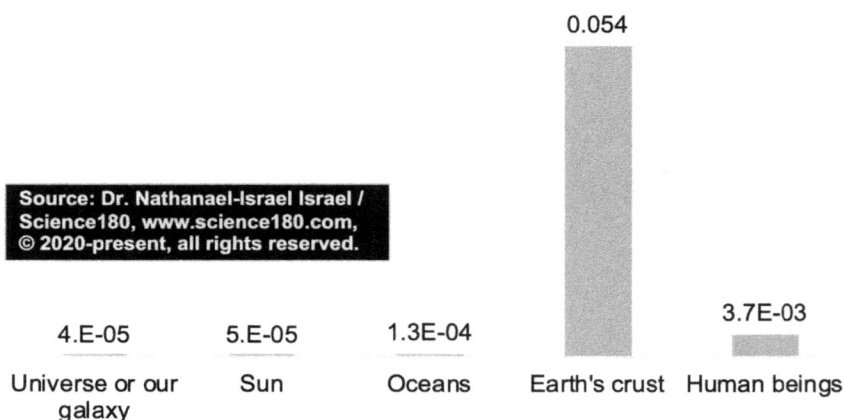

Source: Dr. Nathanael-Israel Israel / Science180, www.science180.com, © 2020-present, all rights reserved.

Universe or our galaxy	Sun	Oceans	Earth's crust	Human beings
4.E-05	5.E-05	1.3E-04	0.054	3.7E-03

23.10. Neon (Ne, Z=10)

The highest abundances of neon are stipulated to be in the universe and in the Sun (Fig. 101). Neon is found in the atmosphere of the 4 innermost planets of the Solar System. While a trace of it is found in the atmosphere of Mercury, in the atmosphere of Venus, Earth, and Mars, its content ranges from 2.5 ppm to 18.8 ppm (Fig. 101), meaning that Neon gas is not abundant at all in the atmosphere of any planet in the Solar System.

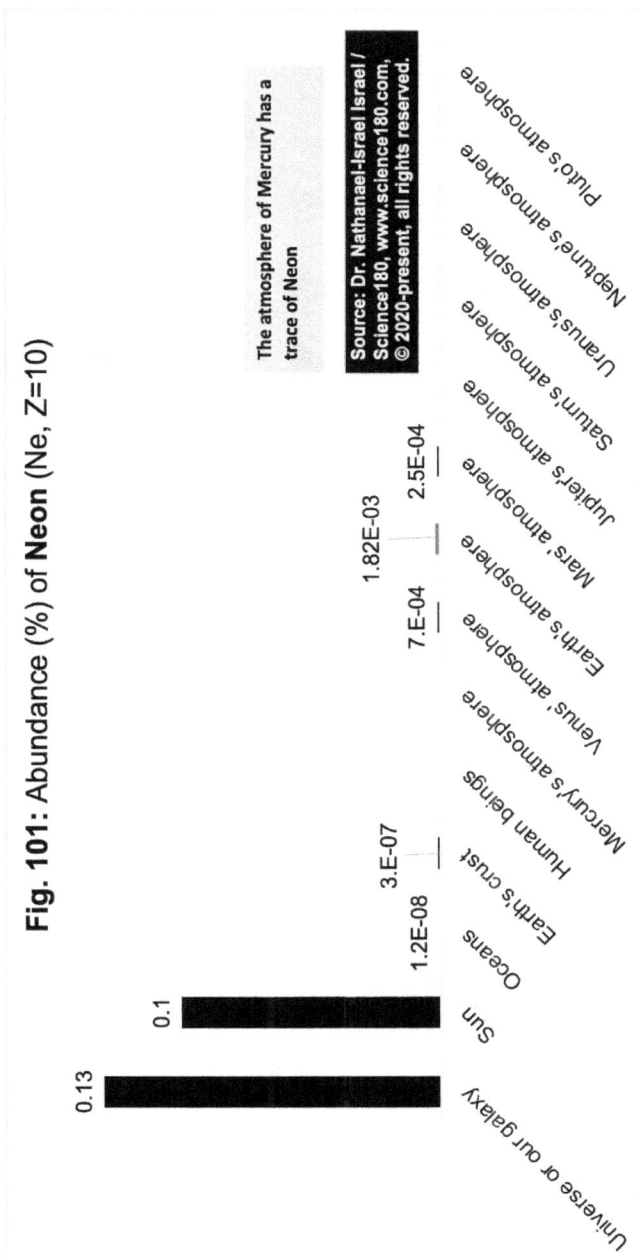

Fig. 101: Abundance (%) of Neon (Ne, Z=10)

The atmosphere of Mercury has a trace of Neon

23.11. Sodium (Na, Z=11)

Sodium is mostly found in the atmosphere of Mercury (29%) (Fig. 102).

Nevertheless, I expect it to be abundant in some crust, at least those of the terrestrial planets. If Sodium (Na, Z=11) is 2.3% of the Earth's crust, it should not be surprising for it to also be abundant in the crust of other planets.

Fig. 102: Abundance (%) of **Sodium** (Na, Z=11)

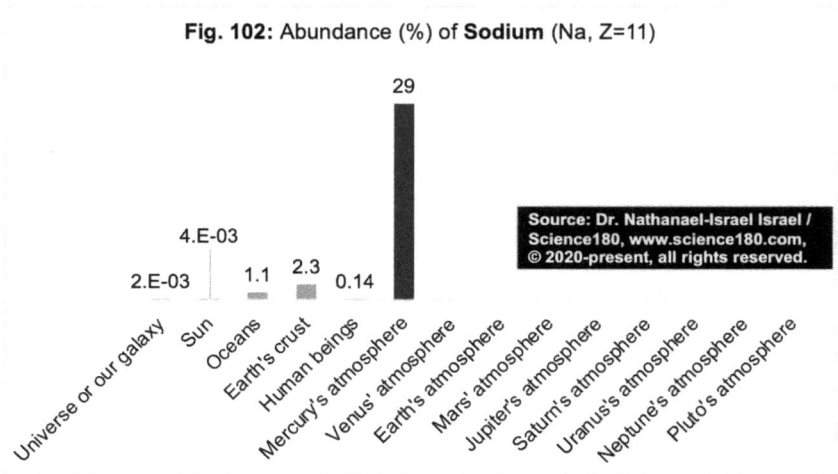

23.12. Magnesium (Mg, Z=12)

The highest abundance of magnesium was in the Earth's crust (Fig. 103).

Fig. 103: Abundance (%) of **Magnesium** (Mg, Z=12)

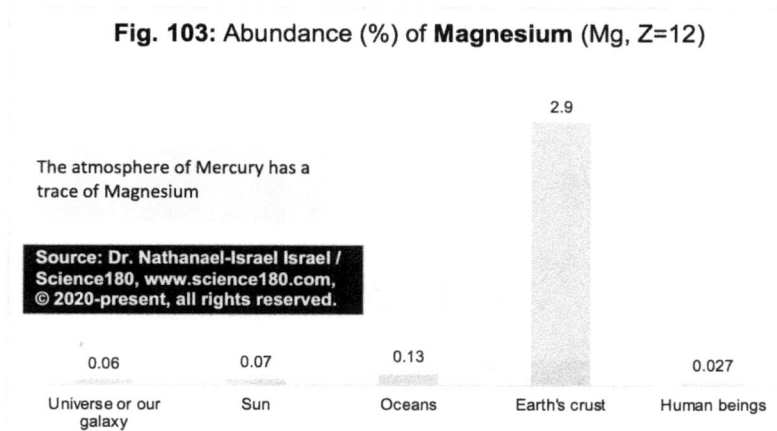

The atmosphere of Mercury has a trace of Magnesium

23.13. Aluminum (Al, Z=13)

Aluminum is abundantly found in the Earth's crust (8.1%) (Fig. 104). Knowing that aluminum is also found in Mercury's atmosphere and that heavier elements could be found in the crust than in the atmosphere, it is highly possible that Mercury's crust would be very rich in aluminum and other metals.

Nathanael-Israel Israel: Has had the honor to be Acknowledged the First Human Being that Scientifically Reconciled Science and Biblical Creation

Fig. 104: Abundance (%) of **Aluminum** (Al, Z=13)

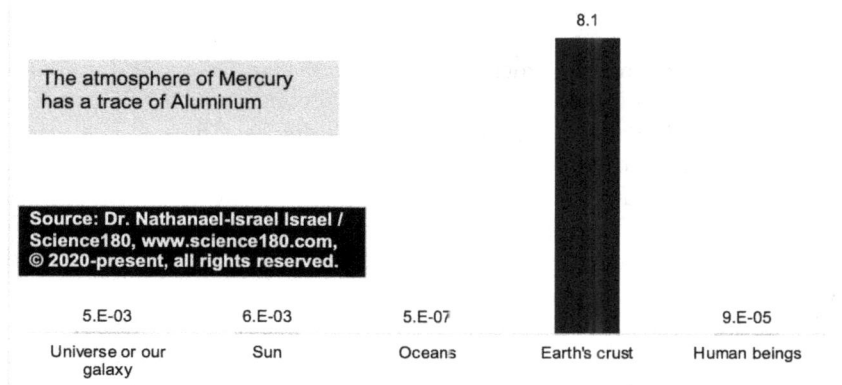

The atmosphere of Mercury
has a trace of Aluminum

5.E-03	6.E-03	5.E-07	8.1	9.E-05
Universe or our galaxy	Sun	Oceans	Earth's crust	Human beings

23.14. Silicon (Si, Z=14)

Although not found in the atmosphere on the planets, silicon is the second most abundant element in the Earth's crust (27%) (Fig. 105). This fact suggested that many chemicals absent in the atmosphere of the planet could be very abundant in the crust. It also pointed to the possibility that conditions in the precursors of the crust and that of the atmosphere could have been different, else the same chemical elements could have been present in both the crust and the atmosphere of the celestial bodies.

The stronger squeezing that the precursor of the Earth's crust could have gone through as compared to that of the precursor of the Earth's atmosphere could explain why conditions in the precursors of the crust favored the formation of silicon (Si, Z=14) as the second most abundant chemical in the Earth's crust. Indeed, silicon is heavier and denser than oxygen. In fact, the density of silicon (2329.6 kg/m3) is 1630.23 times that of Oxygen (O, Z=8). Moreover, the molar mass of silicon (28.09 g/mol) is 1.76 times that of oxygen (16 g/mol). The abundance of silicon in the Earth's crust is smaller than that of oxygen for reasons I gave previously about why heavier and denser elements are usually less abundant than smaller and lighter ones. Not knowing that oxygen and silicon were formed in the same environment and that a major difference in their genesis could have been that the precursor of silicon was bigger and ended up being fragmented and aggregated into a heavier and denser element, some people mistook the usual association of silicon and oxygen in silica and silicates as if silicon and oxygen have some affinity and friendship with one another. Based on what I know about the genesis of the chemical elements, instead of being treated as a matter of affinity, some chemical elements found themselves in association with others in chemical compounds just because they were formed in neighboring environments and consequently have been wrapped together according to the intensity of the force that mixed them and defined the boundaries of their territory or space.

As I close this section on silicon, it is important to keep in mind that despite being the second most abundant chemical element in the Earth's crust, silicon is the 12th most abundant element in human beings, pointing to the fact that human beings selectively eat things that contain elements they need instead of things that contain abundant chemical elements. Life is selective and active and no one should let problems or anything else to cause him to become passive. To put it another way, problems in life are not usually solved by using easy means. If it was not so, plants and animals could have been feeding themselves just on the abundant oxygen and silicon in the Earth's crust and the abundant nitrogen in the atmosphere. We need to learn to work hard in life to find answers for our questions rather than relying on easy ways out. Although silicon is so abundant in most soils on Earth, some plants even lack it and need to be fertilized with silicon-based fertilizers just as other plants need to be fertilized with nitrogen-based fertilizers despite the abundant amount of nitrogen in the atmosphere. When I saw those contrasting similitudes and challenges some plants and animals face, I wondered if some of them may be complaining about their "misery" and "hunger", while they are surrounded by what they need just as some poor human beings complain about their poverty, while they are surrounded by filthy, rich people who don't usually care about the needy but just about how to amass more resources into their bank accounts and reserve!

Fig. 105: Abundance (%) of **Silicon** (Si, Z=14)

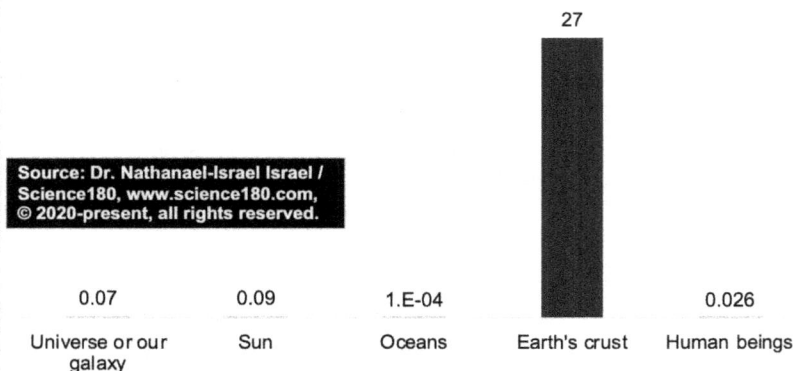

Universe or our galaxy	Sun	Oceans	Earth's crust	Human beings
0.07	0.09	1.E-04	27	0.026

23.15. Phosphorus (P, Z=15)

Not detected in the atmosphere of the planets, phosphorus is more abundant in human beings (1.1%) (Fig. 106), where it plays a key role for instance in the structure of DNA, membranes, bones, transport of energy, etc.

Fig. 106: Abundance (%) of **Phosphorus** (P, Z=15)

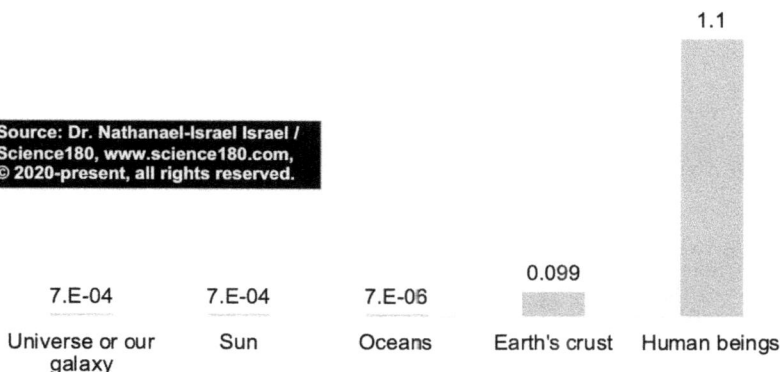

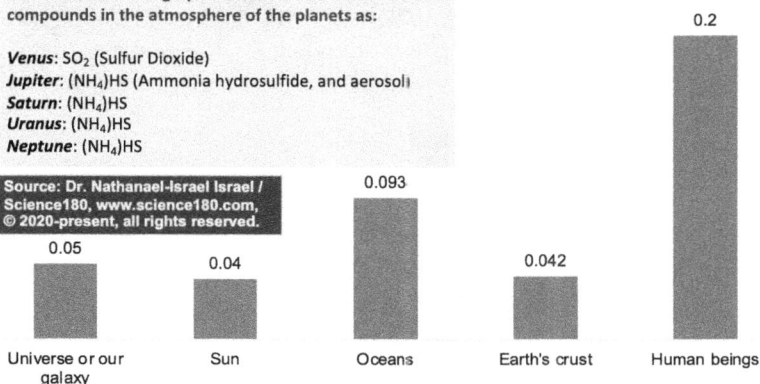

1.1

0.099

| 7.E-04 | 7.E-04 | 7.E-06 | | |
| Universe or our galaxy | Sun | Oceans | Earth's crust | Human beings |

23.16. Sulfur (S, Z=16)

The highest abundance of sulfur is in human beings (Fig. 107). In the atmosphere of the planets in the Solar System, sulfur is present in association with other atoms in compounds such as (Fig. 107):

- Sulfur Dioxide (SO_2)
- Ammonia hydrosulfide ($(NH_4)HS$, which is an aerosol

Venus is the only planet where a significant amount of sulfur dioxide (150 ppm) was found in the atmosphere.

Fig. 107: Abundance (%) of **Sulfur** (S, Z=16)

Sulfur shown in this graph did not include that found in compounds in the atmosphere of the planets as:

Venus: SO_2 (Sulfur Dioxide)
Jupiter: $(NH_4)HS$ (Ammonia hydrosulfide, and aerosol)
Saturn: $(NH_4)HS$
Uranus: $(NH_4)HS$
Neptune: $(NH_4)HS$

0.2

0.093

0.05

0.04

0.042

| 0.05 | 0.04 | 0.093 | 0.042 | 0.2 |
| Universe or our galaxy | Sun | Oceans | Earth's crust | Human beings |

23.17. Chlorine (Cl, Z=17)

Although less dense than oxygen and nitrogen, chlorine is not found in the atmosphere of planets, but is abundant in oceans (Fig. 108). Its radius is higher than that of carbon, oxygen, and nitrogen. The nature of the turbulence that birthed the chemicals in oceans could have favored its abundance there.

Fig. 108: Abundance (%) of **Chlorine** (Cl, Z=17)

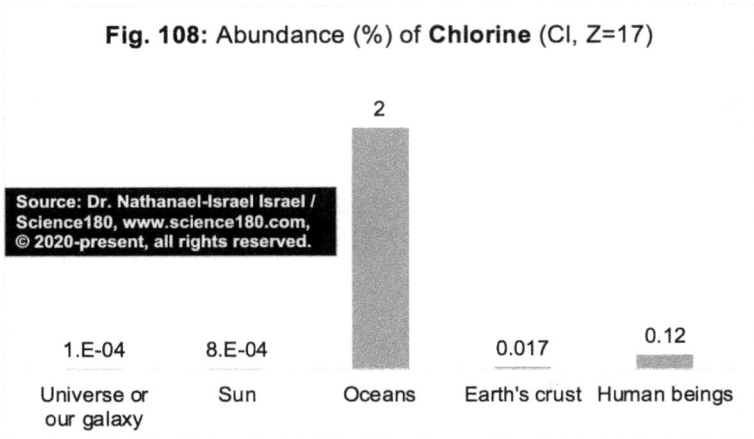

Universe or our galaxy	Sun	Oceans	Earth's crust	Human beings
1.E-04	8.E-04	2	0.017	0.12

23.18. Argon (Ar, Z=18)

Argon may be abundant in the atmosphere of the terrestrial planets (Fig. 109) because conditions in the precursors of the crust of the planets could have not allowed its formation. Its higher density and radius as compared to helium (another noble gas) may explain why it is not found in the atmosphere of the 4 giant planets (Fig. 109) where helium thrives (Fig. 95). Its higher size and density may have better suited the denser conditions of the atmosphere of the terrestrial planets. Just as the precursors of the giant planets did not favor the abundance of argon, neither did the precursors of the stars, which could have been bigger and less dense than planets. The same logic can explain why the noble gases that are denser or heavier than argon are less abundant in the universe than argon. In contrast, helium, which is less than and lighter than argon, is more abundant than any other noble gas.

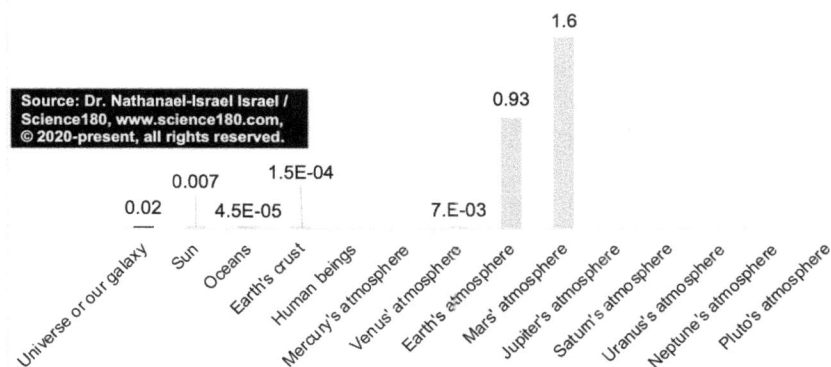

Fig. 109: Abundance (%) of **Argon** (Ar, Z=18)

23.19. Potassium (K, Z=19)

Potassium is more abundant in the Earth's crust than in Mercury's atmosphere (Fig. 110). Yet, it is not present in the Earth's atmosphere. As a metal, its precursor could have had a higher chance to be molded into potassium in a denser environment than in a more loose or voluminous one. Hence, potassium is absent in the atmosphere of the giant planets, and less abundant in the Sun and the universe. The fact that its abundance is higher in the Earth's crust than in the Earth's atmosphere suggests the differentiation of the precursors of atoms into potassium could have required conditions that favored the formation of denser and heavier chemical elements.

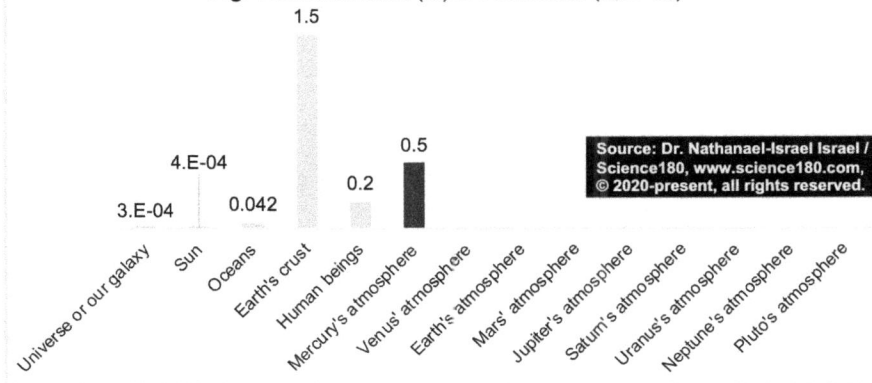

Fig. 110: Abundance (%) of **Potassium** (K, Z=19)

23.20. Calcium (Ca, Z=20)

Of the bodies which chemical contents I studied, the Earth's crust recorded the highest content (5%) of calcium (Fig. 111). Calcium could not be formed in the atmosphere because the conditions in its precursor did not allow the formation of such dense atoms. Similarly, its content in the universe, the Sun or the oceans is small because their precursors would have not been suitable for the massive formation of heavier atoms.

Fig. 111: Abundance (%) of **Calcium** (Ca, Z=20)

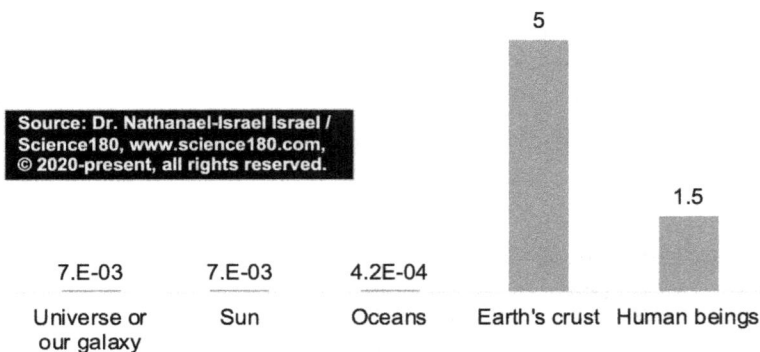

23.21. Titanium (Ti, Z=22)

Titanium is more abundant in the Earth's crust than in the oceans, Sun, or galaxies (Fig. 112) because of the same reasons I explained earlier concerning why heavier elements are less abundant than lighter ones. It should not be surprising that titanium is also found in the crust of some planets, particularly the terrestrial ones.

Fig. 112: Abundance (%) of **Titanium** (Ti, Z=22)

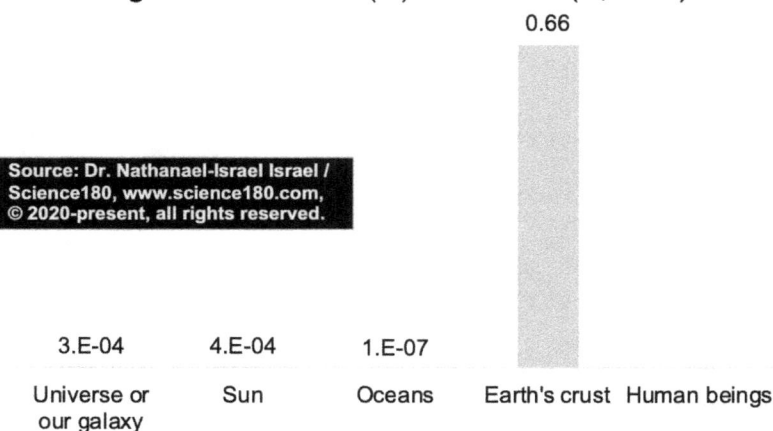

23.22. Manganese (Mn, Z=25)

Based on the data I studied, manganese is more abundant in the Earth's crust than anywhere else (Fig. 113).

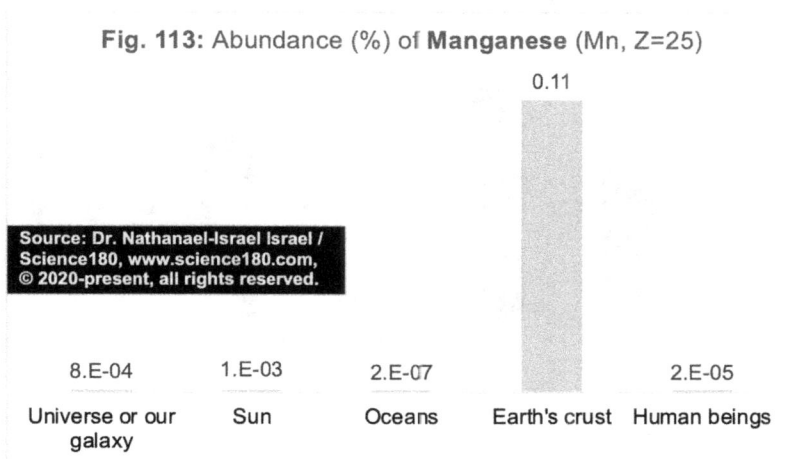

Fig. 113: Abundance (%) of **Manganese** (Mn, Z=25)

Source: Dr. Nathanael-Israel Israel / Science180, www.science180.com, © 2020-present, all rights reserved.

Universe or our galaxy	Sun	Oceans	Earth's crust	Human beings
8.E-04	1.E-03	2.E-07	0.11	2.E-05

23.23. Iron (Fe, Z=26)

Like most metals, iron is more abundant in the Earth's crust than anywhere else (Fig. 114). I would not be surprised if the iron content of the crust of the other terrestrial planets and even of the Moon is very high as well. Although it may be present in the crust of many planets (at least the terrestrial planets), its presence in the atmosphere has been found in traces only on Mercury.

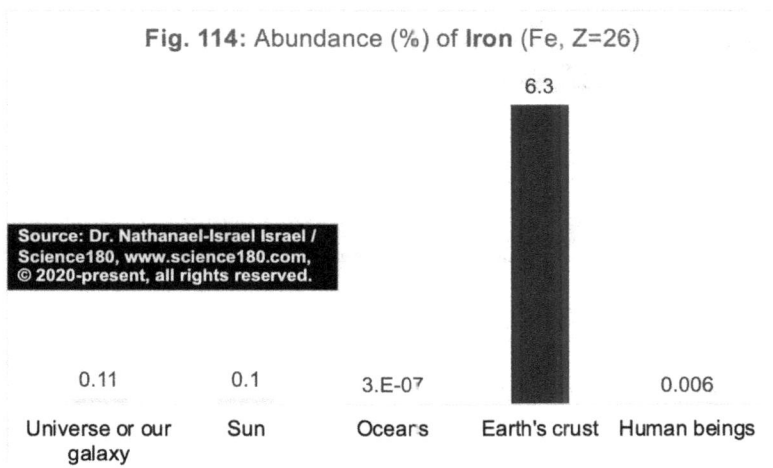

Fig. 114: Abundance (%) of **Iron** (Fe, Z=26)

Source: Dr. Nathanael-Israel Israel / Science180, www.science180.com, © 2020-present, all rights reserved.

Universe or our galaxy	Sun	Oceans	Earth's crust	Human beings
0.11	0.1	3.E-07	6.3	0.006

23.24. Nickel (Ni, Z=28)

Nickel is abundant in the Earth's crust (Fig. 115).

Fig. 115: Abundance (%) of **Nickel** (Ni, Z=28)

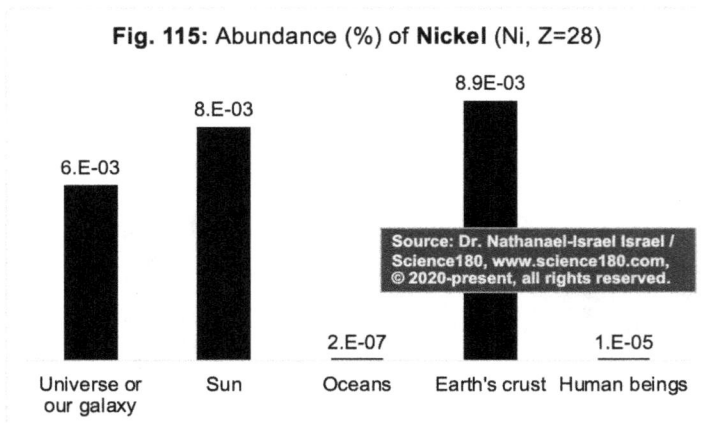

Source: Dr. Nathanael-Israel Israel / Science180, www.science180.com, © 2020-present, all rights reserved.

- 6.E-03 — Universe or our galaxy
- 8.E-03 — Sun
- 2.E-07 — Oceans
- 8.9E-03 — Earth's crust
- 1.E-05 — Human beings

23.25. Zinc (Zn, Z=30)

Zinc is more abundant in the Earth's crust than in human beings or in the universe (Fig. 116).

Fig. 116: Abundance (%) of **Zinc** (Zn, Z=30)

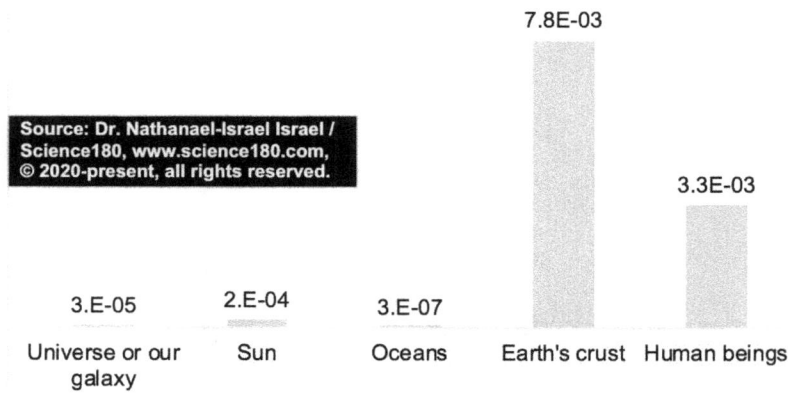

Source: Dr. Nathanael-Israel Israel / Science180, www.science180.com, © 2020-present, all rights reserved.

- 3.E-05 — Universe or our galaxy
- 2.E-04 — Sun
- 3.E-07 — Oceans
- 7.8E-03 — Earth's crust
- 3.3E-03 — Human beings

23.26. Bromine (Br, Z=35)

The highest concentration of bromine is in the oceans (Fig. 117).

Fig. 117: Abundance (%) of **Bromine** (Br, Z=35)

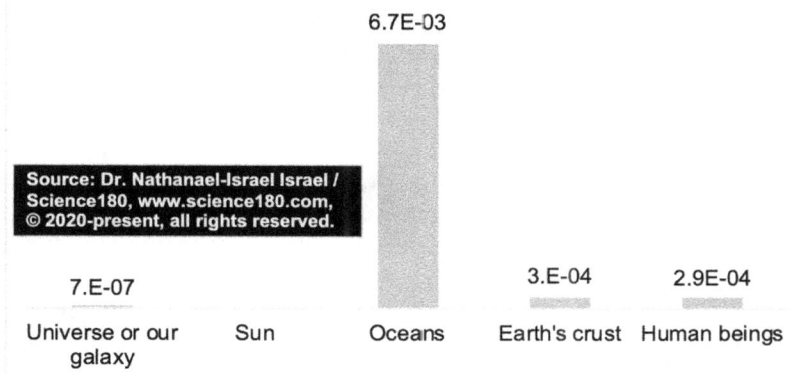

6.7E-03

7.E-07   3.E-04   2.9E-04

Universe or our   Sun   Oceans   Earth's crust Human beings
galaxy

23.27. Strontium (Sr, Z=38)

Finally, strontium is more abundant in the Earth's crust than anywhere else (Fig. 118). It may also be abundant in other terrestrial planets.

Fig. 118: Abundance (%) of **Strontium** (Sr, Z=38)

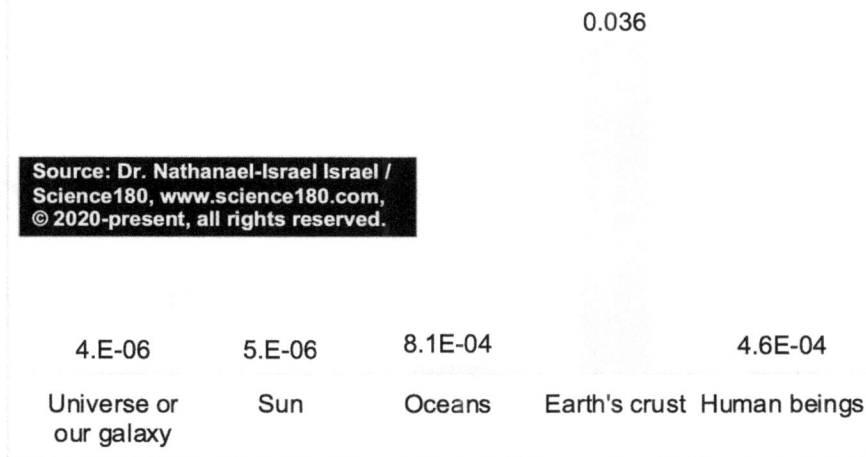

0.036

4.E-06   5.E-06   8.1E-04   4.6E-04

Universe or   Sun   Oceans   Earth's crust Human beings
our galaxy

'Science180 Academy' Success Strategy
HOW TO RAISE RATIONAL CHILDREN IN OUR MODERN WORLD

In our modern secular world, and with the many things that kids are taught at school and over which parents have little control once the kids head to public school, parents have a lot to worry about. But it does not have to be that way. Universe origin and life-origin scientist Dr. Nathanael-Israel Israel has discovered that, more than ever, parents have a crucial responsibility to rationally prepare their kids to have a strong worldview that properly embraces both science and faith, so their kids are not pulled on one side by the secular education and on the other side by religious belief. But how can parents and their children achieve that common goal?

Listen to this Beninese-American scientist and mathematician Dr. Nathanael-Israel Israel to figure it out. Nathanael-Israel is the author of the acclaimed book *"How Baby Universe was Born"*, an easy to understand scientific book primarily written for children age 7-12 years old to help them properly crack the code of the formation of the universe in a language they completely enjoy, and that prepares them to fight any secular or religious theory that may try to rationally drift them away from the reality of everything!

Sample questions that will get answered include the following and many more:

- How can parents use the latest breakthrough about the universe origin to rationally raise their kids?
- How can parents prepare their children from being victims of the danger of wrong theories and dogmas on the origin of life and the universe?
- What can parents do to shield their children from the influence of religious and scientific beliefs that try to enslave them in the name of reason or faith?
- Why is wrong science not the only danger of raising rational children, but wrong belief as well?
- How can we help children to positively navigate the intersection of science and faith?

Learn more at Science180.com/children

CHAPTER 24

UNDERSTAND THE SCIENTIFIC POWER OF THE HISTORIC MACHINERY OF THE FORMATION OF CHEMICAL PARTICLES SO YOU CAN START ASKING THE RIGHT QUESTIONS THAT THE FUTURE NOBEL LAUREATES IN CHEMISTRY COULD WORK ON–I AM NOT KIDDING!

On Tuesday, June 3rd, 2020, as I was deeply and carefully reviewing the data on chemical elements, I noticed some links between the mass of the most abundant chemical elements and some characteristics of the planets where they were formed. One of the challenges in explaining the complex formation of the atoms is that, as they were being formed, the planets and the atmosphere they belonged to were also going through their genesis. This means that the formation of the bodies and systems of bodies in the universe, regardless of their size (e.g. galaxies, stellar systems, stars, planetary systems, planets, satellite systems, asteroids systems, asteroids, atoms, subatomic particles, invisible or crypto particles, etc.) was a dynamic and systemic process, which took place at the same time that their components were also being formed according to their size, position, and other characteristics I detailed throughout my writing. That is one of the reasons when I talked about the precursors of atoms, I also talked about the precursors of the atmosphere and precursors of their planets or planetary systems. In others words, when the atoms were being formed, no definite atmosphere or planet or planetary system was completed yet. Everything formed in the universe was being shaped at the same time until a certain "equilibrium" that I called turbulent equilibrium reached a level whereby movements and certain morphological changes were locked into more dynamically stable stages. I used the term turbulent equilibrium because it involved various scales and intensities of turbulence. I also used the term "dynamically stable stages" because no perfect and stable equilibrium have existed and will never exist in the created or formed world. Everything has been in motion and changing all the time to some extent since then. It is the degree of freedom of these changes and the time it could take before these changes could be perceived, which caused some human beings to think that certain things in nature are static.

As I was pondering on the abundance of the chemical elements in the atmosphere, I felt like as the precursors of atoms were being shaped into the precursors of the atmosphere of each planet, the pathway of the formation of the atoms favored the abundance of lighter atoms than heavier ones. The atomic mass of the most abundant atoms and that of the heaviest atoms in each system depended on the conditions that reigned in the environment where the atoms and the celestial bodies they belonged to were formed. For instance, in the precursor of the atmosphere of Mercury (the innermost planet), conditions have favored the formation and abundance of metallic atoms, while in the precursors of the atmosphere of other planets, such conditions could not have existed, hence they are depleted of metals. Instead, the atmosphere of these planets besides Mercury were filled with nonmetals. To be clear, although the atmosphere of Mercury is also dominated by nonmetals (e.g. O_2), it is the only atmosphere dominated by metals.

Although it may sound like I said certain things over and over, it is important to know that I was dealing with a very complex data and the time and the strategic thinking it took for me to gradually unearth the details and put them into a chronological and meaningful logic cannot be easily comprehended by many, else, the origin of the universe could have been explained for many thousands of years by now. Therefore, because I did not find any better way to gradually explain this complexity, I prefer repeating some facts every once in a while, but I always add different perspectives and insights to them as I understood them from various angles. To put it another way, the process of discovering the real origin of the universe is like peeling the layers of an onion one by one until reaching the core of the onion and then trying to put together these layers to tell a complex story. If it was just like that, the narration of the story could have been easier. But in fact, the peeling of the layers occurred after the onion was first cut into pieces, therefore calling for an extra need to put together the cut layers to make the transition of the story smoother. Therefore, the repeats in the story I am telling are like the connection made between layers found in different pieces of the onion.

Well, in the atmosphere of the other planets besides Mercury, because metals could not be significantly formed, nonmetals dominated. Although as of 2025, metals are said not to be present in the atmosphere of 8 planets out of the 9, I believe that, as more sophisticated means are used to explore the atmosphere of the planets, at least some traces of metals will be discovered in the atmosphere of most planets, at least the terrestrial ones known for their higher density, which can also favor the formation of metals, which among many characteristics are usually dense.

As of 2025, the nonmetals that dominated the atmosphere of each planet depended on the conditions in the precursors of those atmospheres and the system they belong to. As I reached this stage of the writing, I felt like to properly explain the way precursors of atoms were molded into different atoms, I needed to talk a little bit about the physics of fluid fragmentation—more details can be found in *"Turbulent Origin of the Universe"*. Fluid breakup can help to better understand the factors that have acted on the precursors of matter to fragment and aggregate them into atoms constituted of various subatomic particles all of which are embedded in systems of celestial bodies, which at their turn were produced by a major fragmentation and aggregation of matter on a bigger scale. The physics of the fragmentation of aggregation of matter is crucial to understand how the precursors of matter were formed and how turbulent processes affected and shaped them. To address the origin, formation, differentiation or specialization of the compounds and elements that they are made, some of the data I used in the next segments are summarized in Table 11.

Because I detailed these characteristics in other chapters, in the following segments, I just pinpointed some key aspects of them that can help explain their presence or abundance according to the planets. I will start with the metals.

24.1. Metals in the atmosphere of the planets in the Solar System

Four kinds of metals were found in the atmosphere of the planets and they were all observed in the atmosphere of Mercury:

- 1 transition metal: Iron (Fe, Z=26)
- 2 alkali metals:
    - Potassium (K, Z=19)
    - Sodium (Na, Z=11)
- 2 alkaline earth metals:
    - Calcium (Ca, Z=20)
    - Magnesium (Mg, Z=12)
- 1 metal: Aluminum (Al, Z=13)

The most abundant of these metals is Sodium (Na, Z=11) and its abundance in Mercury's atmosphere is 29%. The abundance of sodium is followed by that of Potassium (K, Z=19): 0.5%. The other 4 metals found on Mercury are in traces. Of the 6 metals found in the atmosphere as of 2025, sodium is the lightest, having a molar mass of 22.99 g/mol. In other words, all the metals in the atmosphere of the planets were found in the atmosphere of the innermost planet (Mercury) and of all these metals found in the atmosphere of the planets, the most abundant one is the lightest of them. The high orbital speed of the precursor of Mercury could have played a role in favoring the formation of metals in the atmosphere of that planet.

Although it can be easy to pinpoint that the small size of sodium could have favored its abundance in the atmosphere of Mercury, I did not initially, quickly understand why potassium is the second most abundant metal in the atmosphere of that planet. If the mass of the metals is the only defining factor that explains their abundance, Magnesium (Mg, $Z=12$) and Aluminum (Al, $Z=13$) which are smaller than Potassium (K, $Z=19$) could have been more abundant.

Table 11: Characteristics of the chemical elements in the atmosphere of the planets in the Solar System

Name (Symbol, Atomic Number)	Series or Classification	State at standard temperature and pressure: 273 K	Atomic Number	Atomic mass (g/mol)	Density (kg/m3) at 293 K	Liquid density (kg/m3)	Brinell hardness (MPa)	Mohs hardness	Vickers hardness (MPa)	Bulk modulus (MPa)	Shear modulus (GPa)	Young's modulus (GPa)	Calculated radius (pm)	Covalent radius (pm)	Empirical radius (pm)	Van der Waals radius (pm)
Hydrogen (H, Z=1)	Non-metal	Gas	1	1.008	0.09								53	34	25	120
Deuterium (D, Z=1)	Non-metal	Gas	1	2.014	0.18	162.4										
Helium (He, Z=2)	Noble gas	Gas	2	4.003	0.179								31	32		140
Carbon (C, Z=6)	Non-metal	Solid	6	12.011	2267			0.5		33			67	77	70	170
Nitrogen (N, Z=7)	Non-metal	Gas	7	14.007	1.251								56	75	65	154
Oxygen (O, Z=8)	Non-metal	Gas	8	15.999	1.429								48	73	60	152
Neon (Ne, Z=10)	Noble gas	Gas	10	20.180	0.9								38	69		154
Sodium (Na, Z=11)	Alkali metal	Solid	11	22.990	971	927	0.69	0.5		6.3	3.3	10	190	154	180	227
Magnesium (Mg, Z=12)	Alkaline earth metal	Solid	12	24.305	1738	1584	260	2.5		45	17	45	145	130	150	173
Aluminum (Al, Z=13)	Metal	Solid	13	26.982	2698	2375	245	2.75	167	76	26	70	118	118	125	205
Sulfur (S, Z=16)	Non-metal	Solid	16	32.066	2067	1819		2		7.7			88	102	100	180
Argon (Ar, Z=18)	Noble gas	Gas	18	39.948	1.784								71	97	71	188
Potassium (K, Z=19)	Alkali metal	Solid	19	39.098	862	828	0.363	0.4		3.1	1.3		243	196	220	275
Calcium (Ca, Z=20)	Alkaline earth metal	Solid	20	40.078	1540	1378	167	1.75		17	7.4	20	194	174	180	
Iron (Fe, Z=26)	Transition metal	Solid	26	55.845	7874	6980	490	4	608	170	82	211	156	125	140	
Krypton (Kr, Z=36)	Noble gas	Gas	36	83.800	3.733								88	110		202
Xenon (Xe, Z=54)	Noble gas	Gas	54	131.29	5.887								108	130		216

Designed by Dr. Nathanael-Israel Israel

Table 12: Mass, density, hardness and radius of the 6 metals found in the atmosphere of Mercury

Name (Symbol, Atomic Number)	Atomic number	Atomic Mass (g / mol)	Density @ 293 K (kg/m3)	Brinell hardness (MPa)	Mohs hardness	Calculated radius (pm)	Covalent radius (pm)	Empirical radius (pm)
Sodium (Na, Z=11)	11	22.99	971	0.69	0.5	190	154	180
Magnesium (Mg, Z=12)	12	24.31	1738	260	2.5	145	130	150
Aluminum (Al, Z=13)	13	26.98	2698	245	2.75	118	118	125
Potassium (K, Z=19)	19	39.10	862	0.363	0.4	243	196	220
Calcium (Ca, Z=20)	20	40.08	1540	167	1.75	194	174	180
Iron (Fe, Z=26)	26	55.85	7874	490	4	156	125	140

But because potassium is more abundant than magnesium and aluminum, I suspected that, in addition to mass, another factor must have also affected the abundance of these metals. As I started investigating what that factor may have been, I was drawn to look at the density, hardness, and radius of these elements. In the Table 12, I present the mass, density, hardness, and radius of the 6 metals found in the atmosphere of Mercury.

To present the data into a simpler and easily visualized format, I illustrated them in the graphs below (Fig. 119 - Fig. 124). From these graphs, I noticed that, of the aforementioned 6 metals, potassium is the least dense while sodium is the second less dense. Indeed, the bulk density at 293 K of magnesium (1738 kg/m³), aluminum (2698 kg/m³), calcium (1540 kg/m³), and iron (7874 kg/m³) is higher than that of potassium (862 kg/m³) and of sodium (971 kg/m³) (Fig. 119). In other words, of the 6 metallic elements found in the atmosphere of Mercury, sodium is the smaller and the second less dense. In other words, potassium is less dense than sodium. Yet, potassium (39.1 g/mol) is heavier than sodium (22.99 g/mol). The density of some of the other 4 metals that are found in a trace in Mercury's atmosphere is more than twice that of sodium and potassium. Therefore, I felt like the small density of potassium could have militated in favor of its abundance.

Moreover, the 4 metals found in traces are harder than sodium and potassium. For instance, the Brinell hardness of these 4 metals present in trace is 242 to 710.1 times that of sodium and 460.1 to 1349.9 times that of potassium (Fig. 120). The Mohs hardness of the 4 metals in trace is higher than that of the 2 most abundant metals (Fig. 121). To compress the precursors of atoms into the 4 metals that are in trace, much more energy could have been required. In other words, the energy to "fashion" or "mold" the precursors of atoms into denser and harder atoms could have been much higher than that required to form less dense and less hard or soft atoms. The data of the radius of the 6 metals found on Mercury support this statement. For instance, the calculated radius and the covalent radius of potassium and sodium are among the 3 highest, suggesting that, despite their relatively smaller size, these metals have a bigger radius than their peers present in trace in the atmosphere of Mercury. For instance, although the weight (55.84 kg/mol) of Iron (Fe, Z=26) is more than twice that of sodium (22.99 kg/mol), the calculated radius of sodium (190 pm) is higher than that of iron (156 pm) (Fig. 122). The same trend holds for the covalent radius (Fig. 123). Likewise, the radius of potassium is higher than that of any other metal found in the atmosphere of Mercury. I felt like the radius of the 2 most abundant metals in the atmosphere of Mercury is relatively higher than that of the metals found in trace because the precursors of these 2 elements could have not been squeezed much. In other words, instead of using its limited energy to form heavier, denser, and harder elements, the precursors of the metals in the atmosphere of Mercury were converted into smaller and less dense metallic atoms, and because the force applied to them was not very strong, their radius ended up being among the smallest of the 6 metals.

Fig. 119: Density (kg/m3) at 293 K of the 6 metals found in the atmosphere of Mercury

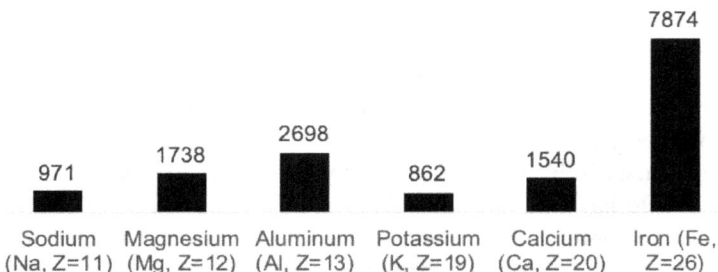

					7874
971	1738	2698	862	1540	
Sodium (Na, Z=11)	Magnesium (Mg, Z=12)	Aluminum (Al, Z=13)	Potassium (K, Z=19)	Calcium (Ca, Z=20)	Iron (Fe, Z=26)

Fig. 120: Brinell Hardness (MPa) of the 6 metals found in the atmosphere of Mercury

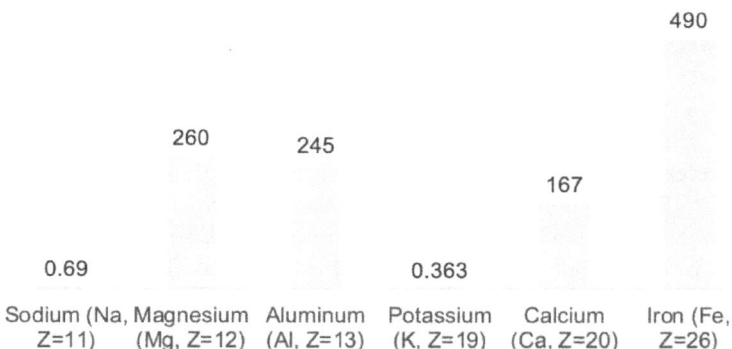

					490
	260	245		167	
0.69			0.363		
Sodium (Na, Z=11)	Magnesium (Mg, Z=12)	Aluminum (Al, Z=13)	Potassium (K, Z=19)	Calcium (Ca, Z=20)	Iron (Fe, Z=26)

Fig. 121: Mohs Hardness of the 6 metals found in the atmosphere of Mercury

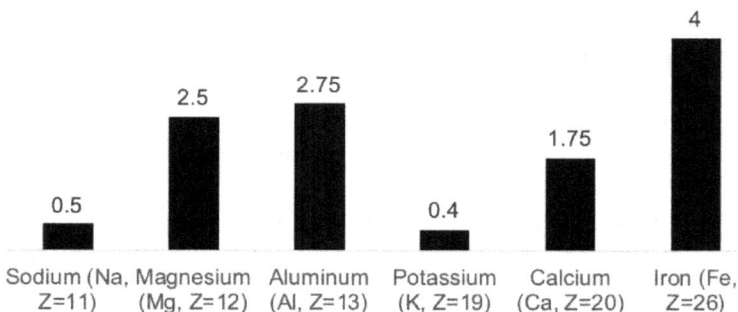

					4
	2.5	2.75		1.75	
0.5			0.4		
Sodium (Na, Z=11)	Magnesium (Mg, Z=12)	Aluminum (Al, Z=13)	Potassium (K, Z=19)	Calcium (Ca, Z=20)	Iron (Fe, Z=26)

Fig. 122: Calculated radius (pm) of the 6 metals found in the atmosphere of Mercury

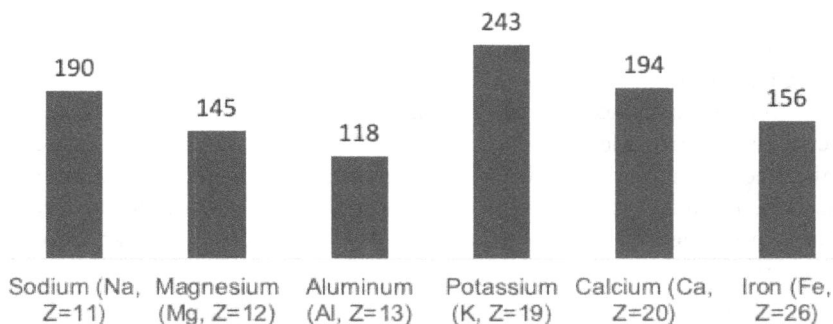

Sodium (Na, Z=11)	Magnesium (Mg, Z=12)	Aluminum (Al, Z=13)	Potassium (K, Z=19)	Calcium (Ca, Z=20)	Iron (Fe, Z=26)
190	145	118	243	194	156

Fig. 123: Covalent radius (pm) of the 6 metals found in the atmosphere of Mercury

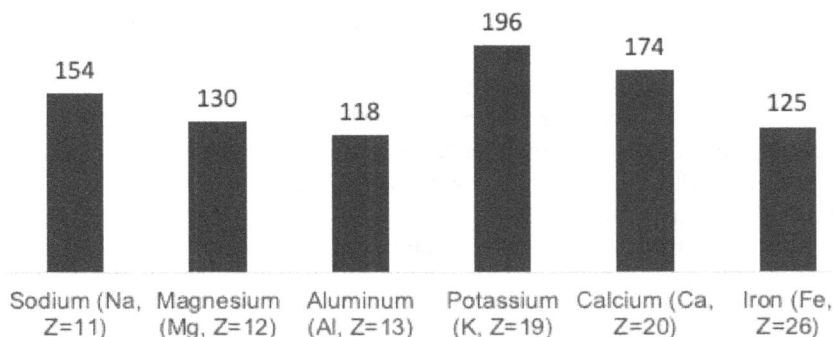

Sodium (Na, Z=11)	Magnesium (Mg, Z=12)	Aluminum (Al, Z=13)	Potassium (K, Z=19)	Calcium (Ca, Z=20)	Iron (Fe, Z=26)
154	130	118	196	174	125

Fig. 124: Atomic Mass (gram / mol) of the 6 metals found in the atmosphere of Mercury

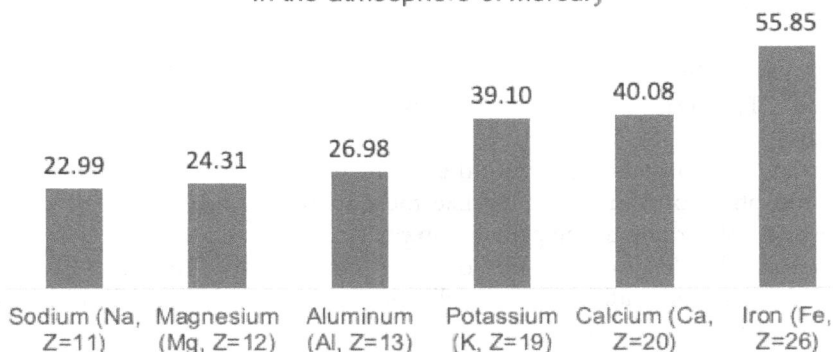

Sodium (Na, Z=11)	Magnesium (Mg, Z=12)	Aluminum (Al, Z=13)	Potassium (K, Z=19)	Calcium (Ca, Z=20)	Iron (Fe, Z=26)
22.99	24.31	26.98	39.10	40.08	55.85

Therefore, considering all I said above, I felt like the 4 metals that are found in traces (Magnesium (Mg, Z=12), Aluminum (Al, Z=13), Calcium (Ca, Z=20), and Iron (Fe, Z=26)) were produced in small amounts because the machinery that converted the precursors of the metallic atoms did not favor their pathways that could have required more energy and resources. In other words, the energy required to transform the precursors of these 4 metallic atoms in the atmosphere of Mercury into denser and harder atoms could have been limited. Consequently, the machinery that molded the precursors of the metallic atoms could have preferred converting these precursors into the smallest and the least dense of the 6 metals. Because sodium which is denser than potassium is more abundant than potassium, I think that the pathway of forming the metals could have preferred molding more metallic atoms into the smallest of the 6 metals than into the least dense of them. For it could have taken more energy to make a big potassium atom than the small sodium one. These trends I observed for the metallic atoms gave me a glimpse into what could explain the machinery that shaped and controlled the formation of all elements in the universe. After this hint of what could have happened during the formation of the metals in the atmosphere, I turned my attention to the nonmetals.

24.2. Nonmetallic atoms found in the atmosphere of the planets

Ten (10) nonmetallic atoms are found in the atmosphere of the planets in the Solar System: 5 typical nonmetals and 5 noble gases. I purposely used the term "typical nonmetal" because some people think that the atoms usually labeled as nonmetals are the only ones that are really nonmetallic. However, as I explained in the chapter on the groups, series, or classification of the atoms, noble gases and halogens are also not metals.

With that being said, while the 4 giant planets are dominated by noble gases, the terrestrial planets are dominated by the most known nonmetals. As I detailed next, although most of the noble gases found in the atmosphere are heavier than the nonmetals, the highest densities were observed with nonmetals. The density of hydrogen and helium, which dominate the atmosphere of the 4 giant planets, is smaller than that of carbon, nitrogen, and oxygen, which dominate the atmosphere of the terrestrial planets. This means that the atmosphere of the 5 terrestrial planets is dominated by denser nonmetals, while that of the 4 giant planets is dominated by the lightest nonmetal (hydrogen) followed by the lightest noble gas (helium). The most abundant noble gas in the atmosphere is the lightest one (helium).

Based on the insight I got from the careful analysis of the data on the metals in the atmosphere of Mercury, I felt like the data on the nonmetals present in the atmosphere according to the planets can provide me with a wealth of information. I therefore, decided to focus on some key characteristics of the atoms such as mass, density, and radius. In the following segments, I detailed some factors, which, during the genesis of the universe and its content, could have played a crucial role in the differentiation of the precursors of atoms into specific atoms known today. Toward that end, I summarized the data pertaining to nonmetals in Table 13.

Table 13: Mass, density, and radius of non metallic elements (noble gases and reactive nonmetals) found in the atmosphere of planets

Name (Symbol, Atomic Number)	Categories of elements	Atomic Number (Z)	Atomic Mass (g/mol)	Density @ 293 K (kg/m3)	Calculated radius (pm)	Covalent radius (pm)	Vanderwaals radius (pm)
Hydrogen (H, Z=1)	Nonmetal	1	1.008	0.08988	53	34	120
Deuterium (D)	Nonmetal	1	2.014	0.18			
Helium (He, Z=2)	Noble gas	2	4.003	0.1785	31	32	140
Carbon (C, Z=6)	Nonmetal	6	12.011	2267	67	77	170
Nitrogen (N, Z=7)	Nonmetal	7	14.007	1.2506	56	75	154
Oxygen (O, Z=8)	Nonmetal	8	15.999	1.429	48	73	152
Neon (Ne, Z=10)	Noble gas	10	20.180	0.8999	38	69	154
Sulfur (S, Z=16)	Non-metal	16	32.066	2067	88	102	180
Argon (Ar, Z=18)	Noble gas	18	39.948	1.7837	71	97	188
Krypton (Kr, Z=36)	Noble gas	36	83.800	3.733	88	110	202
Xenon (Xe, Z=54)	Noble gas	54	131.290	5.887	108	130	216

For more clarity and easier visualization of the trends, I presented the data from Table 13 in the graphs below (Fig. 125-129):

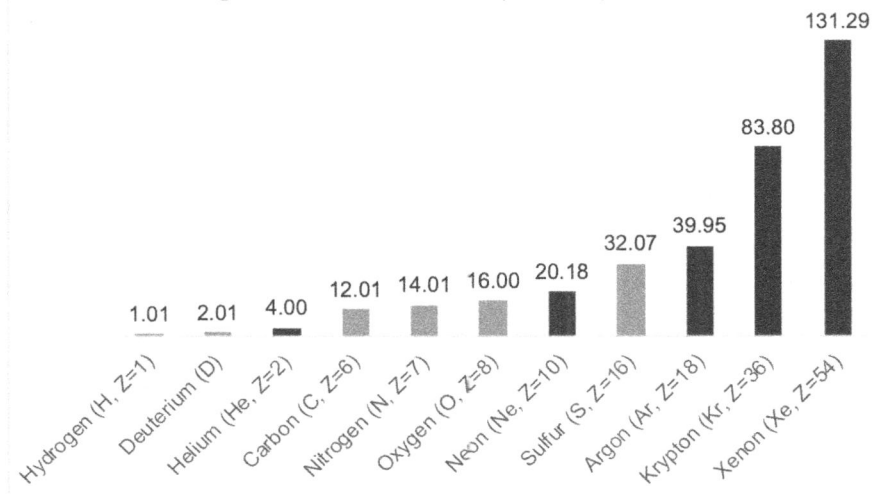

Fig. 125: Atomic Mass (g/mol) of **nonmetals** and **noble gases** found in the atmosphere of planets

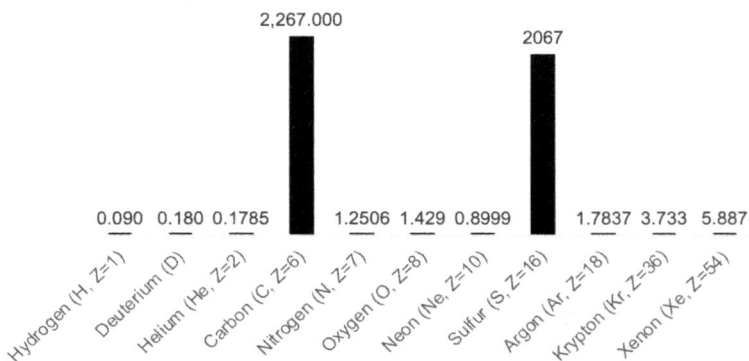

Fig. 126: **Density** (kg/m3) at 293 K of **nonmetals** and **noble gases** found in the atmosphere of planets

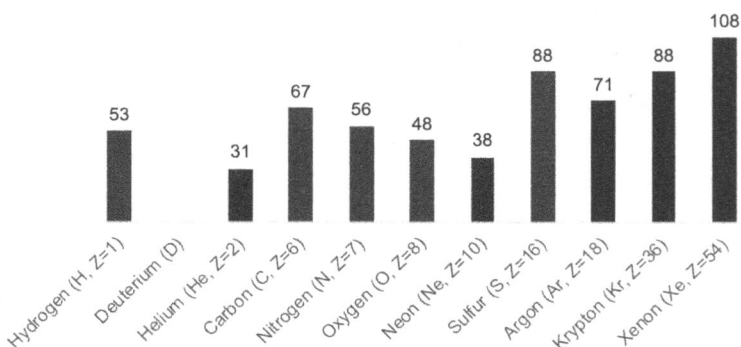

Fig. 127: **Calculated radius** (pm) of **nonmetals** and **noble gases** found in the atmosphere of planets

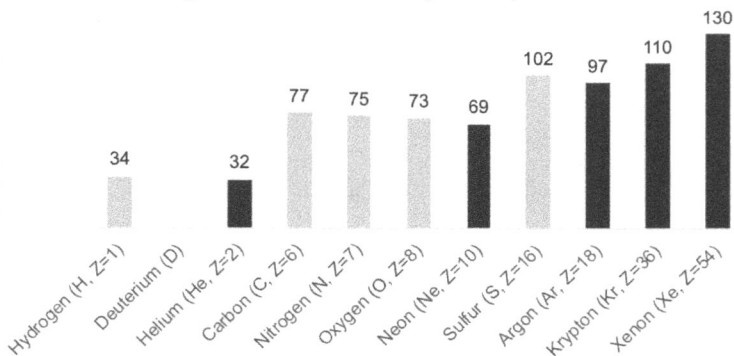

Fig. 128: **Covalent radius** (pm) of **nonmetals** and **noble gases** found in the atmosphere of planets

Nathanael-Israel Israel: Historic Discoverer of the Life Turbulent Origin Formula™

Fig. 129: Vanderwaals radius (pm) of nonmetals and noble gases found in the atmosphere of planets

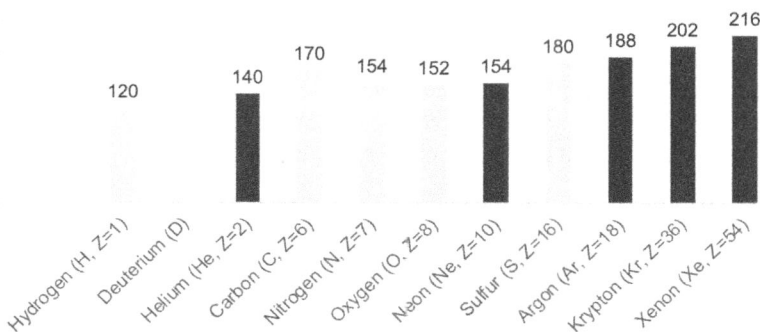

24.2.1. Nonmetals in the atmosphere of the planets

By increasing order of their atomic mass, the 6 nonmetals found in the atmosphere of the planets are:

- Hydrogen (H, Z=1)
- Deuterium (D, Z=1), a hydrogen isotope with 1 neutron, 1 proton, and 1 electron
- Carbon (C, Z=6)
- Nitrogen (N, Z=7)
- Oxygen (O, Z=8)
- Sulfur (S, Z=16)

Four of the nonmetals found in the atmosphere are present in significant amounts in the atmosphere of all the planets. These 4 are:

- Hydrogen (H, Z=1)
- Carbon (C, Z=6)
- Nitrogen (N, Z=7)
- Oxygen (O, Z=8)

In contrast, Deuterium (D, Z=1) is only found in the atmosphere of the 4 giant planets and of Mars. Likewise, Sulfur (S, Z=16) is said to be significantly found only in the atmosphere of the 4 giant planets and of Venus. In most of the writing below, I did not mention deuterium too much, but just the ordinary hydrogen for, besides their atomic mass, they share many features together. Indeed, hydrogen, the lightest nonmetal is the most abundant element in the atmosphere of the giant planets where its abundance varies between 80% and 96.3% of the volume of these atmospheres. In contrast, the heaviest nonmetals were abundant in the atmosphere of the terrestrial planets, meaning the 4 innermost planets and Pluto, the outermost planet. For instance, although Pluto is located far beyond the giant planets, 99% of the volume of its atmosphere is said to be made of nitrogen, a nonmetal 14 times heavier than hydrogen. Therefore, if people do not consider Pluto as a planet or remove it from the analysis, they may mistakenly think that the heavier nonmetals found in the atmosphere are only in the atmosphere of the 4 innermost planets, which can cause them to mistakenly tie the abundance of heavier nonmetals to the semi major or orbital speed of the planets or of their precursors.

Carbon is the second lightest nonmetal found in the atmosphere, but it is the densest of all the nonmetals found in the atmosphere. Its density (2267 kg/m^3) is 1.1 to $25,222.5$ times that of the other nonmetals found in the atmosphere:

- 25222.5 times that of Hydrogen (H, Z=1)
- 1812.7 times that of Nitrogen (N, Z=7)
- 1586.4 times that of Oxygen (O, Z=8)
- 1.1 times that of Sulfur (S, Z=16)

Carbon is even denser than some metals found in the atmosphere of the planets. In fact, of the 16 chemical elements found in the atmosphere of the planets, only 2 metals are denser than carbon:

- Aluminum (Al, Z=13): 2698 (kg/m^3), meaning 1.19 times denser than carbon and
- Iron (Fe, Z=26): 7874 (kg/m^3), meaning 3.47 times denser than carbon.

Both of these 2 metals are more than two times heavier than carbon:

- Aluminum (Al, Z=13) is 2.246 times heavier than carbon and
- Iron (Fe, Z=26) is 4.65 times heavier than carbon.

Moreover, the calculated radius of carbon (67 pm) is 1.2 to 1.4 times that of the other nonmetals found in the atmosphere of the planets except Sulfur (S, Z=16) which radius is 1.31 times that of carbon. In other words, sulfur is the nonmetal which density, mass, and radius are the closest to that of carbon. To put it another way, as the second smallest nonmetal, carbon is smaller and lighter but denser than sulfur. Finally, Oxygen (O, Z=8) is heavier and denser, but smaller than Nitrogen (N, Z=7). The characteristics of these chemical elements gave me a glimpse into how their precursors could have selected or forced to choose which pathways to prioritize to produce each of them according to the dynamic and turbulent conditions, opportunities, and constraints in the environment where they were shaped.

To explain the pathway that defined the formation and abundance of the chemical elements, I studied the planets one by one, starting with those which atmospheres are filled with the simpler chemical elements. To unearth the formation of carbon, I proceeded by first focusing on the elements formed in the atmosphere of the giant planets. Indeed, in the atmosphere of the giant planets, after molecular hydrogen (the smallest nonmetal) and helium (the smallest noble gas), carbon-based compounds and mostly methane (CH_4) are more abundant than oxygen-based, nitrogen based, or sulfur-based compounds in the atmosphere. To put it another way, the few atoms of the heavier nonmetals formed in the atmosphere of the giant planets bound to others to form various compounds containing them. The most abundant of these compounds is methane made by the association of carbon and hydrogen. The least abundant of these nonmetal compounds are in the form of aerosols such as:

- ammonia ice
- water ice
- ammonia hydrosulfide
- methane ice (?)

Before I say anything else, let me recall that when I was looking at the data of the chemical elements in the atmosphere, I could not clearly understand the meaning of the preliminary trends I noticed. As I tried to put my thoughts down, I did not know where to start and where to finish. If it was in the early stage of my writing, I could have left the data to focus on others which inspired me more. But I had to break through the hinderances and unearth some secrets, for I was toward the end of my writing and I needed to wrap everything up. After 3 days of looking at the data while my legs were hurting from standing for hours and typing because I could not sit long for fear of having again some problems with my neck and shoulders, I decided to grab the data from what seemed to be as the simpler angle: the giant planets which are blatantly filled with hydrogen followed by helium. As I finished dealing with these chemical elements, methane (CH_4) appeared to me as the next target to handle as if I was peeling the layers of an onion one by one from a slice constituted of the abundance of chemical elements. In the following segments, I tried to recount how I figured out why and how some nonmetals are found in the atmosphere of the giant planets. Let us continue with methane.

Although it may sound like a simple fact, I would like to recall that a molecule of methane (CH_4) consists of one carbon atom (the second smallest nonmetal) and 4 atoms of hydrogen (the smallest nonmetal). Just by considering the composition of methane and the ranking of its abundance in the atmosphere of the giant planets caused me to believe that the leftover of the precursors of nonmetallic atoms—that were converted into the abundant hydrogen found on the giant planets—could have been preferably molded into carbon. For it could have taken less energy to form carbon out of some precursors of nonmetallic atoms than forming heavier and larger nonmetals. Because carbon is smaller than the other nonmetals found in the atmosphere, its high density made me to believe that its precursors could have been highly compressed or spiraled. And because a lot of hydrogen atoms were formed in the atmosphere of the giant planets, some of them bound to the limited carbon atoms to form methane (CH_4), which, because of the colder environment of Uranus and Neptune, is also found as methane ice in the form of aerosol. Similarly, 2 carbon atoms were bound to one another and to 6 hydrogen atoms to form ethane (C_2H_6) in the atmosphere of some giant planets (e.g. Jupiter, Saturn, and Neptune). And because most of the carbon atoms formed on Uranus were used to abundantly form methane, less and/or none were left to form ethane in the atmosphere of that planet. That is why methane, which is more abundant in the atmosphere of Uranus than in that of any other planet, occupied 2.3% of the volume of Uranus, while ethane is not even reported for its atmosphere. Likewise, because less of their carbon was used to form methane, the atmosphere of Jupiter, Saturn, and Neptune contain less methane but more ethane.

Finally, because forming the other heavier nonmetals (nitrogen, oxygen, and sulfur) could have taken more energy or resources, the pathway of their production was downgraded, hence the amount of these 3 nonmetals in the atmosphere of the 4 giant planets is very small. Consequently, the compounds (e.g. Ammonia (NH_3), Ammonia ice, and Ammonia hydrosulfide (NH_4)HS were found just in traces in the atmosphere of the giant planets. I think that the atmosphere of Venus—which also contains sulfur and where Ammonia hydrosulfide (NH_4)HS could have been made—could have been limited by its low abundance of hydrogen. For the few atoms of hydrogen formed in the atmosphere of Venus could have been quickly trapped by the abundant oxygen atoms to form water, while other oxygen atoms on the same planet (Venus) reacted with the few sulfur atoms to produce Sulfur Dioxide (SO_2). And because oxygen could not be abundantly produced in the atmosphere of the giant planets, the sulfur available on them could not form SO_2 but reacted with the abundant hydrogen and the limited nitrogen to form (NH_4)HS in the atmosphere of the giant planets. That is why Ammonia hydrosulfide (NH_4)HS is present in the atmosphere of the 4 giant planets while Sulfur Dioxide (SO_2) is only present in the atmosphere of Venus.

I could not finish explaining the compounds and chemical elements of the giant planets without addressing deuterium and Hydrogen Deuteride (HD) reported mostly in the atmosphere of the 4 giant planets. Knowing that deuterium is a heavier isotope of hydrogen, I think that as the precursors of the hydrogen atoms were being formed, some were split into and/or gained a neutron to form a nucleus containing a proton and neutron orbited by an electron. Such atoms differentiated into deuterium, which reacted with the abundant protium (ordinary hydrogen atoms) to form the Hydrogen Deuteride (HD), which is a molecule consisting of an atom of protium and an atom of deuterium. Because hydrogen is abundant in the atmosphere of the 4 giant planets, I perceived that the limited amount of Hydrogen Deuteride (HD) in the atmosphere of these planets is due to the limited formation or production of deuterium in those environments. Hydrogen Deuteride (HD) could have been formed also in the atmosphere of Mars and because the precursor of that atmosphere could have been rich in oxygen atoms, which bound to carbon to form CO_2, some of those oxygen atoms on Mars reacted with the available Hydrogen Deuteride (HD) molecules to form Hydrogen Deuterium Oxygen (HDO or D_2O) also called heavy water which is reported only in the atmosphere of Mars. Because many oxygen atoms were formed in the Martian atmosphere, the production of heavy water on Mars' atmosphere could have been limited by the availability of Hydrogen Deuteride. Hydrogen Deuterium Oxygen (HDO or D_2O) could not have been found in the atmosphere of the 4 giant planets despite them containing significant amounts of

CHAPTER 24: HISTORIC DISCOVERY OF THE MACHINERY OF THE FORMATION OF CHEMICAL PARTICLES

Hydrogen Deuteride (HD) because oxygen, one of the elements to form them, could have been limiting. For the few oxygen atoms formed in the hydrogen-dominated atmosphere of the 4 giant planets could have been associated with the abundant hydrogen to form the few molecules of water found in Jupiter only. In other words, ordinary water and heavy water were not found nor reported in the atmosphere of Saturn, Uranus, and Neptune because they lack oxygen, the second heaviest nonmetal found in the atmosphere of the planets.

Unlike the atmospheres of the 4 giant planets which are filled with hydrogen and helium, the atmospheres of the 5 terrestrial planets are poor in hydrogen-containing compounds except Mercury where molecular hydrogen (H_2) is abundantly found (22%), a record not found in the atmosphere of any other terrestrial planet. Some hydrogen related compounds which are abundant in the atmosphere of the giant planets are not even found in the atmospheres of the terrestrial planets. For instance, Hydrogen Deuteride (HD), Ethane (C_2H_6), and Ammonia (NH_3) are not even recorded in the atmosphere of the terrestrial planets, yet they are in a significant amount in the atmosphere of the giant planets. The only hydrogen related compounds that are found in the atmospheres of the terrestrial planets are:

- Hydrogen Deuterium Oxygen (HDO or D_2O) reported only in the atmosphere of Mars in a small amount (abundance = 8.5E-05%),
- Methane (CH_4) recorded only in the atmosphere of Earth and Pluto where the highest abundance (0.5%) was found; and
- Water (H_2O) recorded in the atmosphere of Venus, Earth, and Mars, with the highest abundance reported in the atmosphere of Earth (1%).

Aside from molecular hydrogen which occupied a significant portion of the atmosphere of Mercury, most of the compounds found in the atmosphere of the terrestrial planets are dominated by:

- Carbon (C, Z=6)
- Nitrogen (N, Z=7)
- Oxygen (O, Z=8)

Sulfur (S, Z=16), the heaviest nonmetal found present in the atmosphere of Venus in combination with oxygen to form SO_2 at a relatively small abundance (0.015%). To explain the origin and abundance of these nonmetals in the atmosphere of the terrestrial planets, I turned to their characteristics. Below, I carefully addressed the heavier nonmetals one by one.

24.2.1.1. Sulfur

With a molar mass of 32.07 g/mol, Sulfur (S, Z=16) is the heaviest nonmetal found in the atmosphere of the planets and it is found in the atmosphere of:

- the 4 giant planets in traces as an aerosol in the form of Ammonia hydrosulfide, $(NH_4)HS$ and on
- Venus in the form of Sulfur Dioxide (SO_2) where it is more abundant.

Among the 2 compounds that contain sulfur in the atmosphere of the planets, Sulfur Dioxide (SO_2) is the most abundant: 0.015%. In other words, although sulfur is not the only atom in $(NH_4)HS$, considering the other atoms in that molecule and its abundance as a whole, I think that the abundance of sulfur on Venus may be higher than that in the atmosphere of the giant planets. To put it another way, of the planets which atmosphere contains sulfur, the highest abundance was recorded with Mercury, the innermost planet. The intensity of the turbulence that prevailed during the genesis of the planets and their atmosphere could have caused the precursors of atoms in Venus' atmosphere to favor the formation of denser atoms like sulfur just as the planet itself is also one of the densest. The high orbital speed that the precursor of Venus had could have also favored the formation of heavier elements such as sulfur and oxygen, which in the end associated with one another to form sulfur dioxygen (SO_2).

24.2.1.2. Oxygen

After sulfur, the second heaviest nonmetal present in the atmosphere of the planets is Oxygen (O, Z=8), which mass (15.99 g/mol) is half that of sulfur. The highest concentration of oxygen was found in the atmosphere of the 4 innermost planets, whereas on Pluto, it is found only in the form of Carbon Monoxide (CO) in association with the limited carbon formed there. In the atmosphere of the planets, oxygen is present in 8 compounds or gases:

- CO (carbon monoxide),
- CO_2 (carbon dioxide),
- H_2O (water),
- H_2O ice (water ice as aerosol),
- HDO (hydrogen-deuterium-oxygen also called heavy water),
- NO (nitrogen oxide),
- O_2 (oxygen), and
- SO_2 (sulfur dioxide).

The most abundant of these oxygen-containing materials are:

- CO_2 (carbon dioxide), which accounts for 96.5% of Venus' atmosphere and 95.32% of Mars' atmosphere and
- O_2 (oxygen) which is the most abundant element in the atmosphere of Mercury where it accounts for 42% of the volume. The other planet where oxygen is abundant in the atmosphere is the Earth (20.95%).

Considering the abundance and the formula of the oxygen-containing compounds, I think that oxygen atoms are more abundant in the atmosphere of the planets than carbon ones. Indeed, in the atmosphere of the 4 giant planets, oxygen is generally found in traces, in association with hydrogen to form water ice as an aerosol. Additionally, oxygen is also found in the atmosphere of Jupiter in the form of ordinary water at a higher concentration 4.E-04% of the volume, meaning that of the 4 giant planets, oxygen would be more abundant in the atmosphere of Jupiter, the innermost giant planet. On Pluto, oxygen is found in association with carbon in the form of carbon monoxide, which abundance is 0.05% of the volume of that planet. To put it another way, the abundance of carbon monoxide (CO) on Pluto is 125 times that of the water (H_2O) on Jupiter, water being the main compound that contains oxygen on Jupiter. Comparing that abundance with that of the compounds in which oxygen is found in the atmosphere of the giant planets, it seemed to me that oxygen could be more abundant in the atmosphere of Pluto than in that of the giant planets.

Of the 8 oxygen-containing compounds found in the atmosphere of the terrestrial planets, carbon is absent in 6 of them:

- H_2O (water),
- H_2O ice (water ice as aerosol),
- HDO (hydrogen-deuterium-oxygen also called heavy water),
- NO (nitrogen oxide),
- O_2 (oxygen), and
- SO_2 (sulfur).

And in the compounds where oxygen is associated with carbon, the number of oxygen atoms is equal or higher than that of carbon. For instance, for Carbon Monoxide (CO) to be formed, one atom of carbon must be associated with one atom of oxygen, meaning that the number of carbon atoms and oxygen atoms needed to make molecules of Carbon Monoxide (CO) is the same. However, to make Carbon Dioxide (CO_2) the number of oxygen atoms required must be double that of carbon. Considering what I just said about carbon monoxide and carbon dioxide, the number of oxygen atoms needed to form the Carbon Monoxide (CO) and Carbon Dioxide (CO_2) molecules present in the atmosphere of the planets is higher than that of carbon atoms. Considering the abundance of the oxygen and carbons containing molecules, it appeared to me that oxygen atoms are more abundant in the atmosphere of the planets than carbon atoms. While oxygen is able to react with itself to form O_2, which abundance alone reaches 42% of Mercury's atmosphere and 20.95% of Earth's atmosphere, whereas carbon is not able to react with itself to form any carbon molecule. Of the terrestrial planets, Pluto is the only one where more carbon was formed in the atmosphere than oxygen. In fact, aside from Nitrogen (N_2), the other compounds recorded in the atmosphere of Pluto are dominated by carbon instead of oxygen:

- Ethane (C_2H_6) (%),
- Methane CH_4 (%), and
- Carbon Monoxide (CO).

The formula of those compounds alone is enough to know that more carbon atoms are present on Pluto than oxygen. Considering all I said above as to oxygen and carbon, oxygen is more abundant in the atmosphere of the 4 innermost terrestrial planets than carbon, while on Pluto, the outer terrestrial planet, carbon atoms are more numerous than oxygen atoms.

TURBULENT ORIGIN OF CHEMICAL PARTICLES

Knowing that the innermost terrestrial planets have a higher orbital speed than the outermost planets (see my book on the formation of the universe) and that oxygen is the second heaviest nonmetal found in the atmosphere of the planets, it appeared to me that the high speed that the precursors of nonmetals in the precursors of the atmosphere of the 4 innermost planets could have favored their abundant differentiation into oxygen atoms. Because the conditions could not have favored the abundant differentiation of the atoms into sulfur, most of them were converted into the second heaviest atoms. Hence, the higher abundance of oxygen atoms than carbon ones for the 4 innermost terrestrial planets. However, because Pluto is located far away, the energy and movement of the precursors of the atoms in the precursors of Pluto's atmosphere could have been smaller, hence, it was not able to abundantly produce heavier nonmetals like Sulfur (S, $Z=16$) or Oxygen (O, $Z=8$), but lighter ones like Nitrogen (N, $Z=7$). Hence the abundant production of nitrogen gas on Pluto followed by that of hydrogen and carbon.

Like I already explained in the section on metals, sodium is abundantly produced on Mercury (29%) and its abundance could have reduced the pool of precursors of atoms from which nonmetals could have been produced. Hence, oxygen, which is the second heaviest nonmetal coming right after sodium, is produced in lesser amounts, which in return affected its abundance. In other words, the massive production of sodium (a more energetic and heavier atom) on Mercury is what reduced the abundance of oxygen on that planet. Because the energy in the precursors of atoms could have not been enough to convert all of the remaining into oxygen, some of them were differentiated into hydrogen atoms, which then bound to one another to form the molecular hydrogen abundantly found on Mercury (22%). On Mercury, after as many precursors of atoms were being differentiated into sodium and oxygen, the remaining could not have been molded into nitrogen or carbon because those require more energy, which could have lacked in the environment where those precursors went through changes. Hence, the remaining of the pool of precursors were converted into the lightest atom, hydrogen. Like I explained in the section on metals, a few precursors of atoms on Mercury were converted into metals, but their abundance is low, for the conditions were not met for them to be produced beyond a certain amount.

However, on Venus, Earth, and Mars, because the precursors of atoms could not be molded into sodium atoms, many of them were differentiated into oxygen atoms. However, the abundance of oxygen in the atmosphere of these 3 planets depended also on their conditions and the other atoms that were formed there. For instance, the abundance and composition of the compounds on Venus suggested that after oxygen, the second most abundant chemical element could be carbon followed by nitrogen. In other words, just like in the atmosphere of Mercury, the atmosphere of Venus is dominated by oxygen, but unlike sodium, which is the second most abundant chemical element on Mercury, carbon is the second most abundant chemical element on Venus' atmosphere. As I tried to understand what could explain such a trend for Venus, I noticed that, of the 6

Nathanael-Israel Israel: Historic Discoverer of the Life Turbulent Origin
Formula™

nonmetals found in the atmosphere, carbon is the one that resembles sodium the most at least as far as their density is concerned. Indeed, although the mass of Carbon (C, Z=6) is almost 37.5% that of Sulfur (S, Z=16), the density of Carbon (2267 kg/m^3) is 109% that (2067 kg/m^3) of sulfur, while it is 1812.7 times that of nitrogen and 1586.4 times that of oxygen. Similarly, the calculated radius and the Vanderwaals radius of carbon is closer to that of sodium than that of nitrogen and oxygen. In other words, among the nonmetals found in the atmosphere of the planet, the best replacement for sodium is carbon. Therefore, on Venus, because the conditions did not allow the differentiation of the precursors of atoms into sodium, many of them were molded into carbon as others were being converted into oxygen. As these atoms were being formed, oxygen reacted with some carbon atoms to form Carbon Dioxide (CO_2), which occupies 96.5% of the atmosphere of Venus. The other oxygen atoms formed in the atmosphere of Venus reacted with others of carbon to form Carbon Monoxide (CO), or with sulfur to form Sulfur Dioxide (SO_2) or with hydrogen to form Water (H_2O). Other precursors of atoms in the atmosphere of Venus were converted into nitrogen, the 3rd most abundant chemical element in that atmosphere, where they bound to one another to form Nitrogen (N_2), which is 3.5% of the volume of that atmosphere.

On Earth, because the abundance of none of the two compounds that contain carbon (e.g. Methane (CH_4) and Carbon Dioxide (CO_2)) is not higher than 0.04%, while no sodium is reported in such an atmosphere, I deducted that the conditions in the precursor of the Earth's atmosphere could not have favored the formation of denser atoms which density is in the order of magnitude of that of sodium or carbon. Therefore, while some few precursors of atoms were molded into carbon during the formation of the Earth's atmosphere, most of the precursors of atoms in that atmosphere were differentiated into atoms less dense than sodium and carbon. Besides the few atoms of carbon that were formed, the pathway of nonmetals which were left were that which led to the formation of oxygen, nitrogen, or hydrogen. To explain why in the atmosphere of the precursor of the Earth, the pathway to produce nitrogen was prioritized over that of oxygen, let me first recall some key characteristics of these two atoms.

Nitrogen (N, Z=7) atoms are a little lighter (14.007 kg/mol) than those of Oxygen (O, Z=8) which weight 15.99 kg/mol. Likewise, nitrogen atoms are less dense (1.25 kg/m^3) than those of oxygen (1.429 kg/m^3). In other words, it could take more energy to transform the precursors of atoms into oxygen than into nitrogen. Therefore, the pathway that produced nitrogen atoms were activated to mold most of the precursors of atoms in the precursors of the Earth's atmosphere into nitrogen more than into oxygen. By the time all of the atoms in the Earth's atmosphere were formed, nitrogen atoms dominated. That is why in the Earth's atmosphere, nitrogen gas (N_2) which molecule consists of the association between 2 atoms of nitrogen dominate (78.08%) of the Earth's atmosphere, while oxygen comes next (20.95%). Because oxygen was not as squeezed as nitrogen, the density of oxygen is higher than that of nitrogen. Consequently, the calculated radius of nitrogen (56 pm) is smaller than that of oxygen (48 pm). Likewise, the covalent radius of nitrogen (75 pm) is slightly smaller than that of oxygen (73 pm). It also explains why the Vanderwaals radius of nitrogen (154 pm) is a little higher than that of oxygen (152 pm) as atoms were mixing in the early atmosphere, some of the few carbon atoms bound with hydrogen atoms to form Methane (CH_4) while others bound with oxygen atoms to form Carbon Dioxide (CO_2). Some of the remaining oxygen atoms reacted with hydrogen to form Water (H_2O) which is just 1% of the Earth's atmosphere. The abundance of water in the Earth's atmosphere could have been limited by the lack of hydrogen atoms. According to the abundance of the atoms of the other chemical elements present in the atmosphere, oxygen was bound to certain elements to form different compounds. For instance, in the atmosphere of the Earth and particularly in the lower atmosphere, most of the oxygen bound to each other 2 by 2 to form oxygen gas (O_2), which fortunately is one of the main molecules indispensable for life. In the upper atmosphere of the Earth, the conditions have favored the interaction of 3 oxygen atoms to form the ozone (O_3). In the Earth's atmosphere, one atom of oxygen was associated or combined with 2 atoms of hydrogen to form water (H_2O). In many spots during the genesis of the atmosphere, one atom of carbon was associated with 2 atoms of oxygen to form carbon dioxide (CO_2). The presence of water deep below some impermeable layers beneath the surface of the Earth evidenced that water was available during the formation of the Earth. Despite the abundance of hydrogen atoms in the atmosphere of the giant planets, they do not have much water in their atmosphere probably because of a lack of oxygen, which is needed to react with the hydrogen to form water. To put together everything I have been saying about atoms in the Earth's atmosphere, their abundance was in response to the pathways that favored a better conversion of the energy available in the environment they were formed in accordance to the constraints in that environment. The production of nitrogen was easier than that of oxygen and consequently, nitrogen ended up dominating.

Finally, in the precursors of the atmosphere of Mars, because the conditions could have favored the process that split and aggregated the precursors of carbon and oxygen, these atoms were abundantly produced and in the end Carbon Dioxide (CO_2) is very dominate in the Martian atmosphere (95.32%). From the left over of the precursors of atoms, nitrogen was produced as the second most abundant. As the atoms were mixing in the early Martian atmosphere, some of the oxygen atoms reacted with nitrogen to form Nitrogen Oxide (NO), while others reacted with hydrogen atoms to form Water (H_2O) and others reacted with carbon to form Carbon Monoxide (CO). Finally, some atoms of oxygen associated with others of the same kind to form Oxygen (O_2), which is 0.13% of the Martian atmosphere.

24.2.1.3. Nitrogen

The third heaviest nonmetal found in the atmosphere of the planets is Nitrogen

(N, Z=7). It is mostly found in the atmosphere of Pluto (99%) and of Earth (78.08%) followed by that of Venus (3.5%) and Mars (2.7%). On these terrestrial planets, nitrogen is mostly found in the form of nitrogen gas (N_2). Like I explained in the other sections, nitrogen is in traces on Mercury because on that planet, the precursor of atoms was mostly molded into 3 heavy atoms (oxygen, sodium, and potassium) and most of the rest was molded into hydrogen and helium. To put it another way, by the time the 3 heaviest atoms of Mercury's atmosphere were formed, the leftover of the precursor of atoms could not be molded into nitrogen, but into less smaller atoms such as hydrogen and helium.

In the atmosphere of the 4 giant planets, nitrogen is present in the form of:
- Ammonia hydrosulfide (NH_4)HS (aerosol),
- Ammonia (NH_3), and
- Ammonia ice (aerosol).

Among these 3 forms in which nitrogen is found in the giant planets, ammonia (NH_3) is the most dominant, being 0.026% in Jupiter's atmosphere and 0.0125% of Saturn's atmosphere. While ammonia (NH_3) is not found in Uranus and Neptune, all of the other 3 forms exist on all 4 giant planets. Considering the above, nitrogen could be more abundant in the atmosphere of Jupiter than in that of any other giant planet.

24.2.1.4. Carbon

The 4th heaviest nonmetal found in the atmosphere of the planets is Carbon (C, Z=6) and it is associated with other atoms to form compounds such as:
- C_2H_6 (ethane),
- C_2H_x (hydrocarbons),
- CH_4 (methane),
- CH_4 ice (methane ice in the form of aerosol),
- CO (carbon monoxide), and
- CO_2 (carbon dioxide).

All of these compounds are not present in the atmosphere of all the planets and the abundance of those present varies. The most abundant compounds in which carbon is found in the atmosphere of the planets are:
- CO_2 which abundance is 96.5% on Venus and 95.32% on Mars, and
- CH_4 (methane) most abundant in the atmosphere of the giant planets.

Methane is more abundant in the atmosphere of the 4 giant planets because of the abundance of hydrogen there. However, its abundance in the atmosphere of these giant planets could have been limited by the low amount of carbon atoms formed there. Their abundance on Uranus and Neptune is higher than that on Jupiter and Saturn because on these 2 outer giant planets, the abundance of carbon could have been higher than that of Jupiter and Saturn. The higher abundance of carbon could explain the lower abundance of hydrogen on Uranus and Neptune. For, without carbon, methane could not have been made. In other words, I perceived that, because many precursors of atoms could have been molded into carbon on Uranus and Neptune than on Jupiter and Saturn, the abundance of methane on Uranus and Neptune is slightly higher than that on Jupiter and Saturn.

Methane is not found in the atmosphere of Mercury, Venus, or Mars (Fig. 130). However, on Earth it is just 1.7 ppm of the atmosphere. On the 5 planets beyond Mars, the concentration of methane ranges from 0.25% to 2.3%. The highest concentration is found on Uranus, followed by Neptune. The concentration of methane in Pluto's atmosphere (0.5%) is 2941.2 times that of Earth's atmosphere.

Fig. 130: Abundance (% by volume) of **Methane** (CH$_4$) in the atmosphere of the planets in the Solar System

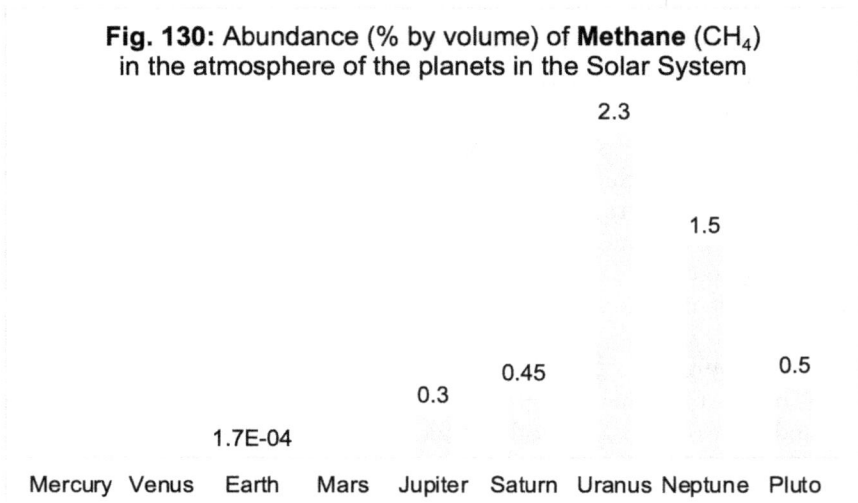

						2.3		
							1.5	
			0.45				0.5
	0.3						
1.7E-04							
Mercury	Venus	Earth	Mars	Jupiter	Saturn	Uranus	Neptune	Pluto

24.2.1.5. Hydrogen

Finally, the lightest nonmetal in the atmosphere of the planets is hydrogen. The two isotopes of hydrogen found in the atmosphere are:

- Protium: the ordinary hydrogen atom which has one proton and one electron and
- Deuterium: the hydrogen isotope which has one proton, one neutron, and one electron.

Deuterium is the least abundant hydrogen isotope in the atmosphere of the planets where it is present in 2 forms:

- Hydrogen Deuteride (HD) found in the atmosphere of the 4 giant planets
- Hydrogen-Deuterium-Oxygen (HDO) also called heavy water only found on Mars

In contrast, the ordinary hydrogen is most dominant and is found in combination with other atoms in the form of 12 molecules:

- C_2H_6 (ethane),
- C_2H_x (hydrocarbons),

- CH_4 (methane),
- CH_4 ice (methane ice: an aerosol),
- H_2 (molecular hydrogen),
- H_2O (water),
- H_2O ice (water ice: an aerosol),
- HDO (hydrogen-deuterium-oxygen also called heavy water),
- Hydrogen Deuteride (HD),
- NH_3 (ammonia),
- NH_3 ice (ammonia ice: an aerosol), and
- $(NH_4)HS$ (ammonia hydrosulfide: an aerosol).

Molecular hydrogen (H_2) is the most dominant form in which hydrogen is found in the atmosphere. For instance, in the atmosphere of the 4 giant planets, molecular hydrogen accounts for 80% to 95.3%. Aside from molecular hydrogen, the other 2 forms in which hydrogen is found in the atmosphere are methane (CH_4) and water (H_2O). In contrast, the least abundant compound in which hydrogen is found is C_2H_x hydrocarbons, which as of 2020 is reported only on Pluto.

Of the 12 compounds that contain hydrogen in the atmosphere, 7 are only found in the atmosphere of the 4 giant planets, meaning they are found on none of the 5 terrestrial planets including Pluto:

- C_2H_6 (ethane),
- CH_4 ice (methane ice: an aerosol),
- H_2O ice (water ice: an aerosol),
- HD (hydrogen deuteride),
- NH_3 (ammonia),
- NH_3 ice (ammonia ice: an aerosol), and
- $(NH_4)HS$ (ammonia hydrosulfide: an aerosol).

Of these 7 compounds mostly found on the atmosphere of the giant planets only, ammonia (NH_3) is the most abundant and is found only on Jupiter and Saturn. More specifically, it is more abundant on Jupiter than on Saturn. In contrast, the least abundant of these 7 compounds found on the giant planets is methane ice, which is postulated to be present only on Uranus and Neptune. I think that these 7 hydrogen-based compounds are found in limited amounts in the atmosphere of the 4 giant planets because the elements other than hydrogen which constitute them are limited in those environments. Moreover, these 7 compounds could not be abundant in the atmosphere of the terrestrial planets because there, the conditions have favored the formation of heavier atoms, therefore reducing the amount of hydrogen present.

24.2.2. Noble gases in the atmosphere of the planets

Noble gases were found in the atmosphere of all the planets except Pluto (Table

14). Of the 6 known noble gases, 5 were found in the atmosphere of planets:
- Helium (He, $Z=2$),
- Neon (Ne, $Z=10$),
- Argon (Ar, $Z=18$),
- Krypton (Kr, $Z=36$), and
- Xenon (Xe, $Z=54$).

Table 14: Characteristics and abundance of noble gas (% by volume) of the planets in the Solar System

Name (Symbol, Atomic Number)	Atomic Number	Atomic mass (g/mol)	Density (kg/m3) at 293 K	Calculated radius (pm)	Covalent radius (pm)	Van der waals radius (pm)	Abundance (% by volume) in the atmosphere of planets									Number of planets where the elements are found in the atmosphere
							Mercury	Venus	Earth	Mars	Jupiter	Saturn	Uranus	Neptune	Pluto	
Helium (He, Z=2)	2	4.003	0.179	31	32	140	6	1.2E-03	5.24E-04		10.2	3.25	15.2	19		7
Neon (Ne, Z=10)	10	20.180	0.9	38	69	154	Trace	7.E-04	1.82E-03	2.5E-04						4
Argon (Ar, Z=18)	18	39.948	1.784	71	97	188	Trace	0.007	0.934	1.6						4
Krypton (Kr, Z=36)	36	83.800	3.733	88	110	202	Trace		1.14E-04	3.E-05						3
Xenon (Xe, Z=54)	54	131.290	5.887	108	130	216	Trace		8.E-06							2

The only noble gas which is not found in the atmosphere of the planets is Radon (Rn, Z=86), the heaviest noble gas. The second heaviest noble gas, Xenon (Xe, Z=54), is present in the atmosphere of 2 planets: Mercury (in traces) and in Mars. The third heaviest noble gas, Krypton (Kr, Z=36), is present in the atmosphere of 3 planets: Mercury, Earth, and Mars. Then, the 4th and the 5th heaviest noble gases, Argon (Ar, Z=18) and Neon (Ne, Z=10), are present in the atmosphere of all 4 innermost planets. In contrast, the lightest noble gas, Helium (He, Z=2), is found in the atmosphere of 7 planets except Mars and Pluto and is the most abundant noble gas in the atmosphere of the planets. To put it another way as illustrated in Fig. 131, the number of planets where the noble gases are found in the atmosphere decreased as the atomic number of those noble gases increased. In other words, the bigger or larger a noble gas, the lesser the number of planets where it can be found in their atmosphere.

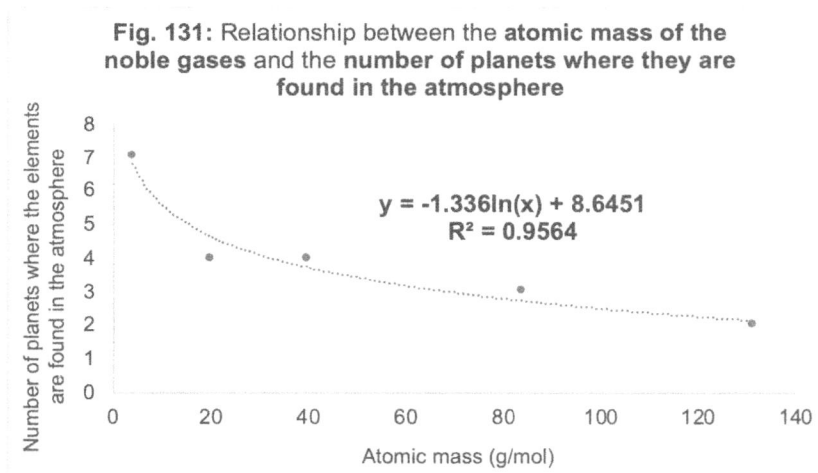

Fig. 131: Relationship between the **atomic mass of the noble gases** and the **number of planets where they are found in the atmosphere**

$$y = -1.336\ln(x) + 8.6451$$
$$R^2 = 0.9564$$

Although noble gases are most abundant in the atmosphere of the giant planets, Helium has an exception. Indeed, the abundance of Helium in the atmosphere of Mercury (6%) is higher than that in the atmosphere of Saturn (3.25%). In other words, the number of planets where the noble gases are found is inversely proportional to their mass. To put it in other words, the heavier a noble gas, the smaller the number of planets where it is found in the atmosphere. Additionally, the lighter noble gases are generally more abundant in the atmosphere than the heavier ones. After Helium (He, Z=2), Argon (Ar, Z=18) is the second most abundant noble gas in the atmosphere of the planets. One may question why, despite its smaller size and density, Neon (Ne, Z=10) is less abundant than Argon (Ar, Z=18). Nevertheless, in general, helium (which is the lightest and least dense noble gas) is the most abundant in the atmosphere, while the heaviest one is not even found in the atmospheres.

To try to understand why the presence and abundance of noble gases in the atmosphere are negatively affected by their mass, I turned to the data on mass, density, and radius of the atoms. As illustrated below, a positive relationship exists between the density and atomic mass of the noble gases found in the atmosphere ($r^2=1$) (Fig. 132). Similarly, density is positively correlated with covalent radius ($r^2=0.97$) (Fig. 133) and the Vanderwaals radius ($r^2=0.97$) (Fig. 134) of the noble gases found in the atmosphere. In other words, as more material was aggregated into the atoms of the noble gases, they became denser and larger. Probably, to make the noble gases bigger and denser, more energy could have been needed. And because energy is usually a limiting factor, the machinery that molded the precursors of noble gases into different kinds preferred making many smaller noble gases than bigger ones. Hence, the smaller ones ended up generally more abundant than the bigger ones. On Pluto, because the conditions were met to abundantly produce nitrogen, no noble gas was formed there. Similarly, in the atmospheres of the giant planets, because of the conditions that reigned during the formation of the atoms, only one type of noble gas (helium) was produced, suggesting that the position (implying semi major of the planets) of the precursor of the atoms could have affected their ability to yield some noble gases.

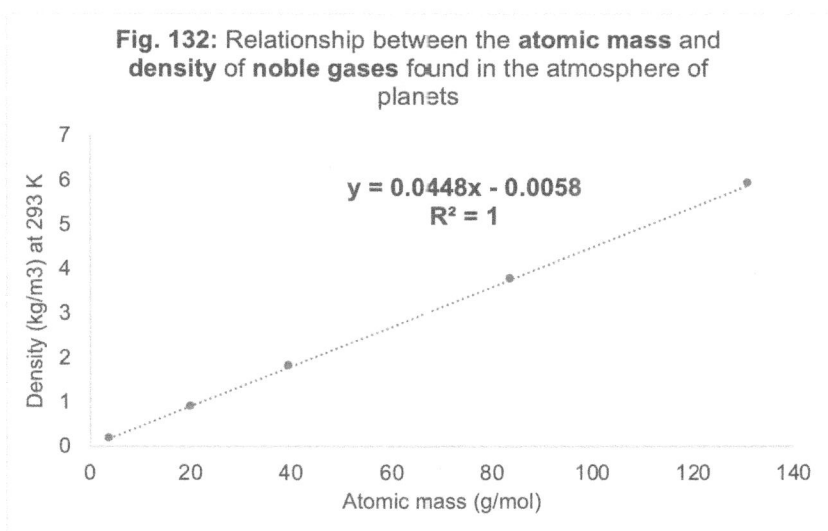

Fig. 132: Relationship between the **atomic mass** and **density** of **noble gases** found in the atmosphere of planets

$$y = 0.0448x - 0.0058$$
$$R^2 = 1$$

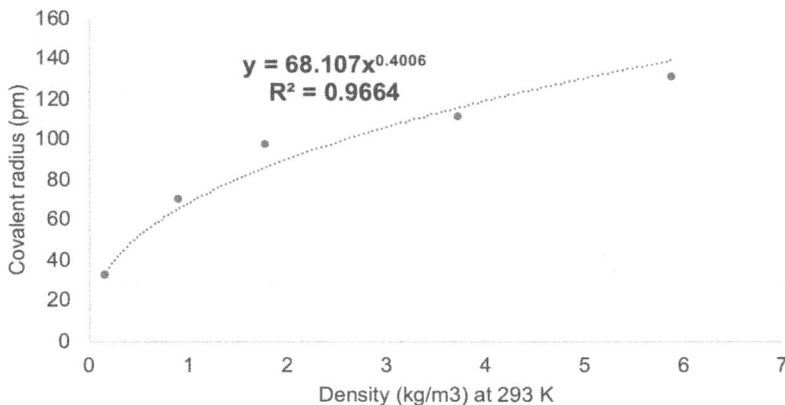

Fig. 133: Relationship between **density** and **covalent radius** of the noble gases found in the atmosphere of planets

$y = 68.107x^{0.4006}$
$R^2 = 0.9664$

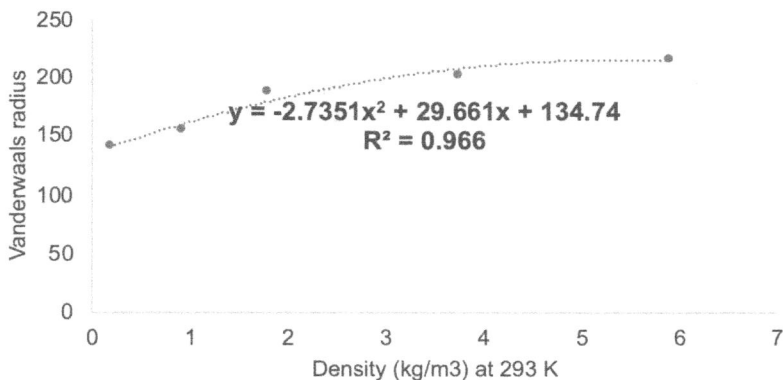

Fig. 134: Relationship between **density and Vanderwaals radius** of the **noble gases** found in the atmosphere of planets

$y = -2.7351x^2 + 29.661x + 134.74$
$R^2 = 0.966$

'Science180 Academy' Success Strategy:
SCIENCE180 MODELS OF THE ORIGIN OF THE UNIVERSE AND ITS CONTENT

Science180 Models consist of all the theories elaborated by Nathanael-Israel Israel regarding his ground breaking discovery on the origin of the universe and its content including all forms of life and chemical particles. Contact me (Israel12.com/contact) about how I can teach you these models:

1. SCIENCE180 MODEL OF COSMOLOGY

2. SCIENCE180 CREATIONISM

3. SCIENCE180 MODEL OF THE ORIGIN OF CHEMICAL PARTICLES

4. SCIENCE180 MODEL FOR THE GENERAL PUBLIC

5. SCIENCE180 MODEL OF LIFE-ORIGIN,

6. SCIENCE180 MODEL FOR CHILDREN,

7. SCIENCE180 MODEL OF PSEUDEPIGRAPHA

8. SCIENCE180 MODEL OF THE PROOF OF THE EXISTENCE OF GOD

9. SCIENCE180 THEORY OF EVERYTHING

Checkout Science180.com to learn more.

CHAPTER 25

THE GENERIC PROCESSES OF THE FORMATION OF ELEMENTARY AND COMPOSITE PARTICLES THAT DEFIED THE STANDARD MODEL OF PARTICLE PHYSICS AND SHOOK THE FOUNDATION OF CHEMISTRY

Although in the previous chapters, I addressed how some chemical elements and compounds were formed, here I will lay the generic processes of the formation of subatomic particles, atoms, molecules, rocks, and other chemical particles. If you did not read the chapter(s) on the abundance of the chemical elements and compounds yet, please do so first before you read this one, else, you may not understand the underlying cause of what I will say in this chapter.

25.1. Formation of the elementary and composite particles

As of today, human beings are organized into families, cities, countries, continents and between and within these structures, organizations of various purposes are established. These organizations and the ways they interact with one another were not formed overnight, but progressively throughout generations, as human beings were multiplying, moving, and interacting with one another (e.g. friendly, hatefully, stealthily, and violently) according to the opportunities and constraints in the world that some of them want to dominate. Likewise, but under completely different conditions, the turbulent prima materia (i.e. the bulk of the precursor of all matters in the universe) was split-gathered into clusters of matters ranging from the invisible ones to those as large as clusters of galaxies, passing by subatomic particles, atoms, molecules, minerals, rocks, and the celestial bodies found in the universe. To explain this process differently, just as water can evaporate, reach different altitudes in the atmosphere, condense and birth diverse kinds of precipitations such as rain, hail, fog, snow, sleet, ice, dew, etc., and just as a kitchen

CHAPTER 25: DECODING THE GENERIC PROCESSES OF THE FORMATION OF CHEMICAL PARTICLES

flour can be used to make many types of meals, and just as the same soil or concrete can be used to construct many types of buildings ranging from huts to the most beautiful architectures, so also, although under special circumstances, the turbulent prima materia (the mother of all matters in the universe) was shaped into different particles and clusters of bodies according to the environments of their precursors.

Just as the interactions between humans and their organizations cannot be modeled just by using a simple mathematic formula, so also the origin and the interactions between particles and others bodies in the universe cannot be properly understood by forcing them to fit simple statistical models. Mathematics itself is limited and has not been sufficient to explain even some simple things in nature. How much more insufficient it is to address the ultimate complexity in nature, which mere human beings cannot fully grasp with their limited mind. Therefore, conscious of these limitations (which are also part of the holistic spectrum of what can be known, explained, and revealed, even if I know a lot), I did not seek to mathematically model a lot of things in nature without understanding the mechanisms behind them, nor to dare to say that I unearthed every secret related to the formation of the universe. Below, in addition to what I have already addressed concerning the formation of celestial bodies, I summarized, according to my current understanding of the data, what I felt could be the best explanation of the origin of subatomic particles, atoms, and the various clusters of compounds they form in the universe which limits we still don't know.

The precursors of the subatomic particles were diverse according to their size, movement, speed, and location. Just like the precursors of the celestial bodies, the precursors of the subatomic particles could have been at one point organized into tiny layers or tiny pockets of matter, which, upon tightly folding, wrapping up, winding up, packing, or spiraling, yielded the plethora of subatomic particles according to their characteristics. Some particles could have resulted from an over and over packing or compacting of their precursors. Subsequently, some particles smaller than the currently known elementary particles were differentiated or gathered together to form elementary particles including leptons, quarks, and bosons. Some leptons are:

- electrons,
- muons,
- tau, and
- neutrinos (e.g. electron neutrino, muon neutrino, and tau neutrino).

Other precursors of particles were differentiated into different types of quarks such as:

- up quark (u),
- down quark (d),
- charm quark (c),
- strange quark (s),

- top quark (t), and
- bottom quark (b).

Other precursors of particles were differentiated into bosons such as:

- scalar boson (e.g. Higgs bosons)
- Gauge bosons which have been classified into 4 groups:
    - photon,
    - W (W+ and W-),
    - Z bosons, and
    - gluons.

The characteristics of these subatomic particles depend on where and how their precursors were shaped. The size, mass, or radius of the subatomic particles depend on the amount of matter in their precursors as they were compressed or gathered into their daughter bodies. As of 2025, the particles I mentioned above (from the electrons until the gluons) are believed to be elementary. Nevertheless, I think that many of them will very soon be proven to consist of other particles.

The precursors of some elementary particles were organized into clusters to form composite particles. Although the precursors of many composite particles could have been split-gathered into the particles constituting them, in other cases, the constitutive particles could have been formed separately elsewhere before the mixing and movement occurring in their environment brought them together and forced them to associate with others to form composite particles. In the same manner, some subatomic particles were arranged into clusters to form various composite particles, some of which are stable, while others are very radioactive. In other words, although some composite particles could have been formed after their elementary particles were shaped at different places and brought together by the processes that mixed everything, in most cases, the precursors of the elementary particles were also components of the precursors of the composite particles and both were molded together. Another way of saying this is that, although composite particles are usually defined as a mixture of 2 or more elementary particles, it can be misleading to say that the elementary particles were formed first before being brought together or combined to form composite particles. Just as the precursors of the planetary systems were split-gathered while the precursor of the stellar systems they belong to were undergoing their own genesis, so also some elementary and composite particles were being split-gathered as the system of matter they belong to (e.g. atoms) were being shaped. In other words, it is very important not to view composite particles as having been formed after their elementary or constitutive particles were formed. I will provide more details very soon.

Because of how their precursors were positioned during their genesis, some elementary particles were clustered into groups, while others were isolated. For instance, as the precursors of the quarks were being split-gathered, some of them were grouped into diverse numbers. The precursors of the 6 types of quarks I mentioned above were associated in different numbers to form different kinds of composite particles, some of which are called hadrons (quark-based particles made of two or more quarks). Composite particles which precursors birthed daughter bodies consisting of 3 quarks are generally termed fermions. While other precursors birthed composite particles consisting of 2 quarks, 4 quarks, 5 quarks, and even more. Some of these particles that ended up having 2 quarks are usually called ordinary mesons, while some of those containing 4 and 5 quarks are called exotic tetraquark or pentaquarks.

The precursors of the fermions (particles generally containing 3 quarks) were clustered into different combinations of quarks. According to their number of quarks and how they spun, most of these composite particles were classified into 6 groups (also called baryons of composite fermions):

- Nucleons (N)
- Delta baryons (Δ): e.g. uuu, ddd
- Lambda baryons (Λ): e.g. uds, udc, udb
- Sigma baryons (Σ): e.g. uus, uds, dds
- Xi baryons (Ξ): e.g. uss, dss, usc
- Omega baryons (Ω): e.g. sss, ssc, ssɔ

Instead of going over the characteristics of these particles that I already described in the early chapters of this book, I will just illustrate their formation with a few examples. For instance, in the case of some nucleons, two up quarks were gathered together with one down quark to form a proton (uud). In other circumstances, two down quarks were gathered together with one up quark to form a neutron (udd). Some protons and neutrons could have been formed outside the nucleus of an atom, but in many cases, these nucleons were formed inside the precursors of system of bodies that later became the nucleus. Some nuclei were formed inside the precursors of clusters of matter that became atoms and others did not. In other words, all nucleons are not found inside atoms. For besides protons and neutrons, many other nucleons were formed, some of which are found inside traditional atoms, while others are found in exotic atoms and even in isolation. Because the conditions of the precursors of the Earth could have favored their formation, nucleons are believed to have a longer lifespan, while the other 5 types of baryons are short-lived although some of them are heavier than nucleons. The short lifespan of some particles may be caused by the fact that they are quickly transformed into other particles fitting conditions on Earth and/or they were quickly expelled from the environments where they were formed on Earth. The speed which some of those particles were launched when formed, for instance in a laboratory setting, can also propel them very fast into space where they may be molded into other particles according to their destination and characteristics.

In other planetary systems and stellar systems beyond the Solar System, conditions could have favored the dominance of other types of baryons including nucleons. Similarly, the characteristics of the elementary particles and the composite particles could depend on the conditions in the systems where they are found. Another way of saying this is that the characteristics of the particles should not be expected to be the same in all planetary systems, stellar systems, galaxies or everywhere in the universe. For although on the galactic supercluster scale, the universe may look the same, the internal conditions of its constituents are not the same everywhere. A simple look at the differences between the celestial bodies in the Solar System and the diversity of their abundance in chemical elements (e.g. in the atmosphere that I extensively studied) suggested to me that particles are not distributed the same in the Solar System. Because this is true in the Solar System, it cannot be ruled out in other stellar systems. Consequently, in other stellar systems, baryons or fermions as defined on Earth could not be the most abundant types of matter. In other words, the theories according to which baryons are the most dominant kind of matter in the universe may not be true for all stellar systems.

Some of those precursors were formed in isolation meaning not embedded in other clusters of subatomic particles, while other clusters were inside bigger ones. Although subatomic particles can be found free and not clustered with others like hadrons, many are associated with others into systems of matter like atoms, atomic nuclei, super atoms, exotic atoms, molecules, and other forms of clusters. While a traditional atom consists of a nucleus (dominated by neutrons and protons and some bosons) "orbited" by electrons, in exotic atoms, one or more subatomic particles are replaced by other particles of the same charge. The electrons in some exotic atoms are replaced by other particles like muon, tau, etc.

The nature and the changes that the precursors of these composite particles went through defined the arrangement, topological and structural changes, number, localization, and proprieties of the particles of their daughter bodies (Fig. 135). Some particles do not orbit others because when they were being formed, they were ejected from the system that split to birth them. For instance, just as some bodies can be found between systems of bodies because they were ejected from the original system formed by other daughter bodies of their mother, so also some composite and elementary particles were ejected from the precursors of some atoms or other clusters of particles and pushed into a motion, which does not allow them to orbit a direct primary body. Just as on the galactical scale, some stars in the galactic arms look as if they are not really orbiting the galactical core, but seem to be getting away from it and may never have a chance to orbit that core, and just as some comets in the Solar System seem to have a very eccentric orbit with orbital period estimated at millions of years, which I doubt is true, so also some particles are "fleeing" what could have been their primary bodies. Photons which radiate as light may be one of those particles which do not really depend on a primary body which they orbit. However, in the Solar System, or at least in the vicinity of the Earth, most of the matter are believed to be made of atoms dominated by protons and neutrons.

Fig. 135: Sketch of the winding up, spiraling, and gathering together of subatomic particles

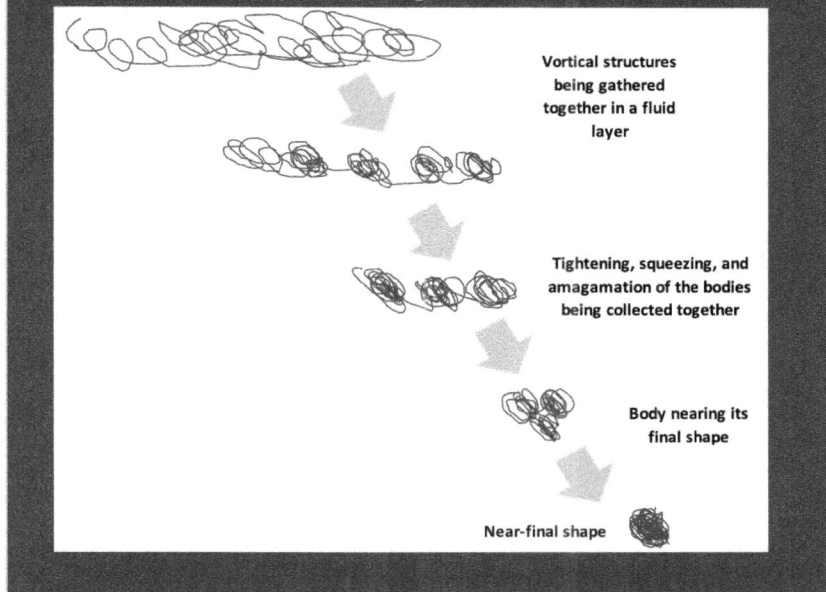

Vortical structures being gathered together in a fluid layer

Tightening, squeezing, and amagamation of the bodies being collected together

Body nearing its final shape

Near-final shape

On the morning of September 1st, 2020, I lost sleep around 4 AM although I went to bed around 1 AM. I tried to fall back asleep, but I could not. As I was laying on my bed, my mind was turned toward the coiling of the DNA. It then appeared to me that the process that coiled the DNA testified of what could have happened to the precursors of some particles on a completely different scale and under different conditions. Indeed, the coiling of DNA is one of the biological evidences of folding, supercoiling, winding, twisting, spiraling, and rolling of systems of matter. Just as DNA is coiled and supercoiled into nucleosomes, and wound up into chromatin and into chromosomes in some organisms, so also the gathering together of some particles (e.g. subatomic particles) could have involved complex processes (Fig. 135), which could have had some similarities with folding, twisting, rolling, tightening, coiling, supercoiling, compacting, or squeezing, etc. We may not be able to see those structures on subatomic particles; because they are very small and much more, no one has ever seen any subatomic particle yet. I am not saying that the gathering together of subatomic particles followed the exact process as the supercoiling of a DNA, but I felt like we can learn something from the supercoiling of DNA to enrich our understanding of what may have happened to the turbulent prima materia as it was being molded into chemical particles. Just as all strands of DNA are not condensed or coiled the same way, and just as others are tightly condensed and consequently their genes are repressed and could not be quickly expressed, and other sequences of DNA are loosely bound and can be easily unwound so that their genes can be expressed, so also to some extent, all chemical particles in nature are not tightly compressed the same way. Furthermore, just as some factors can activate DNA by loosening it, unwinding it, and getting it ready for gene expression, so also certain phenomena can unwind or compress the radius of particles if things change in their system. For instance, the proton radius puzzle can be an example of how the introduction of certain new things in the environment of some particles can affect their characteristics (e.g. the radius). The increase of radius can occur when the concerned particles are kind of unzipped, unwound, or loosened, while the decrease of radius of a particle is like a tightening, squeezing, coiling, and compression. This suggests that subatomic particles and even atoms, larger particles, and celestial bodies may be under some pressure or stress, which they can release to increase their size if the conditions allow so, or which can be increase to tighten them. Therefore, I cannot rule out that the size of all particles or bodies in a system are dynamically in equilibrium with one another. Moreover, the systems in the universe as a whole may even be connected and linked by laws such as if one system is affected, others also can be affected or must adjust their constituents.

Just as nucleosomes are supercoiled structures of DNA containing a specific number of particles, so also some particles found in the nucleus of atoms (e.g. protons, neutrons) are like supercoiled structures of the turbulent prima materia, separated by other subatomic particles. Hence, the density of some subatomic particles is very high. The difference in the size, spiraling, coiling, supercoiling, and twisting of the precursors of the subatomic particles may explain the difference in density and other proprieties. The comparison I just did between particles and DNA should not be taken literally for all particles, but was a way to visualize what may have happened to the precursors of some particles. The day that scientific equipment can allow to better visualize the structures of subatomic particles, more insight will be gotten.

In general, as the precursors of the subatomic particles were being wrapped, different levels of organization took place. During the process that molded the turbulent prima materia, some particles smaller than the subatomic particles could have been first wrapped into some initial structures, which then could have been wrapped again into bigger structures, which at their turn could have been wrapped or coiled again until the final size of the subatomic particles could have been reached. The packing of the subatomic particles could have been differently done from one type of subatomic particles to the other. Similarly, some subatomic particles could have been tightly packed, while others were loosely packed according to the environmental conditions of where the packaging occurred. From one celestial body to the others, the subatomic particles could have been tightly or loosely packed differently. Consequently, the atoms and the compounds made from them could have also been organized differently. Some bosons that are

280

Nathanael-Israel Israel: Known as the #1 Universe-Origin, Life-Origin, and Chemicals-Origin Scientist & Mathematician

known to be very heavy (e.g. Higgs bosons) could have been tightly packed hence their higher density and mass. In contrast, electrons and neutrinos could have been loosely packed hence their smaller mass. On a subatomic level, when particles radiate or are energized, their modifications can change their morphology, structures, and functions, all of which can transform some particles into others.

At the time of the publication of this work, subatomic particles are thought to make up of all forms of matter (e.g. electrons, photons, protons, neutrons, etc.) in the universe. Subatomic particles and electromagnetic waves share some properties, suggesting that it is possible that subatomic particles can be a condensed or compressed electromagnetic waves. Scientific experiments showed that particles are discrete packets of energy which behave like a wave. This suggests that the turbulence that took place during the formation of the universe could have condensed the available energy into small compartments or pockets, which were gathered together into vortex-like clusters. Those packets of energy would have been packaged into different types of matter, which formed the subatomic particles. Once some subatomic particles were formed, they also went through some instability, which condensed or grouped them into packages of various atomic matter. The various combinations of the subatomic particles led to the diversity of atoms. Some subatomic particles were not able to be assembled into atoms, therefore they exist alone and are not found in atoms.

25.2. Formation of atoms and isotopes

In addition to what I described above as the origin of composite particles, here, I will give more details on atoms. Lessons I learned from the turbulence of the precursors of the celestial bodies helped me to understand how the turbulence on the atomic and subatomic levels could have birthed atoms, non-atomic particles, and the subatomic particles constituting them. For instance, just as the turbulence of the precursors of the planetary systems defined the characteristics of their planets and satellites, the diversity of the turbulence that the precursors of the atoms and subatomic particles went through defined the proprieties of their daughter bodies. If data were collected on many stellar systems as they were on the Solar System, I could have used them to explain the formation of atoms and subatomic particles. But because I disposed of more data on the planetary systems, I referred to them to illustrate how atomic systems could have been formed. In other words, my comparing atoms to planetary systems does not imply that atoms are miniature planetary systems, but is a way to relate the two systems of bodies, which of course have significant differences and similarities according to the impact of their scaling.

For instance, just as the precursors of the planetary systems affected their daughter satellites, so also the distribution of electrons around the nucleus of atoms would depend on the characteristics of the precursors of the atoms. As the precursor of the Martian planetary system for instance did not allow the Martian satellites to exist in Zones 2, 3, 4, and 5 and just as the precursor of the Plutonian planetary system did not allow the formation of the satellites in Zones 1, 2, and 5, and just as the precursor of the Earth-Moon system did not allow the formation of satellites in Zones 1, 2, 4, and 5, so also the precursors of some atoms did not allow the formation of certain kinds of electrons. Just as some planets do not have satellites, some particles do not have electrons. I am not saying that the atomic systems are the same as planetary systems, but they share some similarities on different scales because they were all produced by turbulence.

Just as in the outermost turbulence zone (e.g. Zone 5), bodies are usually smaller than those in the most turbulent zone (e.g. Zone 3), when the precursors of the chemical elements matched conditions found in outer zones, many more small atoms meeting characteristics of Zone 5 (at the atomic level) for instance could have been formed. Similarly, when conditions met those in Zone 3 (at the atomic level), many larger atoms could have been formed. Likewise, when many layers of fluids in the precursor of a celestial body met conditions of a certain turbulence zone, the atoms that formed from those layers would have been dominated by those favoring such zones. On the scale of an entire celestial body, certain atoms or particles were abundantly formed because the conditions that prevailed in the precursors of those celestial bodies favored the formation of those chemical elements and particles. The presence of smaller numbers of certain atoms or particles inside the bulk of other atoms can be explained by the intermittence of the fluid layers of the precursors of particles or by the mixing of atoms during the genesis of particles. Consequently, when atoms of the same kind were abundantly formed in larger regions, the purity of the minerals resulting from them can be higher.

Just as smaller celestial bodies are found between bigger ones, so also are smaller atoms found between bigger ones. Likewise, smaller particles are also found between bigger ones. According to the scale of the turbulence that formed chemical elements, smaller clusters of chemical elements of one kind can be found between or inside clusters of another kind. The diversity of particles constituting atoms and other composite particles is the consequence of the diversity of the turbulence that split-gathered their constituents. The similarities between clusters of particles can be traced back to the similarities of the environmental conditions and turbulence that shaped their precursors.

For example, just as some planetary systems are found in the inner zones of the Solar System and others in the outer zones, so also atoms could have been formed in regions having different characteristics. Just as the biggest planetary systems are formed in the most turbulent zone of the system of bodies orbiting the Sun for instance, so also some large atoms could have been formed in regions which, at the atomic level, could have had characteristics of turbulence Zone 3. For instance, the biggest chemical elements could have been present in the most turbulent zone at the atomic level while the densest chemical elements could have been near the inner zone. Small and less dense chemical elements like hydrogen could have been formed in a region which turbulence could not have been strong.

CHAPTER 25: DECODING THE GENERIC PROCESSES OF THE FORMATION OF CHEMICAL PARTICLES

Stable gases such as helium and xenon could have been formed in the least turbulent zone (e.g. Zone 5) and they do not like to associate with others much. Their precursors could not have rotated considerably to increase their density and interactions with others. This also explains why helium is not abundant at the surface of the Earth or in lower altitudes, but farther away in the upper atmosphere where air movement is smaller. In other words, noble gases dislike reacting with other chemical elements just as how bodies in Zones 4 and 5 are farther away from the other zones and have a relatively smaller movement. The densest chemical elements could have been formed in Zones 1, 2, and 3.

Oxygen and other chemical elements that react a lot, could have been formed in zones or environments that favored a strong rotational movement. Large chemical elements like Uranium could have been formed in regions of relatively strong turbulence such as Zone 3 on the scale of atoms. Oxygen is very reactive maybe because its high rotational angular speed and/or its ability to attract atoms. The ozone layers in the upper atmosphere of the Earth may be located at an altitude that favored its formation and sustenance. Knowing the reactivity of oxygen, the ozone layer in the upper Earth's atmosphere protects the Earth by the ability of oxygen to react with or bind to undesirable and harmful particles and radiations that can hurt living organisms including human beings on Earth. As anthropogenic activities are claimed to be making holes in the ozone layer, they are just contributing to exposing living things to dangerous particles including radiation from space. Chlorofluorocarbons are one of the compounds blamed to affect the ozone layers. The decrease of chlorofluorocarbons is said to contribute to the recovery of ozone in the upper stratosphere. Interestingly the conductivity and electronegativity of chlorine and fluorine are almost the same as those of oxygen, suggesting that chlorine and fluorine could be able to do to oxygen what oxygen is supposed to do to other particles or chemical elements: react with them. Therefore, the conductivity and electronegativity can also help sort out how chemical elements were formed.

Besides protons and neutrons, the nucleus of atoms contains other particles such as bosons (e.g. Higgs bosons, photons, gluons, W and Z bosons). Just as with any other body or matter, the particles found inside the nucleus of atoms had their own precursors. The variation of the characteristics of these particles could have been defined by their precursors and how they were gathered together as components of a unified nucleus, considered as the system of primary bodies, orbited by electrons and other leptons in some circumstances.

Particles in the atomic nucleus may seem as if they are bound together tightly, but if the scale of their size is compared with the scale of the celestial bodies, it is as if protons and neutrons are 2 different systems of bodies tied together. Just as the ratio between the size of Pluto and Charon, its biggest satellite, is higher than 0.5, in some binary systems, the ratio of the two biggest bodies can be very much close. At the atomic level, a similar process but on a different scale could have happened for the precursors of protons and neutrons in such a way that they are located toward the center of their system and having masses which are not too much different from one another and being separated by a distance small enough to force them to be tied together by the processes that shaped and wrapped them inside the compartment they were formed and which is known today as the atomic nucleus. By the time the secondary bodies of the precursors of atoms were organized around the nucleus, in a fashion that similar to what happened to the secondary bodies around primary bodies, the nucleus ended up being found at the "center" of most atoms.

Because Pluto is the outermost planet and its orbital speed is among the smallest, the precursor of the Plutonian planetary system could have been less turbulent than the precursors of other planetary systems in the Solar System. The chaotic movement of the binary system formed by Pluto and Charon suggests that when the turbulence of the precursors of bodies decreases to a certain level, its daughter bodies can form a binary system. In other words, under some conditions, certain precursors could not orbit one another but bound together like 2 clusters of primary bodies orbited by secondary bodies in their system. On the atomic scale, this could have been what happened to some neutrons and protons. Either of them could have been a primary body and the others the biggest secondary bodies, but their mass or radius ended up being so close to one another that they were bound together and orbited by the secondary bodies of the system represented by the electrons. Another way of saying this is that particles found inside the nucleus are like clusters of primary bodies and secondary bodies orbited by other clusters of secondary bodies of another level.

If chemists had not simplified the systematics of atoms and if most isotopes are stable and long-lived, each isotope should have been considered a different entity worthy to receive as much attention as that of the conventional 118 chemical elements. Instead of seeing isotopes of each chemical element as descending from the common chemical elements that some people think could have just gained additional neutrons, it is important to view them as descending from different precursors of particles, which, during their split-gathering, yielded atoms with different numbers of neutrons, but the same number of protons. In general, isotopes are not intermediary elements, but atoms that show the multifaceted sides of the combinations of the particles used to form the atoms that constitute the matter in the universe. Some atoms were isotopes, not because they derived from other chemical elements, but because they were formed from the combination of different numbers of nucleons. Below, I used the example of the hydrogen isotopes to illustrate how they could have been naturally formed. For instance, unlike what people believe, I don't think that all hydrogen isotopes naturally come from the hydrogen atoms that have only 1 proton and 1 electron. I think that, during the turbulent process that birthed the chemical elements, the precursors of the so-called hydrogen isotopes had different sizes. As they were being split-gathered into their subatomic particles, these precursors of hydrogen isotopes yielded a nucleus containing different numbers of neutrons. For instance,

Nathanael-Israel Israel: Known as the #1 Universe-Origin, Life-Origin, and Chemicals-Origin Scientist & Mathematician the precursor of the protium was split-gathered into the precursor of a proton and the precursor of an electron, which were molded into a nucleus (primary body) made of one proton orbited by a secondary body, which became an electron. The unified particle formed by that proton and electron is the ordinary hydrogen (^1H). In contrast, as the precursor of the deuterium were being shaped, they yielded a system of primary bodies (consisting of one proton and one neutron), which are orbited by a secondary body, which ended up being an electron. The body formed by these 2 nucleons (1 proton and 1 neutron) "orbited" by an electron yielded the deuterium (^2H). Finally, the precursors of tritium (^3H) also went through a similar process during which they were fragmented into a system of primary bodies (containing 2 neutrons and 1 proton) orbited by an electron, the secondary body. The fact that 99.9% of the hydrogen on Earth is believed to be protium suggested to me that the conditions in the precursor of the Earth could have favored the formation of those types of hydrogen atoms rather than the other ones. In other environments, even in the Solar System, conditions could have favored the formation of other hydrogen isotopes. The same logic can be used to explain the formation of the isotopes of all the chemical elements. Chemists have made a remarkable effort to classify atoms into elements and isotopes depending on their number of protons and neutrons of course, but that effort to group different atoms should not mislead people to think that isotopes of a chemical element descended from that chemical element or from the same precursor. The proprieties of isotopes are more than the simple difference in their number of neutrons that most people seem to focus on. For instance, tritium is very radioactive and used in bombs, while ordinary hydrogen is very stable.

Isotopes of elements which atomic number is greater than 82 (Lead (Pb, Z=82)) are said to be radioactive. This suggests that as the size of atoms increases, the forces that bound the particles together may no longer be enough to bind them, hence, their tendency to break and/or radiate. This also implies that if the turbulence that took place during the formation of the universe was stronger, heavier elements could have been more stable and atoms could have been denser and physical laws could have been different. In contrast, if the turbulence was smaller or quicker, atoms could have been less dense, but more unstable. In other stellar systems, it would not be surprising to find atoms bigger and smaller than those found in the Solar System. Subatomic particles were packed differently and consequently, they were compressed differently inside atoms. Because of the way their subatomic particles were combined, some atoms are very dense while others are less dense. The atoms in which the subatomic particles were more compacted ended up being very dense.

I could not finish talking about atoms and particles in their nucleus without addressing what people think are force connecting them. Considering the data that I analyzed and the trends that I found, I personally think that, it was by misunderstanding that some people think that matter are linked by fundamental forces. Because they are allegedly the heaviest subatomic particles in the most conventional atoms, the Higgs bosons are considered as the carriers of mass. In contrast, the photons, gluons, W and Z bosons are believed to carry different forces. For instance, photons are credited for the electromagnetism, while W and Z bosons are believed to carry the weak force; and finally, the gluons are assumed to carry the strong force. However, when I carefully reviewed the subatomic particles in perspective of all the matter and systems of bodies in the universe, it appeared to me that no particle can be credited for carrying a force or being responsible for the mass of all matter. For each particle has its own mass and was differently gathered together, squeezed, coiled, and wrapped into a system of matter, which ended up having different proprieties, which seem to make people think that forces between them were the underlining factors. As the particles were being squeezed, a force was applied to them of course, but no single particle is responsible for that force. Instead of seeing the connections between the particles as mediated by a force which intensity some people believe is the same for all types of particles, I think that the coming together of subatomic particles was differently caused by how each of them was formed and mixed with others to form specific entities. Likewise, the so-called fundamental forces in nature may depend on the types of bodies. For instance, just as gravity is not the same for all celestial bodies, the forces thought to exist between subatomic particles may also vary according to the size of atoms. Similarly, from one planet to the other, the "forces" between subatomic particles may differ. However, due to the small size of atoms, the variation of these interactions may be hard to discern or measure. As I have extensively demonstrated, because the intensity with which particles were gathered together depended on many factors including their size, the turbulence, the location of their precursors, fundamental forces between these particles should also vary. Reciprocally, if the same atomic nucleus is forced to be orbited by a particle different from that which naturally orbits it, changes will have occurred in the force that gathered the nucleus together and its dimensions can also change. I better elaborated on this is the section below on proton radius puzzle.

The strong force is just an expression or a consequence of the processes involved in the split-gathering of the precursors of atomic nuclei. Just as, on the scale of planets and satellites, gravity is used to explain how their bodies are bound together, so also on the subatomic scale, a certain process was used to gather together the constituents of subatomic particles. In other words, the constituents of each system of bodies were wrapped or collected together in such a way that the force that seems to attach their components varies and depended on their size, speed, or energy with which they were gathered together. Particles or subatomic particles that were gathered more strongly express a higher "binding" force, while those loosely gathered together have a smaller binding force. Likewise, the strength or the energy used to gather together subatomic particles found in the nucleus is translated into their higher density, smaller size, and the strong force said to bind them. In the same manner, just as planets and asteroids have different kinetic energy around their primary star, and just as satellites have different kinetic

286

Nathanael-Israel Israel: Known as the #1 Universe-Origin, Life-Origin, and Chemicals-Origin Scientist & Mathematician

energy as they orbit their primary planet, so also electrons were launched into motion and wrapped around the atomic nucleus with unique or different levels of energy. The space occupied by the electrons and the degree of their tightness or winding up around the nucleus defined the radius, density, and many other properties of their atoms and the degree of their interactions with other particles (including other atoms) in their environment.

The electromagnetic force that is thought to connect electrons to the nucleus may be a consequence of the way that the electrons or their orbitals are organized around the nucleus as part of the components gathered together inside the atoms. The process that gathered the electrons together around the nucleus and balance the electrical and magnetic fields created by their movement can be responsible for the electromagnetic force. Even in conditions outside of atoms, when electrons are moving, they create an electromagnetic field. I wonder why some theorists think that photons are needed to mediate the electromagnetic force they claim exist between the electrons and the nucleus. It is unfortunate that most theorists think about the configurations and the interactions of and between particles (large and small) through the glance of forces that they always try to explain using other particles, including virtual ones, that they think mediate them, while they do not want to properly think about how everything quickly and systematically came into being.

This misperception can also be why some people think that the speed of light is a constant and the highest achievable by any particle in the universe. Indeed, as of 2025, the speed of light is defined as about 299,792,458 m/s, generally approximated at 300,000 km/s. Most scientists consider that speed as a constant. Some believe that the speed of light is constant only in a vacuum, while others believe that it varies according to time and space. I think that the speed of light cannot be constant. Being a particle, a photon cannot traverse space, passing through cold and hot environments and still have a constant speed everywhere. As an energy-rich particle passes through a very cold environment, it can lose energy. This suggests that when photons (the particles of light) travel through space, as their energy level can decrease, their speed also must decrease, similar to a basketball player whose energy level and speed decreases during a full court press. Therefore, at the edge of stellar systems or galaxies and between galaxies, some types of photons can be found which may not be similar to those emitted by the star they came out of. Moreover, it is possible that, as they lose energy and speed, photons can be converted into other types of particles. Most of the particles discovered on Earth are likely affected by the light coming from the Sun. Similarly, it may be difficult for scientists to measure light coming from other stars without interference of light from the Sun. Even at nighttime, the light emitted by the Sun from the other side of the Earth which is in the day, and/or the light reflected by the Earth and the atmosphere during the night, can always interfere with most measurements of light coming from space. In other words, it may be very difficult and impossible to detect certain types of photons from Earth. Those types of photons can be energized by the environment around the Earth and consequently, they can lose their low energy state. Unfortunately, it can be hard to go toward the edge of the Solar System to measure the types of particles present and then compare them to those on Earth. Not properly considering that other types of particles can exist in outer space, and consequently a different type of physics, some scientists keep using the characteristics of particles found on Earth to probe other particles in the universe, therefore making significant errors in their description of the universe. Before I close this section on the formation of atoms and subatomic particles, I would like to recall that many of them can be formed from the disintegration of others or the reactions among others.

Another Book by Nathanael-Israel Israel:
HOW GOD CREATED BABY UNIVERSE

THE FIRST AND ONLY BOOK THAT ACCURATELY EXPLAINS EVERYTHING ABOUT THE FORMATION OF THE UNIVERSE AND LIFE IN A WONDERFUL LANGUAGE THAT ALL CHILDREN AGES 7-12 CAN EASILY, FULLY UNDERSTAND & ENJOY!

As the only universe-origin book that your whole family will like and enjoy together, "*How Baby Universe Was Born*" will set children on the path of success by accurately showing them early in life the formation of the universe, and how to detect errors in theories or stories that would misguide them as they grow up. Therefore, you need to add this great, efficient, trustworthy, and cost-effective book to the strategic journey of children toward their best tomorrow.

With "*How Baby Universe Was Born*", you will:

- Have a peace of mind that children will get accurate, fit, and easy to understand universe-origin information that will produce real results in their life
- Become the leader that captures the heart of children craving for the original explanation of the formation of the universe so you can clear their way for freedom, power, technology, innovation, and breakthroughs of the future (learn more at Science180.com/children)
- Protect yourself and loved ones from wrong theories in the literature and the media by keeping children secured and empowered with the true knowledge of how the universe began
- Explain complicated secrets to children about how to locate mistakes in origin-related theories so you can save time, money, and other resources to improve their lives
- Ultimately boost children's confidence in detecting, confronting, and avoiding wrong theories by knowing the facts and real processes involved in the formation of the universe
- Help children to easily sort out their origin-related questions using strategies that get them to tap into deep secrets that even highly educated people ignore
- Clearly explain to children how to mathematically know without a doubt whether God created the universe as the Bible says or billions of years evolution processes formed it

Accurately explaining the complex formation of the universe and of life to children can be very hard in our modern world, but by getting *"How Baby Universe was Born"*, you will know the proven formula to help children to easily understand their huge universe-origin and life-origin questions with confidence, humor, and joy. They will surely laugh aloud while reading this book and thank you for it! It is time to buy this pragmatic book to help the children in your life today.

Member of the American Association for the Advancement of Science, American Chemical Society, and the American Society for Microbiology, **Dr. Nathanael-Israel Israel** is a Beninese-American scientist and international consultant, who shows the world how to scientifically decode the formation of the universe, of life, and who is known as the creator of the Chemicals Turbulent Origin Formula™, the inventor of the Life Turbulent Origin Formula™, and the discoverer of the Universe Creation Formula™. Learn more at Israel120.com.

25.3. Formation of bonds between atoms

As the atoms were being formed, the position of their subatomic particles with respect to one another was changing due to their movement and rearrangement until they were "locked" into interactions, which limited their movements. In the process, the subatomic particles were differently associated with others. In other words, as the particles were interacting with one another, they lost some of their freedom of motion, which was translated into their ability to interact with others differently. For, as the formation of the particles was progressing, changes that occurred in their micro-structural rearrangements also affected their movements. At the subatomic and atomic levels, rearrangements gradually occurred between the precursors of the particles (subatomic and atoms) until they were shaped, "locked up", or settled into their "final" configuration defined by their relationships with one another. Some of these relationships are qualified today as bonds illustrated by ionic bonds, covalent bonds, polar bonds, metallic bonds, etc., defined as (Wikipedia, 2022b):

- ionic bonds (formed by an "attraction" of oppositely charged atoms such as the case of sodium chloride, $NaCl$),
- metallic bonds (in which the electrons are believed not to be strongly held by the atoms, but delocalized, meaning that the involved atoms donate electrons to a pool of electrons existing between the metal atoms)
- covalent bonds (formed by the sharing of electrons between atoms with similar electronegativities); some covalent bonds are simple (e.g. a bond between hydrogen atoms in H_2), double (e.g. between the carbon of ethylene, C_2H_4), triple, etc.
- polar covalent bonds (by which electron density is believed to be shifted

towards the atom with the largest electronegativity, meaning the electrons are unequally shared between the atoms involved),

- network covalent bonds (by which the same molecules repeat the same structure over and over in the entire material) such as the covalent bonds between the carbons of a diamond.

- coordinate covalent bonds or dative bonds (formed when one atom donates both of the electrons to form a single covalent bond) such as the case of the ammonium ion (NH_4^+) in which nitrogen is believed to contribute 2 electrons to form the ion.

Because no one has ever physically seen an electron, it is important not to rely too much on the above description of chemical bonds, for they can be wrong. For it is impossible for human beings to properly describe or characterize things they have never seen. Just as the size of particles can change according to the environment, so also the strength, length, or nature of the chemical bonds can vary according to the environment although it may be hard to notice or measure these variations.

As of today, the cohesion between molecules refers to what is believed to be "attractive" forces between molecules of the same kind. In contrast adhesion is used to refer to what is believed to be "attractive" forces between molecules of different kinds. Cohesive forces between molecules are believed to cause the surface of liquids to contract to the smallest possible surface area, therefore causing the surface tension. Molecules on the surface of a liquid are believed to be pulled inward by cohesive forces, therefore reducing the surface area. Consequently, I think that, as their bonds were forming, the relationship between the particles were also changing. For instance, the cohesion between molecules of the same kinds could also be explained by the way they were compressed or gathered together. Similar processes could also explain the adhesion of some molecules of different types. To some extent, it is possible that the surface of the precursors of the celestial bodies could have been wired in a way that helped pull inward some particles at the surface. Consequently, the air that is in contact with that surface was pulled inward.

25.4. Explanation and solution to the proton radius puzzle

Early in this book, I introduced the notion of the proton radius puzzle. Now, I will give details on it. Indeed, if they can properly measure the size of the subatomic particles of different atoms, scientists may realize that the characteristics of the subatomic particles may depend on the atoms they belong to. In other words, the size of the protons in some smaller atoms can be different than the size of the protons in some bigger atoms. Similarly, within the same atom, it is possible that the size of the protons differs. The configuration or the organization of particles is not static, but dynamic and can be affected by the environment. When different subatomic particles are placed in the vicinity of other particles, they can change the configuration or the organization of one another.

TURBULENT ORIGIN OF CHEMICAL PARTICLES

Some of these changes are responsible for some chemical reactions. Similarly, on the nuclear scale and scales even smaller than the nucleus, when different particles are forced into the environments of other particles, the nature and characteristics can change depending on the amplitude of the induced changes. If some particles found on Earth can be moved to other planets, their structure and characteristics will likely change instantly or progressively depending on the level of the differences between these environments. For the forces and factors that are causing a celestial body to maintain its composition or characteristics can also affect the chemical elements of the particles present in that body. Without gravity and the factors that were involved in its establishment for instance, the Earth could not exist or maintain its form. Consequently, if the gravity of the Earth was significantly different, the characteristics of particles present on Earth could also be significantly different. This means that the gravity, atmospheric pressure, and composition of the Earth for instance significantly impact the maintenance of the particles present on Earth and its vicinity, although these particles were not formed after gravity was established. In other words, gravity and the forces that characterize the particles present on a celestial body are not completely independent. This impact can also be true for other variables characterizing celestial bodies.

Moreover, just as secondary bodies in a system of bodies were not placed by chance, but according to the turbulence of their precursors and the characteristics (e.g. energy) that their mother bodies imparted onto them, so also electrons orbiting a nucleus were not placed by chance, but according to how the turbulence that split-gathered their precursors into electrons differently positioned them. For instance, considering the distribution of the mass of the satellites around their primary planet on one hand, and on the other hand the distribution of the planets around the Sun, I showed that the position of the biggest bodies respects a law. When a primary body is big, its biggest secondary body orbiting it is also big, and when a primary body is small, its biggest secondary body is also small. In other words, a relationship exists between the size of a primary body and the maximum size of a secondary body which can orbit it. Likewise, the position of the secondary bodies with respect to their primary body follows a law that I explained in *"Turbulent Origin of the Universe"*. In that book, I also showed that the size, the rotational angular speed, the positioning of the bodies in a system of bodies are all calibrated according to the turbulence that borne these bodies. When these relationships are broken, the structure of the primary bodies and the secondary bodies forced to orbit them can change. It is a similar problem that protons, muons, and electrons are facing in the proton radius puzzle. And because the physics of the formation of the universe and its constituents has not been understood before my discoveries, the error has never been understood and corrected.

Indeed, the distribution of bodies orbiting a nucleus cannot just be placed any old way. The size and the characteristics of the electrons around a nucleus must respect a law. When an electron is replaced by a muon which is 100 times heavier than an electron, the characteristics of the atom must change to account for the changes of the law. The revolution of a muon around a proton must change the way the proton is wound up, for, the electron and proton constituting a protium for instance were being gathered together as the protium itself was being formed. Meaning that the action of the movement of the electron around a proton is equilibrated with the characteristics of the proton. When the electron is replaced by a muon, the proton can be squeezed more, hence the decrease in the radius of the proton orbited by a muon. For those who positioned the muon around nucleons did not even consider the relationship between the size, positioning, and movement of secondary bodies around specific primary bodies. The same secondary body cannot be positioned around different primary bodies and expect the system of the secondary and primary bodies to always stay the same.

Unfortunately, as soon as the proton radius puzzle was found, some scientists started reducing the size of the protons in their work thinking that the radius is really smaller in all conditions as what was previously used. The fact is that one day, scientists will find out that the radius of the proton and of any matter depends on the environment. This is because physical laws are not the same everywhere. The ecologists can testify to the fact that the size of living organisms (e.g. plants and animals) is not the same when the individuals are grown alone in a big area or when they are in a community. For instance, plants which are isolated and grown alone are bigger than those which are growing in a community of others plants similar to them. Even human beings behave differently when they are in a community than when they are living alone. To some extent, it is the same law but applied to different things. The way living and nonliving matter interact with one another depends on their environment and their characteristics. Unfortunately, instead of considering the proton radius puzzle as a wake-up call to fix their model of the formation of the universe and its particles, the scientific community did not take the proton radius puzzle seriously as it seems to have already claimed that they solved it, not knowing that they are just patching old mistakes with new ones.

25.5. Formation of molecules

As the precursors of chemical particles were being formed, clusters of precursors of atoms were produced. Some of these clusters of precursors of atoms became precursors of molecules. These precursors developed into molecules as the precursors of their atoms developed into matured atoms by the turbulence that subdued them. In other words, most molecules were formed as clusters of precursors of atoms were split-gathered meaning split into different atoms which were molded or gathered together according to their environmental conditions. In the process, some atoms were mixed with others in their vicinity and even far away after being transported by the current flow of fluids of the precursors. Chemical combinations were made between atoms of the same kind (e.g. chemical elements), while others are formed by atoms of different kinds. The degree of the mixing can also explain the diversity, purity, and complexity of chemical compounds. The nature of the composite compounds depends on the type of atoms mixed and the conditions that reigned during their formation. The ways the atoms were combined affected the nature and the characteristics of the compounds formed. Because the precursors of molecules were moved by the turbulence in their environment, they did not land in the same position all the time. A difference in the rearrangement of the atoms can lead to the formation of different molecules or chemical compounds. For instance, graphite and diamond are all made of carbon atoms, but they differ because the arrangement of the carbon atoms is different. As the atoms were moving, some ended up being forced to associate with others because they were pushed by others.

According to the purity of the atoms mixed to form molecules, some molecules are pure, while others are very complex. Molecules were compressed into different compartments that give the impression of them being under the influence of forces believed to be mediated by intermolecular bonds. Some of these so-called intermolecular bonds are (Wikipedia, 2020k):

- cation–pi interaction,
- dipole-dipole interactions,
- hydrogen bond,
- London dispersion force,
- van der Waals bonds, etc.

For instance, a hydrogen bond is said to be an electromagnetic "attractive" interaction between polar molecules in which hydrogen is bound to a highly electronegative atom (e.g. nitrogen, oxygen, or fluorine). Different atoms and molecules are tied together and some of the connections between them are termed as bonds. Consequently, molecules are usually believed to be connected to one another by intermolecular bonds, whereas atoms constituting a molecule are believed to be also tightly connected. At the same time, intermolecular bonds are usually believed to be weaker than intramolecular bonds. For instance, although stronger than a van der Waals interaction, the hydrogen bond is believed to be weaker than covalent bonds or ionic bonds. But in reality, these so-called bonds are the consequence of the processes that configured the particles and forced them to stay together. The so-called forces that bind components of particles together could have varied according to their size. In other words, what is termed chemical bonds today are the consequence of how the precursors of chemical particles were squeezed and gathered together by the turbulent processes that formed the bodies in the universe. The way different atoms associated with others to form molecules depended on how the clusters of precursors of atoms in each environment were molded into specific clusters of atoms according to the characteristics of the precursor of the celestial bodies or environments hosting them, and according to the abundance of the atoms formed. Hence, in some environments, certain atoms can associate with certain types of atoms, while in other environments, the same atoms can be associated with different kinds of atoms. It is not just a matter or atoms preferring to the associated with a certain type of atoms always, but a matter of what kinds of atoms were made in the environment and how these atoms were forced to bind and stay together with others by the turbulent processes that gathered all the matters together to form the bodies. To learn more about the formation of molecules, and how the environmental conditions of the precursors of the celestial bodies defined the formation and abundance of the chemicals, I would like to refer you to the chapter devoted to the abundance of chemicals in the atmosphere where I extensively explained the formation of molecules such as:

- Ammonia (NH_3)
- Ammonia hydrosulfide (NH_4)HS
- Carbon dioxide (CO_2)
- Carbon monoxide (CO)
- Ethane (C_2H_6)
- Hydrogen deuteride (HD)
- Hydrogen-Deuterium-Oxygen (HDO), also called heavy water
- Methane (CH_4)
- Molecular hydrogen (H_2)
- Nitrogen oxide (NO)
- Oxygen (O_2)
- Sulfur dioxide (SO_2)
- Water (H_2O)
- Etc.

I explained how the abundance of each of these chemical compounds depended on the conditions that prevailed during the formation of the celestial bodies or environments hosting them. It is not by chance that some molecules are abundant in some environments, but absent in others.

25.6. Formation of minerals and rocks

In the precursors of the celestial bodies, larger clusters of atoms were split-gathered into aggregates of one or more compounds held together by bonds. Some of these compound clusters are called minerals. Although many minerals were formed from precursors that went through changes as their constitutive atoms were also being shaped, I also think that in some cases, the atoms and compounds that constitute some minerals could have been formed in different places and then brought together by the mixing of particles during the genesis of the celestial bodies. In other words, as the precursors of minerals were being formed, some atoms were moved from one place to another place just to be mixed and incorporated into minerals which precursors they were not initially part of. Just as the precursors of different planetary systems were formed as the precursor of their stellar system was being shaped, so also, to some extent, the precursors of various minerals were formed as different atom clusters were split-gathered from a bulk of clusters of precursors of atoms.

Minerals are not usually pure substances, but they are "contaminated" by other elements in the environment they were formed. We may think that minerals are contaminated, but in reality, the precursors of most minerals are made of atoms that evolved almost together during their formation, and it was not that minerals were purely made first before other atoms were brought in or were mixed with the pure ones to change their composition. It is humans' needs to deal with so-called pure minerals that caused them to think about contamination. The purity of a mineral depends on the homogeneity of its precursors in terms of the atoms born from it. Various combinations of atoms are found in minerals according to the presence and abundance of these atoms in the environment where the minerals were formed. To form minerals, the atoms in the precursors of the minerals were formed and came together and associated with one another. It was not that the atoms were first made and then brought together, but as they were being made, they were brought together. Just as a sauce is made by mixing different ingredients together and tasting them until the mixture is ready, so also as the precursors of minerals were being molded, the precursors of the atoms constituting them were being shaped and mixed together until all the constituent of the mineral was ready and set. The degree of the mixture of the precursors of a mineral can also affect how hard it can be to split its constituents. Because minerals are usually mixed with others, it takes efforts and resources to remove impurities from them or to separate them into more purer compounds.

Below, I used the case of minerals found in the Earth's crust to illustrate how some silicate minerals could have been formed. Indeed, because silicon (Si) and oxygen (O) are the most dominant atoms in the Earth's crust, silica (compound containing silicon and oxygen) is used to classify minerals on Earth. In other words, silicate minerals contain silica, while non-silicate minerals lack silica. This also implies that silicon and oxygen are not present everywhere on Earth. The places where they are not present could be where non-silicate minerals abound. While in some cases, the formation of minerals was a matter of the precursor of secondary bodies escaping the precursor of their primary body, in other cases, it was different individual clusters of atoms that were collected and forced to stay together by the process which molded the celestial bodies of the environments these minerals were formed. The chemical composition of the minerals can hint at how they were formed. Below, I illustrated how some minerals in the Earth's crust could have been formed.

25.6.1. Silicate minerals

The base of silicate minerals is a silica tetrahedron: $[SiO_4]^{4-}$, suggesting that silica could have descended from a precursor of compounds which birthed a silicon as a kind of primary body surrounded by 4 atoms of oxygen as the secondary bodies. In other words, the precursor of the silica was split gathered into the precursor of silicon (the kind of primary body) and the precursor of 4 oxygen atoms (the secondary bodies). It is possible that the precursors of the 4 oxygen atoms escaped the precursor of the silicon and were then split into individual oxygen atoms, which in the end, surround the silicon. This was an example of split-gathering at the mineral level, which occurred at many locations in the precursor of the Earth. The conditions in the precursors of the Earth's crust could have favored the formation of such silica, which them mixed with other atoms in their vicinity to form different silicate minerals. Silicate minerals were formed by the polymerization of the silica tetrahedron. Again, it is not that silica were necessarily formed individually and then polymerized to form silicate minerals, but it is more likely that the precursors of silicate minerals were split-gathered into the precursors of their constituents, which, as they were being formed, were collected together to form these minerals. However, because the precursors of the constituents of the silicate minerals were rearranged differently, the silica tetrahedra were polymerized differently and associated with diverse atoms. Some silicates were made of just one silica tetrahedron, whereas others contain many. All of these depended on the amount of precursors that were split and gathered together. In some cases, the tetrahedra were organized into sheets, chains, rings, and even networks. Therefore, according to the difference in the polymerization of tetrahedra, 6 subclasses of silicate minerals were formed (Dyar and Gunter, 2008; Chesterman and Lowe, 2008):

- orthosilicates: have no polymers (e.g. andalusite, (Al_2SiO_5), pyrope $(Mg_3Al_2(SiO_4)_3)$, almandine $(Fe_3Al_2(SiO_4)_3)$, Zircon $(ZrSiO_4)$, and topaz $(Al_2SiO_4(F, OH)_2)$
- disilicates or sorosilicates: have two tetrahedra with a base structure being $[Si_2O_7]^{6-}$ (e.g. epidote $Ca_2Al_2(Fe^{3+}, Al)(SiO_4)(Si_2O_7)O(OH)$
- cyclosilicates: rings of tetrahedra (base structure is $[Si_6O_{18}]^{12-}$); example: beryl, toumaline
- inosilicates: chains of tetrahedra (e.g. pyroxenes (e.g. $XY(Si_2O_6)$, where X

and Y are variable chemicals and amphiboles (have backbone of $[Si_8O_{22}]^{12-}$)

- phyllosilicates: sheets of tetrahedra (e.g. mica, chlorite, and kaolinite-serpentine)
- tectosilicates: three-dimensional network of tetrahedra (e.g. quartz, feldspars, feldspathoids, and zeolites). Quartz (SiO_2) has many polymorphs (e.g. tridymite, cristobalite, α-quartz, β-quartz). The base unit for feldspars is $[AlSi_3O_8]^-$ or $[Al_2Si_2O_8]^{2-}$. Depending on their content (e.g. in potassium and sodium), the feldspars are divided into many groups.

All these structures testified of the arrangement that the precursors of these silicates have gone through and the destiny of their fluid layers.

25.6.2. Non-silicate minerals

Because silica as not part of their precursors, some minerals lack silica, therefore they are termed non-silicate minerals. The precursors of non-silicate minerals were formed in environments which did not favor the formation of silicon, but other minerals such as carbon, halogens (e.g. fluorine, chlorine, iodine, and bromine), gold, silver, copper, iron, nickel, nitrogen, sulfur, phosphorous, etc. In fact, as the bulk of the precursors of some celestial bodies was being formed, it contained countless clusters of precursors of minerals, some of which ended up forming non-silicate-minerals dominated by the chemical elements that bear their names. For example, gold minerals were formed in an environment that favored the abundant formation of gold atoms. Likewise minerals dominated by copper, silver, iron, etc., were formed in places where the precursors of these minerals were abundantly molded and where silicon could not be formed. The diversity of these atoms in the precursors of the minerals led to the formation of various kinds of non-silicate minerals. According to their composition, non-silicate minerals are divided into categories (Dyar and Gunter, 2008; Chesterman and Lowe, 2008) such as:

- Carbonates: main anionic is carbonate $[CO_3]^{2-}$, e.g. polymorphs calcite and aragonite ($CaCO_3$), and dolomite ($CaMg(CO_3)_2$).
- Halides: main anion is a halogen (e.g. fluorine, chlorine, iodine, and bromine), e.g. halite ($NaCl$), sylvite (KCl), fluorite (CaF_2).
- Native elements: e.g. gold, silver, copper, iron, nickel, arsenic, graphite, and diamond.
- Organic minerals: contain organic carbons (e.g. whewellite, $CaC_2O_4·H_2O$).
- Oxides: e.g. cuprite (Cu_2O), corundum (Al_2O_3), hematite (Fe_2O_3) bauxites (aluminium ore), chromite ($FeCr_2O_4$), and magnetite (Fe_3O_4).
- Phosphates: have a tetrahedral $[PO_4]^{3-}$ unit; e.g. fluorapatite ($Ca_5(PO_4)_3F$), chlorapatite ($Ca_5(PO_4)_3Cl$), hydroxylapatite ($Ca_5(PO_4)_3(OH)$).

- Sulfates: have a sulfate anion, $[SO_4]^{2-}$, e.g. gypsum ($CaSO_4 \cdot 2H_2O$), barite ($BaSO_4$), and celestine ($SrSO_4$).
- Sulfides: e.g. pyrite (FeS_2), sphalerite (ZnS), ore of lead, ore of mercury, and ore of molybdenum.
- Others (e.g. borates and nitrates)

25.6.3. Formation of allotropes and polymorphs

Before illustrating the allotropes and polymorphs, I will first explain the crystal structure of the chemical elements. Solid atoms were arranged in different crystal structures according to the position of their atoms when their formation was completed or when the degree of freedom of their precursors was significantly reduced enough to prevent them from moving in certain directions. The crystal structure of the chemical elements can be classified into 7 groups among which the most dominant is cubic. Below is the number of chemical elements for each type of crystal structure:

- Cubic: 43 chemical elements
- Hexagonal: 36
- Monoclinic: 3
- Orthorhombic: 8
- Rhombohedral: 6
- Tetragonal: 2
- Unknown: 20

Except helium which has a hexagonal crystal structure, all of the other noble gases have a cubic one. The literature also mentioned other crystal structures:

- body-centered cubic,
- face-centered cubic, and
- monoclinic.

The crystal structure greatly influences the physical properties of minerals. As the minerals were being formed, they were packed differently according to the turbulent processes that squeezed their precursors. Some minerals were strongly squeezed, while others were not. The way the minerals were (weakly or strongly) stacked or connected to one another affected their cleavage or breakage along some planes or directions. For instance, depending on how the minerals were packed, some are very hard, while others are not. For instance, talc is one of the least hard mineral, whereas diamond is considered the hardest mineral or one of the hardest. The softness of talc may be one of the reasons some pharmaceutical companies have been using talc powder on babies before they found out it can cause cancer.

As atoms were being formed and mixed with others, some of the same chemical elements bonded to one another, forming different types of structures known as allotropes. The environment where these allotropes were formed favored the conversion of the precursors of the chemical elements into the same type. Therefore, a significant amount of the same chemical was formed at some locations and depending on how they were arranged, they yielded different allotropes. According to their spatial arrangements, these allotropes can have different properties. Polymorphs are groups of minerals sharing the same chemical formula but having a different structure. A crystal structure depends on the arrangement of the atoms involved in the mineral. For instance, the three minerals (kyanite, andalusite, and sillimanite) constituting the group of aluminosilicates share the same chemical formula Al_2SiO_5 but differ by their structure. While kyanite is triclinic, andalusite and sillimanite are orthorhombic. Similarly, diamonds and graphite are both carbon allotropes. Another example of polymorphs are diamonds and graphite, both carbon polymorphs but different by the ways their carbons are bonded to one another. Indeed, while diamond (the hardest natural substance) has isometric crystal, graphite is very soft and has a hexagonal structure. In diamond, each carbon is said to be covalently bonded to four neighbors in a tetrahedral fashion whereas in graphite the carbons are aligned in sheets where each carbon is bonded covalently to three others. In general, according to how carbon atoms bonded to one another, different types of carbon allotropes were formed such as:

- diamond (whose carbons are arranged in a tetrahedral structure),
- graphite (whose carbons atoms are stacked in a hexagonal structure),
- graphene (which is a single strong layer of graphite),
- fullerenes (which carbon atoms have almost spherical shapes), and
- carbon nanotubes (whose carbons are tubes with a hexagonal structure).

Like I explained that how the precursors of the celestial bodies were split-gathered affected the distribution of their daughter bodies, it is also important to keep in mind that such phenomenon also existed at the microscopic level. However, due to the minuscule scale of subatomic particles, atoms, and molecules, it is more difficult to explain how the breakup of their precursors affected some of their structures. Nevertheless, I would like to mention that the particles in the fluid of the precursors could have rearranged their structure as they interreacted with their neighbors, formed networks, jammed, and unjammed as their fluids were flowing. The sheet structure found in some minerals and rocks today could have been caused by a special arrangement of the particles of their precursors. According to the stress applied to them and their movement, some precursors of particles could have polymerized and aggregated into diverse structures and orientations. The viscosity of the fluids could have also affected those rearrangements and orientations of the particles. Just as on the astronomical level, precursors of celestial bodies were tilted, inclined, elongated, stretched, giving rise to celestial bodies having different axial tilt, orbital inclination, eccentricity, and other characteristics, so also at the level of subatomic particles, atoms, molecules, minerals, and rocks, precursors and their daughter bodies were stretched, tilted, oriented differently, and consequently leading to the diversity of the structures and characteristics of these types of particles.

CHAPTER 25: DECODING THE GENERIC PROCESSES OF THE FORMATION OF CHEMICAL PARTICLES

For example, minerals can interact with one another for instance through a process called twinning defined as the intergrowth of crystal of a single mineral species. Many types of twins have been observed in nature such as: contact twins, geniculated twins, penetration twins, reticulated twins, cyclic twins, and polysynthetic twins depending on the planes of their juncture. For instance, contact twins are believed to be joined at a plane. Geniculated twins have a curve in their middle. Penetration twins look like crystals that have grown into each other. Reticulated twins appear as interlocked crystals similar to netting. Cyclic twins look as deriving from a repeated twinning around one or many rotation axes and, depending on the number of axis, they are called threelings (3 axes), fourlings (4 axes), fivelings (5 axes), sixlings (6 axes), and eightlings (8 axes). Polysynthetic twins are like cyclic twins but they arise along parallel planes instead of around rotational axis (Dyar and Gunter, 2008; Chesterman and Lowe, 2008). The different kinds of mineral twinning are also a proof that when minerals were being formed, their precursors could have moved around, netting with others, rotated, and were compressed.

25.6.4. Formation of rocks

The split-gathering of one or more minerals or mineraloids yielded different rocks according to their precursors. Rocks are a mixture of minerals and non-mineral materials. Below, I used rocks found on Earth to illustrate their formation. Indeed, during the formation of the planets, the precursor of the Earth reached a point where it was a magma-like or lava-like fluid, some of which was cooled down, solidified to birth some igneous rocks. The precursor of the Earth itself went through turbulent changes before part of it became a lava or magma. Chemical particles deep inside the Earth's center may be even different from those found in lava or magma erupting from volcanoes. In other words, because the interior of the Earth is far from the surface, where the temperature is cooler, the Earth's interior was not cooled down and molded into the same chemicals as its exterior. In contrast, the Earth's crust solidified as it cooled down. Some igneous rocks termed plutonic rocks (e.g. granite) were formed when the magma cooled and crystallized within the Earth's crust. In some cases, the magma came out of the interior of the Earth (as well as the precursor of the Earth) and after being projected at the surface of the Earth, it cooled down to form volcanic rocks (e.g. basalt). As of today, when some rocks are subdued to a very high temperature, they can melt and become a fluid like a magma or lava, therefore confirming their magmatic origin. Because their precursors contained different quantity of materials that became silica, igneous rocks do not have the same silica content. Consequently, according to their silica content, igneous rocks are divided into:

- Felsic igneous rocks (e.g. granite) are mostly dominated by silica (SiO_2) and alumina (Al_2O_3);
- Intermediate igneous rocks (e.g. andesite and dacite) contain less SiO_2 than felsic igneous rocks;

- Mafic igneous rocks (e.g. gabbro and basalt) contain less silica than intermediate igneous; and
- Ultramafic rock igneous rocks (e.g. komatiite and peridotite) have less silica.

Under conditions (e.g. higher temperature and pressure) different from those in which they were originally formed, some rocks could have metamorphosed into new ones called metamorphic rocks. This process also underlines the fact that the change of environmental conditions can affect how a same precursor could have yielded different rocks. In other words, during the formation of rocks, temperature and pressure of their environment would have played a role in the shaping of the destiny of the precursor of the rocks. As they were formed, some metamorphic rocks had a texture, while others did not. Consequently, metamorphic rocks can be divided into 2 categories: the foliated and the non-foliated depending on whether they have a texture or not. Some foliate rocks are schist and gneiss, whereas some no-foliated rocks are marble and serpentine. The texture, the foliation, and many other proprieties of rocks are imparted onto them by the processes that fashioned their precursors. Unlike what geological timescales claim, I think that the metamorphism of rocks could have been very quick. In my book on the formation of the universe, I showed that some celestial bodies (e.g. Earth, Moon, and Sun) were formed within days. In the next segment, I will elaborate on that topic.

Finally, sedimentary rocks are believed to be formed at the surface of the Earth when some organisms and minerals settled or chemically precipitated from a solution, accumulated, and cemented. Some eroded rocks are believed to have been transported to become sedimentary rocks. Sandstones, carbonate rocks, and limestones are some of the most abundant sedimentary rocks on Earth. Although some sedimentary rocks could have been formed in the conditions I just mentioned, I think that many of them were formed as they are today from the beginning. For instance, covering most of the Earth's surface, sedimentary rocks are usually found in almost horizontal layers or strata, which I think are the consequences of the fluid layers that they birthed them, or the consequences of their sequential deposition. I think that some sedimentary rocks could have been formed after some particles formed in the atmosphere fell to the ground. According to their composition and structure, sedimentary rocks were divided into:

- Carbonate sedimentary rocks (e.g. limestone and dolostone): composed of calcite;
- Evaporite sedimentary rocks (e.g. chlorides and sulfates): composed of minerals believed to be formed from the evaporation of water;
- Iron-rich sedimentary rocks (e.g. ironstones): composed of iron;
- Phosphatic sedimentary rocks (e.g. phosphate nodules and phosphatic mud rocks): composed of phosphate minerals;
- Siliceous sedimentary rocks (e.g. opal, chalcedony): composed of silica;

302

and

- Siliciclastic sedimentary rocks (e.g. sandstone and mud rocks): dominated by silicate minerals.

Depending on the strength of the forces that locally acted on their precursors, some rocks were folded, while others were faulted. Types of folding include antiforms (e.g. anticlines) and synforms, (e.g. synclines) according to the direction of the fold. Some of these folded structures were overturned leading to the formation of overturned anticlines and synclines.

25.7. Duration of the formation and timing of the differentiation of chemical particles

An interesting and important question worth asking is "how long did it take for the formation of chemical particles to be completed?" Many theories claimed that it could have taken millions and even billions of years for the universe to form, but in my books on the formation of the universe, I demonstrated that the Solar System and its content for instance were formed within a few days. Indeed, using scientific data, I showed that the Earth was formed 3 days after the beginning of the formation of the Solar System, while the Moon and the Sun were formed on the 4th day. I calculated this timeline after studying hundreds of variables, and decoding the processes that were behind the formation of the universe. Because chemical particles are components of the celestial bodies, I think that those found on the Earth, the Moon, and the Sun could have been formed within days as well.

I relied on escape velocity, orbital speed, radius, and semi major axis of the celestial bodies to calculate the duration of their formation. Because a lot is not known about the position, speed, and size of most chemical particles, it is difficult to calculate the exact time it could have taken for them to be formed and positioned at their current places. For instance, if I focus on an electron around the nucleus of an atom, it could have taken a time for the precursor of that electron to escape the precursor of the nucleus of that atom, then move to the current orbital of that electron. Because the probability is usually used to mathematically address the position of electrons in atoms, it seems to me that probability may also be needed to handle the duration of the formation of the atoms and the subatomic particles. Nevertheless, considering what I already wrote on the origin of the universe, the duration of the formation of chemical particles could not have been longer than that of the celestial bodies.

While the distance between celestial bodies is massive, usually in the order of millions of kilometers, and even astronomical units (such as light year) for some, in the case of chemical particles (e.g. subatomic particles and atoms), the distance is much smaller than a millimeter. Yet, for those microscopic particles to be formed, processes were involved to split-gather the precursors of their atoms into the precursors of the nucleons (e.g. protons and neutrons) and the precursors of the electrons. Due to the expansion of the universe, the distance separating some electrons from the nucleons of their atoms could have changed since the origin of the universe. Also, the current speed of some chemical particles (e.g. subatomic particles and atoms) could be smaller than that of the fluid flow that carried their precursors. The distance between chemical particles could also have been affected by the diffusion of their precursors. Indeed, diffusion relates to the ability of a fluid to diffuse, move from one place to its outskirts, like from a center to a periphery, or from a region of high concentration to a region of low concentration. The precursor of the viscosity of the precursors of the particles could have affected their diffusion and consequently their positioning in their system. The precursor of the viscosity of the precursors of the particles could have probably had a negative impact on their diffusion and consequently on the distance separating the particles to their primary particles. The variation of the ability of the precursors to diffuse could explain why some particles ended up occupying more space (e.g. fluids) than others. The turbulence and the force that separated the precursors of the particles away from their mother could also have played a role in their sizing. This implies that the distance between adjacent electron layers for instance can vary within the same atom and across many atoms.

To expound on the duration of the formation of chemical particles, I will now explain (1) why all atoms and subatomic particles could not have been differentiated before the precursors of the celestial bodies were split and (2) why the chemical elements found in the atmosphere could not have been just produced by an evaporation from the crust. Indeed, some chemical elements that are not found in the atmosphere are very abundant in the crust of the planets. For instance, Silicon (Si, $Z=14$), which is not found in the atmosphere of the planets in the Solar System, is the second most abundant element in the Earth's crust and probably in the crust of other terrestrial planets. The truth I unearthed concerning the formation of the atmosphere of the planets also gave me a glimpse at what could have happened during the formation of the crust of the planets. To put it another way, considering the trends I found with the formation of the atoms in the atmosphere, I felt like environmental conditions in the precursor of the crust of the planets could have also affected their chemical composition. It appeared to me that all atoms and subatomic particles might not have been completely differentiated before the precursors of the celestial bodies and their clusters were split. For instance, if the atoms in the planetary systems of the Solar System were established before the precursor of the Solar System was split into the precursors of the planets, the atmosphere of the giant planets could not have been dominated by hydrogen and helium which is in very small amounts in the atmosphere of the terrestrial planets. Similarly, the atmosphere of the terrestrial planets could not have been dominated by heavier elements, while that of the 4 giant planets are dominated by lighter elements. The lack of oxygen in the atmosphere of the 4 giant planets, the Sun, or in other parts of the universe may have been limited by the conditions that subdued the precursors of atoms in those bodies. The connection between the factors that shaped the celestial bodies in the universe and their atmosphere better explain the small density of the biggest celestial bodies and their dominance by lighter chemical elements, while the smallest celestial bodies are usually denser and filled by denser elements. The processes that dominated the gathering together of the precursors of celestial bodies have also significantly influenced the differentiation, development, and characteristics of their chemical elements. The same logic can explain why the chemical composition of the Sun is more similar to that of the giant planets than to that of the terrestrial planets.

Likewise, the environment where the crust was formed was different than that of the atmosphere. Consequently, for the same planets, the abundance and nature of the chemicals formed in the crust are different than those in the atmosphere. For the Earth for instance, the crust is dominated by oxygen and silicon while the atmosphere is dominated by nitrogen and oxygen. Yet, nitrogen is just a trace in the crust while silicon is barely present in (if not absent from) the atmosphere. In general, nonmetals (noble gases, halogens, and the conventional nonmetals) are more abundant in the atmosphere and waters than in the crust of the Earth. In contrast, metals and metalloids are more abundant in the crust than in waters and atmospheres. The difference of the environment conditions where turbulence shaped the precursors of the particles in the crust, ocean, and atmosphere could explain the difference, presence, and dominance of certain groups or types of chemical elements according to their locations.

The negative correlation between the abundance of the chemical elements and their atomic mass suggested to me that, according to the environment, it could have been harder to form complex compounds than simple ones. More energy could have been required to convert the precursors of heavier chemical elements into heavy atoms than the precursors of lighter atoms in light elements. That is why many precursors were converted into lighter or simpler chemicals than into heavier or more complex chemicals. In the same manner, the energy in the precursors of bodies (small and large) after they split from their mother could have also affected how their shaping and characteristics were "finalized". The scope of the changes that the daughter bodies went through after their split (from other bodies) can explain the similarities and differences between their peers born from the same mother precursor.

For instance, heaviest chemical elements are more abundant in the atmosphere of the innermost planets because such atmospheres favored the formation of dense particles. For the conditions that compressed the planets to increase their density also affected the nature and density of the precursors of the atoms present in their atmosphere. It is also possible that some compounds in the atmosphere of the planets might have volatilized from the crust after their formation. However, some elements in the atmosphere are more likely directly formed there rather than just having migrated from somewhere else (e.g. the crust) to the atmosphere. In other words, some elements found in the atmosphere could have been formed there and did not just migrate there after being formed somewhere else. Some atoms in the atmosphere could have been formed under conditions different from those in the crust. For instance, while nitrogen is 78% of Earth's atmosphere, in the Earth's crust it is just 0.02%, implying that conditions in the precursor of the Earth's crust could not have favored the formation of nitrogen as conditions in the precursor of the Earth's atmosphere did. If all the nitrogen atoms in the atmosphere were formed in the crust before migrating to the atmosphere, the abundance of nitrogen in the crust could have been higher and more traces of nitrogen could have been found in the crust. Yet, nitrogen is less abundant in the Earth's crust than many lighter gases.

Another way to explain this is that, as the celestial bodies were forming, in addition to some chemicals that were formed in the atmosphere, others might have escaped the crust and/or evaporated to position themselves above the crust or the more condensed part of the outer of these bodies. Those that evaporated could have later cooled down while others might have fallen to the ground due to gravity and other processes which brought them down. Others could have never fallen, but could have stayed in the atmosphere. Some chemicals could have been ejected into the atmosphere as the celestial bodies were being wrapped into their shape. Other chemicals in the atmosphere could have been formed in situ after some of the precursors of their peers could have been condensed and incorporated into the main bodies in their vicinity and ended up forming structures like rings.

The things I said so far also explain why is the chemical composition of the Moon and Earth are said to be similar. Indeed, according to scientific evidences, not only do the Moon and the Earth seem to have a similar bulk composition, but they also virtually share the same isotopic fingerprint. This similarity can be explained by the fact that the Moon and the Earth descended from a common precursor: the precursor of the Earth-Moon system. After the precursor of the Earth-Moon system was split into the precursor of the Earth and the precursor of the Moon, the precursors of their atoms and subatomic particles may not have gone through significant enough changes that could erase the footprint of their common origin and characteristics already imparted into them. Probably, many chemical elements in the Earth and Moon would have been completed or reached a significant stage of differentiation before the precursors of these celestial bodies were split. Or the differentiation of the precursor of their particles could have reached a level of similarity that could not be reversed by the processes that finalized the formation of each of these celestial bodies. Consequently, although some differences exist between the chemical composition of the Earth and the Moon, the core of their characteristics and isotope composition is said to be almost the same. In other words, the similarities of the chemical composition and isotopic fingerprint of the Moon and the Earth can be explained by the impact of their descendance from a common precursor (which was the precursor of the Earth-Moon system) and the changes that their corresponding precursors went through after separating from their mother.

Furthermore, similar reasons explain why the chemical composition of Jupiter is said to be globally close to that of the Sun. Indeed, although the turbulence in the precursor of Jupiter could have been different from that of the precursor of the Sun, they shared some similarities. As the largest body orbiting the Sun and the second largest body in the Solar System, just after the Sun, Jupiter could have gone through some similar changes as the Sun. In other words, because of its size, the precursor of Jupiter shared some resemblances with that of the Sun more than the precursor of any other planet in the Solar System. For instance, the density of Jupiter and that of the Sun are in the same order of magnitude, suggesting that the intensity of the compression, squeezing, or gathering together of their precursors could have had things in common. In the end, these similarities affected some of the proprieties of the Sun and Jupiter, therefore making Jupiter the planet which chemical composition (e.g. hydrogen and helium composition) and other characteristics are more similar to the Sun than other bodies in the Solar System. However, because of its much smaller size, the precursor of Jupiter formed a planet, while that of the Sun formed a star. In the end, the size and the environmental conditions of the precursor of the Jovian planetary system did not allow it to be a miniature Solar System.

'Science180 Academy' Success Strategy:
SCIENCE180 ACADEMY PROGRAMS

Owned by Science180, Science180 Academy is a training, speaking, consulting, and mentoring program specialized in everything universe-origin, life-origin, chemicals-origin, and anything at the intersection of reason and faith, or science and religion.

Science180 Academy deals with different subjects according to the needs of its members or target groups. When people register to Science180 Academy, they must choose the program(s) they want to focus on so their training can be properly personalized accordingly. This is similar to how people register to a university, and take classes in a specific department matching their needs!

Science180's breakthroughs are so complex and dense that it is not realistic or good to try to explain all in just one academy, else people will be overwhelmed, disinterested, and confused by the plethora of data to handle. In other words, Science180 Academy offers a wide range of origin-related training in various domains strategically designed to allow people to choose the most suitable for their needs so that, regardless of their background or field of expertise, people can equip themselves, align their mindset, improve lives today and forever using the accurate explanation of the origin of the universe, of life, and of chemicals. Science180 Academy curriculum is based on 12 years of deep unconventional research that culminated with the publication of many much-admired books on the formation of the universe and its content:

The content of each Science180 Academy is strategically crafted by Dr. Nathanael-Israel Israel (who is acknowledged as the internationally-acclaimed world's authority in origin-related issues) to suit both scientists and nonscientists, religious and nonreligious people, leaders as well as followers, so they can fully decode the proofs of the formation of the universe, of life, and of chemicals they have been wanting to demonstrate or grasp.

Please visit Science180Academy.com to sign up today or to contact me so I can discuss that with you.

The current programs of Science180 Academy are:

- **1. SCIENCE180 ACADEMY OF COSMOLOGY** (Designed for all scientists who want to scientifically study cosmology, the science of the origin and fate of the universe)

- **2. SCIENCE180 ACADEMY OF TURBULENCE** (This is a perfect fit for scientists and other experts interested in studying abiotic turbulence).

- **3. SCIENCE180 ACADEMY OF LIFE SCIENCES** (Tailored to those who want to study biotic turbulence)

- **4. SCIENCE180 ACADEMY OF CHEMISTRY** (Designed for chemists, biochemists, scientists, and other educated people who want to understand the origin of chemical particles)

- **5. SCIENCE180 ACADEMY FOR LAYPEOPLE OR THE GENERAL PUBLIC** (Very fit for any layperson or "less" educated people who wants to learn (in a simple language) deep insights that even those who went to university for years were unable to decrypt by themselves ...

- **6. SCIENCE180 ACADEMY FOR CHILDREN** (This Academy breaks down origin key topics into language that children can fully understand). This is the only Science180 Academy that your whole family will like and enjoy together, and which will set children on the path of success by accurately showing them early in life the formation of the universe, and how to detect errors in theories or stories that would misguide them as they grow up.

- **7. SCIENCE180 ACADEMY OF THE PSEUDEPIGRAPHA AND SPIRITUAL WORLD**

- **8. SCIENCE180 ACADEMY OF CREATIONISM** (Science180 Creationism is a scientific theory spearheaded by the groundbreaking discoveries of Nathanael-Israel Israel, that scientifically explained the origin of the universe, life, and chemicals using turbulence, and that mathematically reconciled science and the Biblical account of creation for the first time in history. Science180 is different from all existing creationist theories known before 2025. Science180 Creationism reconciled science with the Biblical account of creation, including scientifically proving that the Earth was formed on Day 3, while the Moon and the Sun were formed on Day 4 of creation!). As you attend "Science180 Academy of Creationism", you will receive accurate answers to all your questions concerning the creation of the universe).

- **9. SCIENCE180 ACADEMY FOR FREETHINKERS & ALL ANTI-CREATIONISTS** (This Science180 Academy is designed for evolutionists, anti-creationists, and all other types of unbelievers seeking to rationally explore and understand alternative arguments for creation or formation or origin of the universe, life, and chemicals from a fresh, scientific perspective).

o **10. SCIENCE180 ACADEMY OF LEADERSHIP**-(Also called "Science180 Academy for Leaders", this program will enlighten leaders of organizations on how to solve their people problems, process problems, and profit problems related to the origin of the universe, of life, and chemicals according to their domain of expertise). With "Science180 Academy of Leadership", leaders will gain new insights so they can cast new visions and avoid focusing on screwed-up processes, products, and services related to universe-origin initiatives that need to be fixed, faced, or dealt with. Science180 Academy of Leadership will also equip leaders to address process problems related to inefficiency, gaps, missed opportunities, wasted time and efforts, too many steps, bureaucracy, useless layers between organization and customers concerning the innovation, research methodology, research, product development...

Science180 breakthroughs so that they can sell more often at full price, avoid regrets in the end, open new markets focusing on real solutions, expand their products and services lines, cut useless costs and research, stop wasting time on useless products that will yield nothing, start focusing on the real money-making problems, reach and convert more prospects into profitable and loyal customers, speed up time to market, avoid spending resources on unprofitable projects but on profitable ones, take their organization to a higher level, open new groundbreaking doors, ...

11. SCIENCE180 ACADEMY FOR GOVERNMENTAL AGENCIES (Do you want to know how and why most nations and governments are wasting millions of dollars on universe-origin and life-origin researches they don't need ... and how to avoid it? What if your nation or institution can reduce wasteful spending on universe-origin research and life-origin research, as well as your dependency of wrong theories on the origin of the universe and life?

"Science180 Academy for governmental agencies" will show you how to use the latest scientific breakthrough to better understand the origin of the universe without wasting money on what is already known or what we think we don't know, but that most scientists ignore. Having spent years accurately decoding the origin of the universe, of life, and of chemicals, Dr. Nathanael-Israel Israel delivers science-backed insight to properly understand all the processes connected to the universe formation–so you don't waste more money and time on trying to research the beginning of the cosmos, but to focus on reducing budget of spatial agencies, focus on real science, cutting-edge research, and things that inevitably lead to discovery and innovation).

12. OTHER SCIENCE180 ACADEMY: If you did not relate with any of the Science180 Academies mentioned above, but you are still interested in learning something specific about the origin of the universe, life, and chemicals that better fits your needs, please visit Science180Academy.com to contact us so we can discuss that with you.

CHAPTER 26

HOW PHYSICS AND CHEMISTRY MEET TO EXPLAIN THE UNDERLYING CAUSES OF THE DIVERSITY OF CELESTIAL BODIES IN THE UNIVERSE

In *"Turbulent Origin of the Universe"*, I extensively explained the processes involved in the formation of celestial bodies including rings. But in this chapter, I will highlight a few factors explaining the particle composition of the celestial bodies according to their types.

Rocks played an important role in the finishing of the formation of celestial bodies. For instance, in the case of the Earth and other terrestrial planets, rocks were piled or heaped together to form the celestial bodies. While the outer portion of some precursors were cooled down to form a crust, the inner portion kept the fluid state, therefore, they erupted sometimes into volcanos. Some solid rocks were joined together while others were deposited on top of others. The fluids in the precursors also acted as a solvent to glue some rocks and their constituents. While the interior of most planets would contain fluids, in the case of some small asteroids and satellites no fluid could be found, but stacks of rocks.

The size of the precursors of the celestial bodies affected the destiny of their internal particles and consequently of their overall state. For instance, because most of the precursors of their particles did not form very dense particles and because of their huge size, the precursors of stars were not able to have their outer surface solidified as a crust. Lacking an outer solid shell, these stars therefore are believed to be mostly made of hydrogen and helium. At the planetary level, the precursors of some planets were able to form dense bodies and a crust which covers their outer part. That is the case for the terrestrial planets such as Mercury, Venus, Earth, Mars, and Pluto. In contrast, the precursors of some giant planets such as Jupiter and Saturn were not able to form many dense particles which put

Nathanael-Israel Israel: Author of "How Baby Universe Was Born" together could have formed a dense crust. Instead, they formed giant gaseous planets. In contrast, the precursors of Uranus and Neptune ended up forming icy planets. Between all of those planets, asteroids of various chemical compositions were formed. Depending on their position, some asteroids are believed to have a chemical composition intermediate between the metal rich terrestrial planets and the volatile-rich outer bodies. Moreover, because of the circumstances of their precursor, the main belt asteroids were not gathered together into a single asteroid or planet by piling together its rocks, but instead, the rocks were scattered and became individual asteroids organized in a belt. In other words, in the case of the formation of the terrestrial planets, the rocks were not scattered into individual celestial bodies, but they were heaped together and bound by forces, including gravity and/or its precursor. The characteristics in the main belt asteroids could have also been affected or caused by the precursor of the Jovian planetary system which fluid layer was at one point located just below the fluid layer of the main belt asteroids. Beyond the asteroid main belt, because conditions may have not favored the formation of a solid crust, the giant planets are dominated by gases. It is possible that some of the giant planets have a crust surrounded by a thick atmosphere. However, as of 2025, to my knowledge, no mission has ever collected materials at the surface of the giant planets yet. Unlike the precursor of the main belt asteroids where at least rocks were formed without being able to gather together, for the precursors of the giant planets, rocks could not have even been able to form as was the case of the crust of terrestrial planets. Instead, the particles that were supposed to be packed together into the rocks were loosened and spread out over a bigger portion of space, therefore giving the giant planet a large size.

Because it was very big, even bigger than the precursor of any other body in the Solar System, the precursor of the Sun was not much compressed or packed into a solid mass by the forces that shaped it. Instead, most of its constitutive particles were less dense and were not as compressed as the precursors of some other bodies in the Solar System. The precursor of the Sun could have been like a big light or a very hot body filled with energy. The nature of the precursor of the Solar System explains why the center of the earth and of most planets could contain lava. As the precursors of the celestial bodies moved farther from the Sun, their fluids were cooled down by the environment, at least at their surface. This explains why the exterior of terrestrial planets are hard or crusty, whereas, their interior could likely be hotter than lava or magma, which sometimes erupt into volcanos.

Because of its massive form and huge light, the Sun emits a lot of heat into its surroundings and never experienced a night but has been very much hot and not affected by the cold of the space around it. In other words, the precursors of the other bodies in the Solar System were also initially very hot like the precursor of the Sun, but, because of their smaller size and how their turbulence affected them, their outer surface could not preserve its state of matter and temperature, but was cooled down by the cold and darkness of the space they were projected into.

Consequently, the outermost surface of some bodies was solidified, yielding a crust positioned on top of a likely fluid interior. In contrast, the outer surface of some celestial bodies was not solidified, but formed gases, while others formed icy surfaces. For instance, the environment and the nature of the turbulence in the precursors of Jupiter and Saturn explain the gaseous state of their matter (at least in their atmosphere) as the precursors of their particles could have been highly agitated and more dispersed rather than packed to form a solid planet. Furthermore, the precursors of the outer bodies in the Solar System had to overcome colder environments, which took away a significant amount of their energy. Consequently, the surface of some of them was frozen. The precursors of the innermost bodies could have lost less energy for the heat of the precursor of the Sun could have increased the temperature of their environment. In other words, the environment where the outer bodies in the Solar System were formed could have sucked a lot of energy from the precursor of the outer bodies. Even until today, some bodies could be losing more energy because of their location. In other words, in addition to the turbulence that defined the nature of particles in the precursor of the bodies, the temperature of their surrounding also played a role.

However, I think that in some locations of the galaxies, it is possible to find stars which are smaller than some planets in the Solar System. As long as the environment is very hot and clustered with many stars, its size would not matter. For if the environment is very hot and cannot allow the cold to cool down the surface of a precursor of a star, it will form a star regardless of its mass. In the galactical core where many stars are clustered, very small stars can be found, meaning that size alone cannot justify the formation of star.

Photons (which are the particles of light) are matter that was contracted, compacted, and packed together to form big clusters of photons. Therefore, photons freely move at a high speed. Because the particles in the Sun are not tightly tied together, some of them, including photons are easily released. However, planets and satellites had their constitutive particles highly tightened and do not escape or move out easily. Because the precursors of the chemical particles in celestial bodies that are not stars were combined with others to form particles more complex than photons, they do not exist in the form of light. The energy in particles which are not photons somehow bear witness of their original formation based on something similar to what was used to form light as well. The big planets like Jupiter and Saturn were not like the Sun because during their formation, their particles were transformed into particles more complex than the photons which form light.

Just as the chemical composition is not the same at every location on Earth, so also the chemical composition of the Sun and of all other celestial bodies cannot be expected to be the same in all of their locations. For instance, although the Sun is dominated by hydrogen and helium, it also contains heavy atoms although in smaller amounts. Heavy elements are also present in the Sun because, when the precursors of matter in the Sun were being molded into atoms, the environmental conditions in some spots or pockets allowed their split-gathering into heavy elements instead of the smaller and lighter hydrogen and helium. The light and other forms of radiation that the Sun emits at a specific zone can depend on the type of chemical elements present in that zone. Some Sunspots could be dominated by heavy chemical elements, which therefore decreased the brightness of the Sun at their location. The darker sunspots may contain denser chemical elements than the brighter spots of the Sun. The variation and the random appearance of the localization of a sunspot may be due to the movement of the plasma in the Sun. As the Sun is moving, the matters that constitute it keep moving as well.

To sum up, stars are like naked celestial bodies, which do not have the crust cover that some celestial bodies (e.g. satellites, terrestrial planets, and asteroids) have. On another hand, celestial bodies that have a crust are like clothed bodies which nakedness (fluids in their interior) is covered by a crust (the cloth). Lava inside the earth and inside other planets can prove that their precursors were very hot and were cooled down to form a crust outside. The lava or magma inside some planets and satellites is a witness to the fact that all their constituents did not go through some changes that could have solidified them. But as seen with volcanic activities, the lava or magma can solidify. The Sun has no crust because its heat burns everything and could not allow a solid crust to be formed.

The topographic range is the difference in elevation between the highest and lowest points on a celestial body's surface. The topographic range is not known for all of the planets in the Solar System. For instance, as of 2025, the surface of the giant planets is still unknown, so is their topographic range. However, the topographic range of the 4 terrestrial planets (Mercury, Venus, Earth, and Mars) varies between 7 km and 30 km (Fig. 136). Mercury has the lowest topographic range whereas Mars has the highest.

Fig. 136: Topographic Range (km) of the Planets in the Solar System

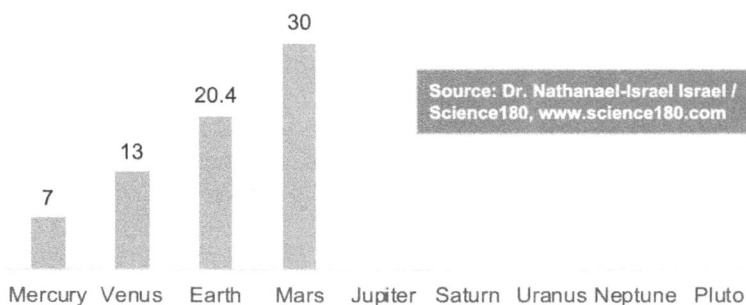

| 30 |
| 20.4 |
| 13 |
| 7 |

Source: Dr. Nathanael-Israel Israel / Science180, www.science180.com

Mercury Venus Earth Mars Jupiter Saturn Uranus Neptune Pluto

Science180: The Only One Formula Accurate Enough to Explain the Formation of Chemicals

TURBULENT ORIGIN OF CHEMICAL PARTICLES

The topography of the celestial bodies which have a crust was affected by their constituents and the force that compressed or packed these bodies. Hence, the topographic range and the semi major of the satellites are positively correlated (r2= 0.995). Aphelion, perihelion, and semi major axis of the planets have a positive effect on their topographic range, meaning the farther a planet is from the Sun, the higher its topographic range. At one point, I wondered why topographic range and the semi major are so highly correlated positively. Then, I felt like, probably, as the semi major of the planets increased, the force that squeezed the planets to yield their final shape caused the topography range to be proportionally affected. Moreover, volcanic activity on the planets may also explain their topographic range. As of 2025, scientific data showed that Mars has more volcanic activities than Earth, Venus, and Mercury. Probably there may be less volcanos on Mercury and that may be why its topographic range is so low. Io, the innermost Jovian satellite in Zone 3 is believed to be the most volcanic body in the Solar System. The crust of Io may not be strong enough to contain the magma or lava inside of it. Hence, it erupts most of the time to yield volcanos. Beside Io, the other volcanic celestial bodies in the Solar System that are said to be volcanically active are Triton, Venus, and Earth. In contrast, Europa is believed to have no volcano and its surface is said to be very reflective, making people think it is made of ice at the surface. It is believed to have an undersurface ocean. Many planets, asteroids, and even satellites contain a lot of fluids inside that are sealed up by a crust at the surface. I presented the topographic range of the planets here to illustrate how the same process that caused the precursors of some celestial bodies to be more squeezed than others, yielding uneven surfaces for the celestial bodies, also caused (at the microscopic level) some chemical particles to be more squeezed than others, leading to their diverse density, hardness, and modulus. In other words, it is not by chance that chemical particles have different densities.

Indeed, as I wondered why the chemical particles have different densities, and why some are denser than others, some similarities with respect to celestial bodies came to my mind. For instance, in *"Turbulent Origin of the Universe"*, I proved that, as the layers of the precursors of particles started collecting together to shape their daughter bodies, their fluids were spiraled, wrapped together as vortexes were being organized inside of them. The speed of the fluid layers affected how their daughter bodies were squeezed. The precursor of particles that did not rotate fast enough could have reduced the ability of their daughter particles to squeeze their content, meaning that their density could have decreased. The data I studied on celestial bodies suggested that the densest bodies are those for which the orbital speed and the rotational speed are balanced and relatively high. For instance, although the densest celestial bodies are usually found in Zones 1, 2, and 3, the densest ones are not usually in Zone 1. In *"Turbulent Origin of the Universe"*, I established that a higher orbital speed of the precursors balanced with the ability of their daughter to rotate faster affected the density of the daughter bodies. Thus, although the terrestrial planets are the densest, the Earth is denser than all of them because its precursor had a relatively higher balance between its orbital speed and rotational speed. In contrast, although Mercury and Venus have a higher orbital speed, their density is smaller than that of Earth because their precursor could not rotate very fast, finishing a single rotation in months, while Earth rotates once every day. Similarly, although the giant planets have a higher rotational speed, their densities are not the highest because their orbital speed is small due to the fact that the position of the fluid layers in their precursors. I have established that the turbulence that the precursors of these bodies went through is at the root of their rotational movement.

'Science180 Academy' Success Strategy:
SCIENCE180 BOOKS THAT WILL HELP YOU!

I, Nathanael-Israel Israel, broke down my discovery about the formation of the universe into many books so that you, the readers, can pick the ones that correspond to your needs and interests without disappointing you or wasting your precious time. These books come in many versions (e.g. scientific version, public version, chemical version, biological version, biblical or prophetic version, pseudepigraphic version, and a children's version) targeting people according to their expertise, educational background, and interests as briefed below:

1. *"TURBULENT ORIGIN OF THE UNIVERSE"* (This is the scientific version of my book tailored to scientists and anyone interested in the detailed scientific demonstration of the universe formation). In this book I used the "mother of all turbulences" to scientifically demonstrate the formation of the universe so that scientists can understand and reorient the course of their research, teaching, and publishing, and accept the truth to better live today and forever. Get *"Turbulent Origin of the Universe"* today to begin an incredible journey of accurately decoding the universe and change your life forever! Learn more at Science180.com/scientific

2. *"RECONCILING SCIENCE AND CREATION ACCURATELY"* (this is the book that I called the "Biblical or prophetic version of my book on the universe-origin, and it targets Christians and anyone interested in knowing the Biblical perspective of the creation of the universe). This important book accurately demonstrates the marvelous creation and formation of the universe by God in six consecutive 24-hour-days, and answers many questions about the universe creation, so that after acknowledging Him (who deserves all the glory now and forever), human beings can choose life and avoid the terrible judgment awaiting the unbelievers in the world to come. Get this thoughtful book now to figure out what happened at the beginning, what is coming up, and why it is time to urgently rethink everything you have been told about the universe-origin so you don't eventually regret! Don't say I did not tell you! Learn more at Science180.com/biblical

3. *"TURBULENT ORIGIN OF CHEMICAL PARTICLES"*
(Called the "chemical version" of my book on the universe-origin,
this elegant book targets chemists, biochemists, and anyone
interested in chemistry). With *"Turbulent Origin of Chemical Particles"*,
the accurate decrypting and understanding of the formation of
chemicals has never been profitable and easy. Hence this great
book is THE ultimate how-to guide for great people wanting to
correctly decode the origin of the chemicals and positively
transform their lives. Get this celebrated book today. Learn more at
Science180.com/chemical

4. *"ORIGIN OF THE SPIRITUAL WORLD"* (This book
is what I called the pseudepigraphic or hidden version of my books
on the universe-origin, and it is meant for believers who want to
tap into a higher level of scriptural secrets that most people may
not believe). This book draws the attention of the world toward the
pseudepigrapha (a collection of hidden and rejected books, yet
filled with deep secrets still valuable today) and explaining how,
since thousands of years, God has already revealed deep details
about the supernatural origin of the universe, but people (including
those who believe or claim to believe in Him) have just refused to
literally accept God's mysterious story of creation, which can never
be understood by just sticking with conventional science. If you
believe in God, have some origin-related questions which answers
you cannot find anywhere, not even in the Bible, and if you want to
tap into historically neglected revelations to answer fundamental
universe and life questions, then be sure to get a copy of *"Origin of
the Spiritual World"* today. More at Science180.com/pseudepigrapha

5. *"FROM SCIENCE TO BIBLE'S CONCLUSIONS"* (I called this
book the "public version" of *"Turbulent Origin of the Universe"* and it
is tailored for the general public, and it is a great summary of the
scientific version from a perspective that laypeople will fully
understand). In this book, I, Nathanael-Israel Israel, broke down
the complicated (scientific, philosophical including religious) data
about the origin of the universe in a simple language that the
general public can fully understand, and know in order to live
happily forever. Quickly grab and read this scientifically verifiable,
bestselling book to finally get the accurate, jaw-dropping answer
that has been rationally shaking both believers, skeptics, and all
freethinkers. Don't wait! Learn more at Science180.com/public

6. **"TURBULENT ORIGIN OF LIFE"** (This is the biological or life version of *"Turbulent Origin of the Universe"*). It is meant to suit both scientists, nonscientists, and all kids of laypeople, and it decodes the origin of all forms of life so human beings can understand and better live. As of 2025, this book is my only book devoted to the origin of all forms of life, and it will help you to grasp in a simple language what is needed to fully understand the formation of all forms of life. Whether you are a scientist or a layperson, a believer, or a skeptic, you cannot afford to ignore the greater, better, faster, simpler, cheaper, easier, and accurate formula unlocked in this important book that successfully decoded the origin of life. Get *"Turbulence Origin of Life"* today and change lives. Don't wait. Learn more at Science180.com/life

7. **"HOW BABY UNIVERSE WAS BORN"** (How was the universe formed? Did God really form it like some people believe, or did it come out of some long processes? How can we scientifically prove and break down this difficult mystery in a language that children will fully understand and like?) Get the answers as you read this book that I called the "children version" of *"Turbulent Origin of the Universe"* and life. Accurately explaining the complex formation of the universe and of life to children can be very hard in our modern world, but by getting *"How Baby Universe was Born"*, you will know the proven formula to help children to easily understand their huge universe-origin and life-origin questions with confidence, humor, and joy. They will surely belly laugh and thank you for it! It is time to buy this pragmatic book and offer it to the children in your life today. Learn more at Science180.com/children

8. **"HOW GOD CREATED BABY UNIVERSE"**. The most difficult part of writing scientific things to children is how to break down complex technical concepts into simple words that they and even anyone who can read and clearly understand (without losing the accurate details and facts). When the topic to address is about the origin of the universe, the task is even more challenging for most people, but not for Nathanael-Israel Israel. As long as you can read, you will find this amazing book extremely helpful to grasp all complicated concepts needed to properly crack the origin of the universe in a language that even children ages 7-12 and anyone who did not go very far in school can fully comprehend.

9. *"SCIENCE180 ACCURATE SCIENTIFIC PROOF OF GOD"* (Whether you are a believer, an unbeliever, a freethinker, administrator, politician, curriculum designer, curriculum specialist, education policymaker, librarian, school board member, parent, researcher, student, teacher, clergy, or a layperson, as long as you are really seeking to scientifically understand the rational proof of the existence of God, *"Science180 Accurate Scientific Proof of God"* is the much-admired book written for great people just like you). As long as you are interested in the first and the only scientific book that talks to anti-creationists, evolutionists, big bang proponents, atheists, and all other freethinkers and rationalists about the universe formation and they bigly beg to know more about God, the creator, that they mistakenly deny; then this book is for you. As long as you are really seeking to scientifically understand the rational proof of the existence of God, *"Science180 Accurate Scientific Proof of God"* is the much-admired book written for great people just like you. Grab it today and start reading it. Don't wait any longer! Learn more at Science180.com/godproof

If you want to have the entire big picture of my discovery of the origin of the universe, life, and chemicals, and to enlighten your life and career, then plan to get all or some of these books that best suit your needs and interests. For more details, visit Science180.com/books

CHAPTER 27

DISCOVER THE SHOCKING STORY BEHIND HOW THE FORMATION OF THE PLANETARY RING SYSTEMS INCLUDING THE SATURNIAN RINGS WAS FINALLY DECODED USING CHEMISTRY AND TURBULENCE!

In *"Turbulent Origin of the Universe"*, I addressed how rings of planets and satellites were formed. In this chapter, I just focused on a few features of the particles in rings. Indeed, the size, nature, and chemical composition of the particles and rocks found in the rings are not the same, but vary according to the types of rings. Because a lot of uncertainty still exists about the chemical composition of some rings, I did not focus on that, but on their size.

Indeed, unlike the rings of the other 3 giant planetary systems, the Uranian rings are said to consist of large bodies 20 cm to 20 m in diameter (Ockert et al., 1987). The bigger size of the particles in the Uranian rings may be connected to the nature of the turbulence of their precursors. In contrast, the Jovian rings are mostly made of dusts (Smith et al, 1979; Showalter et al., 1987). While the size of particles in the Halo ring (the innermost Jovian ring) is estimated to be in the order of submicrometers, particles in all of the Jovian rings are estimated to have of radius about 15 μm (Throop et al., 2004). In contrast, the Saturnian rings are said to consist mostly of water ice and trace amounts of rock, and the size of its particles varies between micrometers to meters (Porco, 2012). Saturnian rings are also believed to be very rich in water ice (Esposito, 2002). Finally, the Neptunian rings are very dusty and the densest regions are said to be like the less-dense Saturnian rings. Both Uranian and Neptunian rings are very dark and are believed to contain organic compounds (Smith et al., 1989). As long as size and faintness are concerned, Neptunian and Jovian rings are said to look similar, both consisting of faint, narrow, dusty ringlets, and even fainter broad dusty rings (Burns et al.,

2001). Finally, unlike what many people think, "rings are not a series of tiny ringlets, but are more of a disk with varying density" (Tiscareno, 2013).

As I explained in *"Turbulent Origin of the Universe"*, the trail of dusts in which some planets orbiting the Sun could have come from some left over of the precursor of their planetary system which was not gathered together with either their planet or satellites. As the layers of fluids in the precursor of the secondary bodies in the precursor of a planetary system started going through turbulence, they were not always able to gather together all of their particles into a single body and form a single satellite that has no trail of dust around it. Instead, some particles were gathered together into rocks of various sizes, while others were gathered together into particles of various sizes. The particles that were gathered together into larger bodies became the satellites, while particles not incorporated into the precursor of the satellites become or formed the rings. In other words, because the particles in the fluid layers of some precursors of secondary bodies could not be gathered together into satellites, they were scattered as they were collected into large particles and/or rocks constituting the rings. Hence, some rings do not contain satellites, while others do. In other words, not all layers were gathered together into satellites. Those which were gathered into satellites were unable to collect together all their particles. The layers that were not gathered together at all and the left-over particles from the layers that were gathered formed the particles and rocks present in the rings. Another way of explaining this is that, during the formation of the rings and satellites, there was a point where the precursors of the rings and satellites were particles which were split-gathered into the precursor of the rings and the precursors of the satellites as their turbulence was underway. As turbulence was developing and fluids were moving, mixing, rotating, spiraling, and packing, some fluids were packed into bigger clusters, which yielded the satellites, while other fluids did not rotate enough to be amass into bigger bodies, but smaller particles spread over the space occupied by their mother precursors. The daughter bodies that were not amassed into satellites became the particles constituting the rings. In contrast, the gathering together of the precursors of particles into denser bodies led to the formation of the satellites. Hence some satellites are embedded inside rings. Additionally, the size of the particles in the rings depends on the type of planetary system.

For instance, no satellite was found inside the Halo ring, the innermost Jovian ring. While the Halo ring is very thick, the other Jovian rings are faint. The fact that Halo ring is the vertically thickest Jovian ring suggests to me that its precursor was not able to be gathered into a satellite, but into particles packed into a thick torus. In other words, the Halo ring is a good example of fluid layers of a secondary body in a planetary system that were not split-gathered into a satellite. The other Jovian rings are faint because part of the precursors of the particles that were present in their neighborhood were gathered together into some of the satellites embedded into those rings and/or the layers of fluids that birthed those rings could not be collected into a satellite. The dust and other smaller particles in the precursors of some secondary bodies diffused after being unable to form a satellite.

The same trend is observed for Saturnian rings as well. Upstream of the innermost Saturnian satellite, at least 7 seven rings are found:

- D Ring
- C Ring
- Titan ringlet
- Maxwell gap/ringlet
- 1.470 Rs Ringlet
- 1.495 Rs Ringlet
- B Ring

Of these inner Saturnian rings, B Ring which is located just upstream of the S/2009 S1, the innermost Saturnian satellite, is the Saturn's opaquest ring. The opacity of B Ring is not by chance but can be explained by the amount of materials gathered into its precursor. The precursor of B Ring could have been dense, but its gathering did not yield a satellite. However, downstream of the precursor of B Ring, conditions were met and allowed the gathering together of the precursors of the secondary bodies into S/2009 S1, the innermost Saturnian satellite known as of 2025. Together with the C Ring (which is upstream or inward of B Ring), the A and B Ring (which are separated by the Cassini Division), are generally considered as the main Saturnian rings. The main rings are the densest rings and they contain larger particles, suggesting that their precursors failed to amass into a single satellite but were gathered into particles constituting the rings. Some of the precursors of the secondary bodies in the vicinity of the main rings which were favorable were gathered together into the innermost Saturnian satellite (S/2009 S1), located inside the main ring. Part of the particles or rocks that were located in the gap between the A ring and the B ring could have also been gathered together to form the innermost Saturnian satellite. In contrast, because the precursors of the bodies downstream of the main rings were able to split-gather together into satellites, their remaining particles which were not incorporated into those satellites became the rings. Hence those rings are usually dusty, narrow, very faint, and are composed of much smaller rocks and particles (NASA, 2018).

As for the Uranian rings, the main Uranian rings are also located inward of the innermost Uranian satellite. This also corroborates the location of the main Jovian rings and the main Saturnian rings: inward of the innermost satellite of their primary planet. Outward of the innermost Uranian satellite, the rings are faint and consists of smaller particles and generally lack dust, suggesting that outward of the innermost Uranian satellite, more precursors of the particles or rocks which could have ended up in the rings were mostly incorporated into the satellites.

Some of the empty spaces or gaps between rings are due to demarcation of the space between the fluid layers of the precursors of the secondary bodies (rings and satellites) and/or due to the incorporation of the precursors of those particles into satellites. For instance, the Saturnian satellite called Pan is located inside the Encke gap, while Daphnis (another Saturnian satellite) is located inside the Keeler gap. Some of the materials used to form these satellites could have been particles that would have otherwise filled these gaps within the rings. In other words, the more the particles in the precursors of the secondary bodies were gathered together into satellites, the less materials were left to form the rings, hence a gap or a faint ring. As the layers of fluids of the precursors of the secondary bodies were separated, distances were also born between them. The density and brightness of the rings could have been determined by the number of particles or amount of matter in the precursors of bodies that were not gathered together into the satellites, but which were left out to become part of the rings. The brightness and density of the rings may depend on the amount and nature of their particles. The variation of the size of the rocks and particles constituting the rings could depend on the size of the vorticial or spiraling clusters of matter in their precursors. In other words, as the precursors of the rings were being shaped into their daughter bodies, different amounts of matter were gathered along the way according to the nature of the turbulence they went through and the factors that split them. The vortical movement and spiraling that occurred inside the precursors of the rings and satellites can explain why some spiral patterns and spiral waves are still observed inside some rings until today. Because most of the precursors of the secondary bodies in Zone 3 were gathered together into the satellites in that zone, rings are not usually found in or outward of Zone 3 and when they are found (like in the case of some Saturnian rings ranging from Methone ring to Phoebe ring), they are usually faint because they consist of tiny, smaller particles or rocks, or just dust. This also suggests that most of the rocks, particles, and dusts in rings are not just escapees from the big satellites or from any other satellites. In contrast, rocks have been falling from the rings onto neighboring satellites or onto their primary planets.

To sum it up, most of the particles found in rings were not transported to their current location by other celestial bodies, but were born from the precursor of the secondary bodies in their planetary system. The law of split-gathering, which defined the formation and size of matter on different scales, allowed that in regions where rings are found, small rocks or clusters of particles were formed, incapable to be amassed into larger celestial bodies such as satellites or incorporated into the bodies of their neighboring satellites (Fig. 137 and Fig. 138). As the precursors of the rings could not have been gathered together into bigger bodies, they were spread all over the domain near the trajectory of their orbit, which could have been cleared or which looked like a gap in these precursors incorporated into satellites. The color, density, brightness, thickness, and other characteristics of the rings depend on their precursors and how their environmental conditions shaped them. The inclination of the rings is a product of the inclination of the precursor of their planetary system and/or precursors of the secondary bodies. The variation of the density, faintness, or brightness of the rings in a ring system can be explained by the variation of the proportion of the precursors of the secondary bodies gathered together into satellites or left over to be incorporated into rings and how this left over was scattered throughout (Fig. 137 and Fig. 138).

Although the precursors of Venus, Mercury and Earth could have been dense, all of their particles were not gathered together without leaving out some which formed co-orbital trails of dust or rings. For, as they were being gathered together, all particles or matter in the precursors of the celestial bodies rarely collected themselves without leaving some out. Just as a gas torus exists around some satellites, for all the matter in the precursor of the secondary bodies in their neighborhood were not gathered together into the satellite but some were left to form the torus, so also, all of the particles in the precursors of the planets or planetary systems were not gathered together but some of them were collected into small asteroids (including some trojans), while other leftover particles became trails of dusts or rings surrounding some planets or in the path of their orbit.

To some extent, the main belt asteroids are for the Solar System what the rings are for the planetary systems. For what explains the formation of the main belt asteroids can also explain the formation of rocks in the rings. To put it another way, what explains why the asteroids in the main belt did not stick together to form a planet or a unique and bigger asteroid can also, to some extent, explain why the particles and rocks in the rings were not gathered together to form other satellites or to increase the size of the satellites they surround. Just as the densest planets are inward of the main belt asteroids and the largest planets are outward of the main belt asteroids, so also the densest, thickest, and opaquest rings are inward of the innermost satellites, and to a larger extent, the ring systems in the planetary systems are generally located inward or upstream of the largest satellites. All of this is the same law of split-gathering affecting the outcome of the precursors according to their location. Even on the scale of atoms and subatomic particles, the same law exists and could also explain why electrons around a nucleus could be organized into clouds instead of being very dense and in orbit.

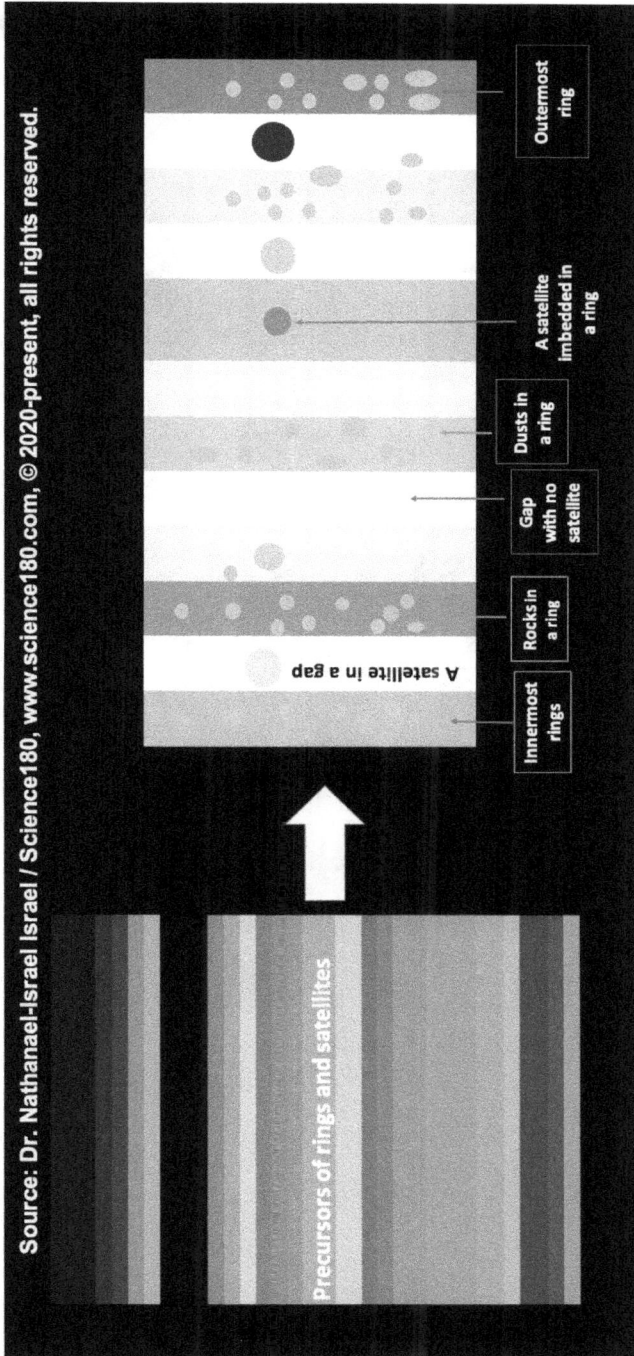

Fig. 137: Layers of fluids in the precursor of a secondary bodies split-gathered into rings and satellites

Science180: The Premier Organization that Scientifically Decoded the Origin of Chemicals Accurately

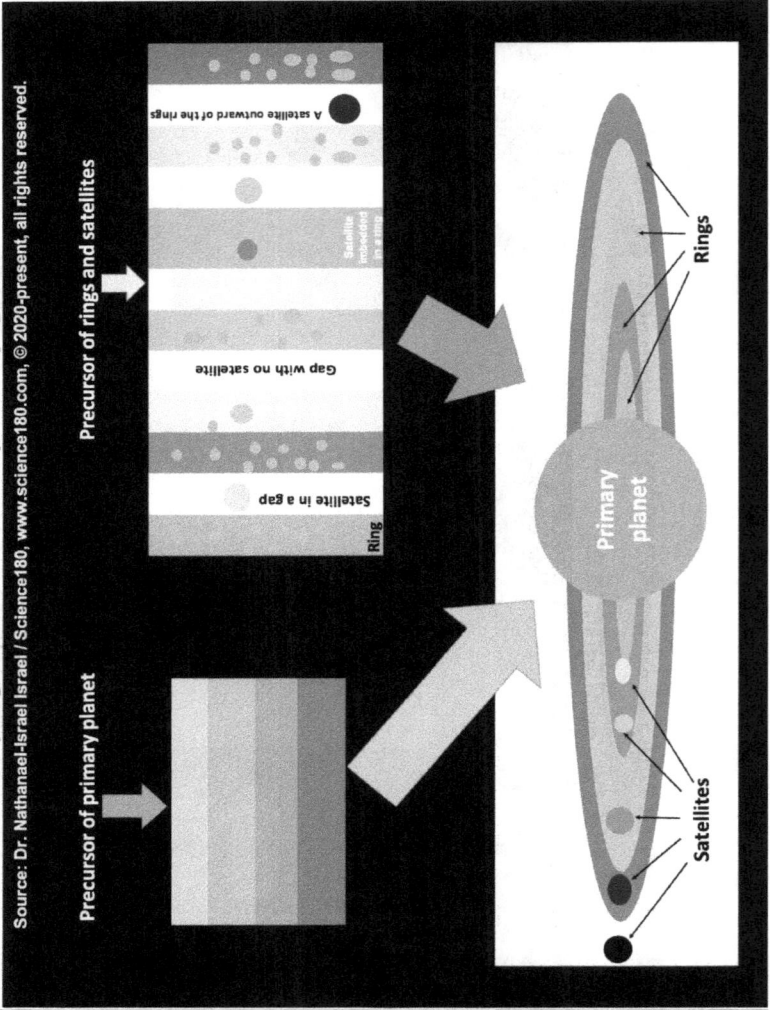

Fig. 138: Layers of fluids in the precursor of the primary planet, satellites and rings split-gathered to form the planeraty system

Nathanael-Israel Israel: Author of "From Science to Bible's Conclusions"

Another Book by Nathanael-Israel Israel:
TURBULENT ORIGIN OF THE UNIVERSE

THE FIRST AND ONLY SCIENTIFIC BOOK THAT ACCURATELY EXPLAINS EVERYTHING YOU NEED TO UNCONVENTIONALLY, EASILY, AFFORDABLY, AND ENJOYABLY DECODE THE UNIVERSE FORMATION

In *"Turbulent Origin of the Universe"*, filled with great diagrams and digestible scientific facts, you will discover, learn, or get:

- The all-in-one, proven & uncomplicated scientific formula that accurately decoded the formation of the universe, and that explained the birthdate of the stars, planets, satellites, asteroids, and all other celestial bodies in the universe, so you can position yourself to stay on top of your competitors, avoid repeating crucial mistakes that many people have ignorantly made at their own perils

- Extraordinary, unprecedented, accurate insights into the first factors (e.g. early universe physics) that defined the history and formation of the universe so you can tap into deep scientific secrets you ignore, and set yourself apart from others

- The new physics that will revolutionize science forever and land you into a zone of original ideas that improve lives nonstop regardless of your expertise

- The 4 simple things without which it is impossible for anyone to ever understand the formation of the universe, think accurately, work differently, achieve, or perform better for superior results

- The verified key to move the cosmological mountains of misunderstanding, so you can confidently free your mind from doubts, improve your health, and prevent you from any danger connected with sticking with wrong assumptions

- Save time and money, and enjoy your life once you remove errors holding your true understanding of the universe-origin captive

- Historic scientific proof of whether a planet was formed in 2.82 days, whether a satellite was formed in 3.32 days, and whether a star was formed in 3.69 days after the beginning of the universe; so you can creatively produce and address a broader work spectrum by learning how to effectively communicate with and establish unusual connections between otherwise disconnected and disparate scientific data

- The scientific formula that successfully tested the existence of God in a way that shocked believers, skeptics, and all other freethinkers

- Why the scientific community has failed to sufficiently explain the origin of the universe; and understand how existing theories have missed and undefined central ideas, and imposed limits on the vision of scientists
  - Specific in-depth knowledge, up-to-the-minute information, and ideas so you can expand your market, cut useless costs, stop wasting time on inadequate projects, and start focusing on the profitable solutions (Science180.com/scientific)
  - How Science180 Academy can strategically enlighten you, guide you to navigate and filter the massive data collected on the universe, so you can answer the world's most challenging questions, remove any scientific and philosophical cataracts that may be blocking you, and bring you many steps closer to your best life
  - How to better resonate with your target market that is craving something original that breaks wrong explanations of the universe-origin

Get *"Turbulent Origin of the Universe"* today to begin an incredible journey of accurately decoding the universe and change your life forever!

Dr. Nathanael-Israel Israel is told by people that he is the #1 Universe-origin, Life-origin, and Chemicals-origin Expert. He is the founder of Science180 and the author of many books on the origin of the universe and its content. Visit his personal site (www.Israel120.com) today to get some help.

CHAPTER 28

REMARKABLE SCIENTIFIC ADVANCEMENT THAT CHANGED THE RATIONAL EXPLANATION OF THE FORMATION OF GALAXIES, GLOBULAR CLUSTERS, STARS, AND THEIR CHEMICAL PARTICLES

Although in *"Turbulent Origin of the Universe"* I spent dozens of pages explaining the formation of galaxies and stars, I felt like in the next segments, it would be appropriate to pinpoint a few facts about some of their characteristics pertaining to the genesis of chemical particles. For the variation of the characteristics of celestial bodies hides a code of the way their chemical particles were organized. Hence, I felt uncomfortable addressing the formation of chemical particles without bringing up some key issues about the formation of some celestial bodies such as galaxies and stars. Most of this chapter is an extract from *"Turbulent Origin of the Universe"*.

Indeed, in 2005, it was believed that the observable universe contains 200 billion galaxies (Gott III et al., 2005). However, a 2016 review of that number suggested that the observable universe contains more than 2 trillion (2×10^{12}) galaxies consisting of more than 10^{24} stars, meaning there are more stars in the universe than all the grains of sand on planet Earth (Mackie, 2002; ESA Staff, 2019). While some galaxies called solitary galaxies, isolated galaxies, or field galaxies are isolated and seem not to be interacting with other galaxies, most galaxies are organized into structures called groups, clusters, superclusters, and cluster of superclusters. Clusters of galaxies are said to be habitually dominated by a giant elliptical galaxy, called the brightest cluster galaxy (Dubinski, 1998) The biggest galaxy in a cluster may be what could have been the primary body or primary galaxy if I can compare clusters of galaxies to planetary systems in the

TURBULENT ORIGIN OF CHEMICAL PARTICLES

Solar System. Superclusters of galaxies are believed to have tens of thousands of galaxies, and on the supercluster scale, "galaxies are organized into sheets and filaments surrounding vast empty voids" (Bahcall, 1988).

The structure of the galaxies is defined by 3 components: spheroidal component, disk component, and spiral arms. In *"Turbulent Origin of the Universe"*, I established that some galaxies could have been rolled over, tilted, or inclined during their formation and consequently, their rotational axis changed. Galaxies which have a prolate spheroid could have been rolled over just as some celestial bodies having a higher axial tilt or higher orbital inclination in the Solar System seemed rolled over. I also explained why it should not be surprising that some galaxies lack a spheroidal component. I showed how the spheroidal component is like the primary system of the galaxies, while the arms and disks of galaxies are like a system of secondary bodies. Just as some celestial bodies in the Solar System do not have a system of primary and secondary bodies, some galaxies also may lack some components.

Because the galactic disks are believed to emit most of the brightness of their galaxies, I explained that it does so because, as being the outer bodies in those galaxies, the stars in the disk were not as compressed, highly wound up, or condensed during their formation like those located in or near the spheroidal component. The small density or compaction of the stars in the disk implies that they could be less dense particles in them and consequently, they are brighter and are responsible for most of the brightness of the galaxies that have disk.

Finally, when I looked at the spiral arms of galaxies, it seemed to me as if their precursors were ejected from the precursor of the spherical component as the precursor of the entire galaxy they belonged to were being split-gathered to birth all their daughter bodies embedded in the galaxies. Unlike systems of celestial bodies in the Solar System, where the primary bodies are clearly separated from their secondary bodies, on the galactic scale, the gathering together of the precursor of some galaxies could not have been strong enough to release or separate the precursor of the galactic arms farther away from the precursor of the spheroidal component. The size and the degree to which the precursors of the spiral galaxies were able to push away, eject, or release the precursor of their secondary stars from the precursor of the primary stars (which constitute the spheroidal component) could have determined the properties of the galactic arms such as the size of the pitch angles, the angle of the extension around the center, the closeness or openness, and the number of complete rotations around the center of the arms. The precursors of the spiral galaxies (which have long arms extended around the center many times) could have flown over a long distance before its arms were wrapped or spiraled around the center of the galaxies. In contrast, the precursors of the galaxies that lack long arms or that are more spherical could not have been spread over a long distance before being spiraled into its "final" shape.

Based on their visual morphology, galaxies are classified into 3 main types:

CHAPTER 28: FORMATION OF GALAXIES, GLOBULAR CLUSTERS, STARS, AND THEIR PARTICLES

elliptical, spiral, and irregular. I demonstrated how the elliptical galaxies may have been inclined and those located toward the central region may have been more squeezed or compressed during their formation than the outer elliptical galaxies, which "orbit" may have been tilted or elongated more than the innermost. I also showed that the elliptical galaxies are darker or less bright than the other types of galaxies because their stars may have been more compressed and/or contained denser particles, hence a lower brightness. Elliptical galaxies could be for clusters of galaxies what terrestrial planets are for the Solar System.

I also found that the precursors of normal spiral galaxies could have had less "difficulty" ejecting and split-gathering their secondary stars than barred spirals. For the stars that are in the galactic arm are like secondary bodies, while those in the galactic core are like the primary bodies. The precursors of barred spiral galaxies could not have been able to eject their secondary stars away directly from the primary stars, but they could have flown for a while before spiraling. At one point, the precursor of the barred region of the barred spirals could have been almost in the middle of the arms organized as if one is leading and the other trailing. As the precursor of the barred spirals was winding up, because it could not wind up all of its stars to form an elliptical galaxy, the leading of the trailing arms were tilted or wound up according to the direction or sense of the movement of the precursor. The central part of the stars could not have been compressed a lot or wound up much because of its thickness, resistance to winding up, and the lack of energy that could have been required to compress it or spiral it.

I also demonstrated that the stars in the disk and arms of spiral galaxies are not as dense as those in the central areas, hence they give a blue color meaning they shine more or release more energy than the central ones, which are red, implying that they radiate less energy, which I link to their denser composition or the presence of denser particles in them which consequently reduced their brightness.

I think that some irregular galaxies are just remote galaxies that current tools are failing to properly resolve. They come in many shapes: lenticular galaxies, ring galaxies, interacting galaxies, etc. The diversity of the turbulence that dominated the universe where some galaxies were formed can explain the nature of the various irregular galaxies. If you are interested in knowing more about these galaxies, please refer to *"Turbulent Origin of the Universe"*. In that book, I also explained why some globular clusters may have been ejected from the precursors of the galaxies they belong to and others may have been born by the way the precursors of the galaxies around which they are found were split from one another, causing the intermittence of stars between galaxies or star clusters. In the same book, I also handle how, under some circumstances, the precursors of globular clusters could have not spun or spiraled enough to cause their daughter stars to orbit the primary star or primary clusters of stars as some stars go around the center of some galaxies. Some precursors of globular clusters could have exploded without enabling their daughter stars to form a well-defined system of secondary stars orbiting a well-defined primary core. Another way of saying this is

Science180: The Undeniable Scientific Challenge to All Metaphorical, Figurative, Loose, Liberal, or Vague Explanations of Chemicals-Origin that the precursors of globular clusters may have also lacked the ability to gather their daughter stars into a unique and condensed cluster, but instead released or scattered them without making a primary star dominating them. All this can be boiled down to the way the precursors of the clusters of stars were split-gathered into their daughter bodies. In the same book, I also delved into why, during their formation, the precursors of the spherical clusters of galaxies could have been more tightly wound, spiraled, or compressed, hence their daughter galaxies are denser than those of the irregular or spiral clusters. I also showed how the distribution of galaxy clusters hint at the location of their mother precursors and potentially their genesis. For instance, denser spherical galaxies could likely be located toward the center of the clusters of spherical galaxies because such a location could have favored the spiraling of their precursors into dense and more compacted galaxies.

As the precursors of galaxies were split-gathering through a sort of cascade, stars were being formed until a point when the split could no longer occur or continue to occur due to the size of the precursors and the environmental conditions. At that moment, the formation of new stars could have stopped. Each of the precursors of the stars were wound up to yield its individual star. According to their environments, some stars were tightly wound up while others were loose. The degree of the tightness and the turbulence of the precursors of the stars could have defined the internal composition and consequently the intensity of the brightness of their daughter stars. Consequently, some stars that contain denser constituents ended up being redder than those which constitutive particles were less dense and which appear bluer.

Finally, although it can be impossible to know exactly the proportion of the universe that is void or not, a look at the night sky suggests that a lot of void spaces exist between the bodies in the sky. The "empty" spaces as seen from Earth are not necessarily void, but they contain things that most human beings are unable to see with the naked eye, or with advanced telescopes. However, no matter the size of the content of space, it is clear that there are more voids than materials in space. In other words, matter is not present everywhere in the universe. Something happened to allow matter to be present in some locations and absent in others. In other words, there is a reason for matter to be distributed unevenly throughout the universe and not present everywhere. During the formation of the universe, it is likely that, in addition to the forces deployed to fragmentate the precursors, the light present in the precursors of the celestial bodies being formed might have also contributed to increasing the distance between the bodies. For light travels at a high speed and has the ability to move things. Therefore, light emanating from the precursors of the bodies could have also contributed to pushing away some bodies in its vicinity. For, in the beginning, when the celestial bodies were being formed, they were initially full of energy, which can also take the form of light, hence, we can see some of them at night time. In addition to the force which scattered the precursors of the matter across the universe, the radiation emanating from those bodies could have increased the distance between them until an equilibrium was reached from which point, the energy of those lights could not significantly affect the location of the neighboring bodies much.

Even on the atomic level, a distance is found between the subatomic particles. A distance exists between electrons, protons, and neutrons. On scales smaller than the subatomic level, a distance exists between everything although current scientific equipment fails to measure them. This is because during the formation of the matter in the universe, things were moved around and the initial energy inside the precursors of all known matter contributed to distancing the types of matter being formed. Even at the microscopic level, distance exists. Some scientists have tried to explain the laws governing the distance between the celestial bodies. Unfortunately, they thought that celestial bodies were attracting one another and that the force of this attraction is correlated with their radius and the distance between them.

During my work on the origin of the universe, I realized that the processes that explain the formation and evolution of vortices can also explain the formation of galaxies. Indeed, when I first looked at the picture of some galaxies, I saw some similarities with pictures of hurricanes. It later appeared to me that, just as a fluid filament can curl to form a vortex, so also the precursors of galaxies could have initially been astronomical fluid layers, which, due to their turbulence, were "curled" or "wrapped" into astronomical 3D bodies. The same thing could have happened to many celestial bodies, meaning that their precursors could have been a layer, which was "curled" to yield them. The shape or the configuration of the celestial bodies could give a clue as to how the curling or wrapping or gathering together of the fluid layers could have been. Just as on the microscopic scale some vortices can be spherical and very dense, so also on the astronomical scale of the galaxies, the precursors of some galaxies were able to be tightly collected together into spherical bodies forming for instance elliptical galaxies (Fig. 139). In other cases, the fluid layers of the precursors of some galaxies were unable to compact into an elliptical galaxy, but instead into a spiral galaxy. In other words, the precursors of spiral galaxies were unable to tightly wrap all their fluid layers around their central or spherical bodies. The same is true for barred spiral galaxies, but here, the wrapping around or the spiraling occurred around a longer central body, implying that the plane of a significant portion of the fluid layer of the precursor was not bent. Hence the presence of a bar in the "middle" or center of the barred spirals (Fig. 139). The precursors of the arms of spiraled galaxies could have just been layers of fluid that were loosely wound around a primary portion of their precursor. The central part, the core, or the spheroidal component of spiral galaxies could have been where the fluid layers of their precursors were most stuck during the processes that gathered them together. As the gigantic fluid layers of the precursors of the galaxies were being gathered together, stellar systems were formed inside of them (Fig. 140). As those stellar systems were being formed, their planetary systems and asteroid systems were also formed. The formation of planetary systems also implies the formation of their planets and satellites. In other words, as the precursors of galaxies were being formed, on different scales and levels of hierarchy, stars, planets, satellites, asteroids, and microscopic particles (e.g. atoms, subatomic particles) were also being formed in the universe as components of the bodies found in those galaxies. On a scale larger than galaxies, larger fluid layers could have been split-gathered so that galaxies could be formed and so on and so forth until the highest level of clusters of galaxies were formed. Just as stars are orbited by their planets and asteroids, so also the core of the spherical components of galaxies are orbited by stars and stellar systems. Just as rings are found orbiting some planets, so also on the galactical scale, the galactic disc contains many stars and planets imbedded in them. Just as in a fluid flow, fluid elements or vortices are pulled and pushed by others in the flow, so also, in the fluid layers of a gigantic precursor of galaxies, some precursors of stars could have been pushed or pulled by those adjacent to them. It is possible that some stars could have been titled or inclined in a certain direction and angle because of the pulling or pushing of their precursors by the precursors of adjacent stars. Likewise, the orientation of the plane of some galaxies could have been caused by the pulling or pushing of the precursors of the galaxies adjacent to them and/or caused by other events associated with their breakup up from others. The galaxies in the universe are positioned in different planes and some of them cannot be fully seen from telescopes placed on Earth because of the position that their precursors landed during their formation after breaking up from others. That is why, while some galaxies can be clearly seen from Earth, only a tiny line of light is seen for others, which likely lay in the same plane as the Milky Way, the galaxy that the Sun belongs to. The characteristic of the fluid layers or their stratification could have affected their turbulence and how their fluids were mixed. As the fluids were mixing, some from different layers could have been mixed and the vortices could have been differently flipped over, inclined, and even overturned for in the vicinity of the interface of two adjacent layers. The shearing of the fluid flow due to the fluid layers moving at different speeds could also have contributed to inclining, flipping over, and even overturning some stars, stellar systems, and even galaxies (depending on the scale). Sometimes, layers of clouds in the atmosphere lead to the formation of billows, some of which have structures resembling spiral galaxies (but on smaller scales). Just as a long vortex which is stretched can spin up at a rate that can depend on the size of the larger bodies controlling the stretching, so also the size, rotational speed, and rotational period of galaxies, stars, planets, satellites, and even microscopic particles could have been affected by the characteristics of their vortices and the size of the fluid layers of their precursors. In the end, the chemical composition of the galaxies in the universe cannot be expected to the same. More information about the formation of galaxies and stars can be found in my books on the origin of the universe. I highly encourage you to check them out.

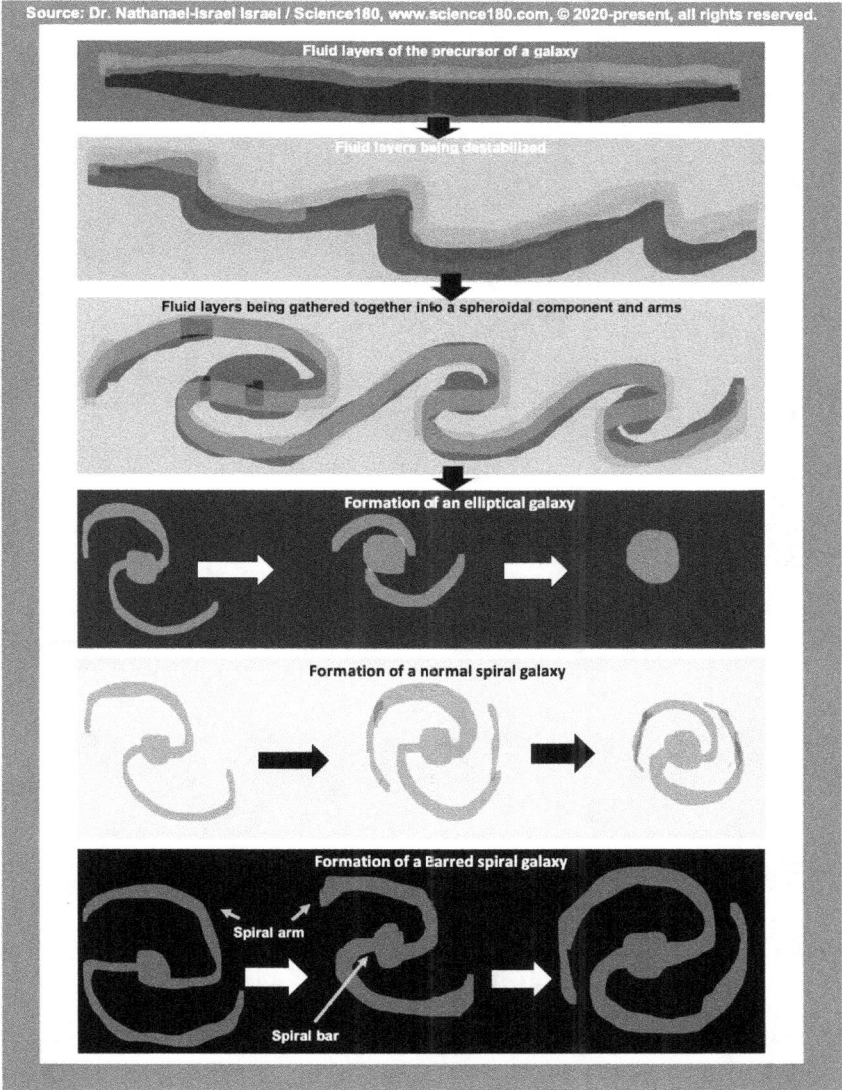

Fig. 139: Formation of elliptical, normal spiral and barred spiral galaxies

Science180: The Undeniable Scientific Challenge to All Metaphorical, Figurative, Loose, Liberal, or Vague Explanations of Chemicals-Origin

Fig. 140: Stars spread throughout the arms and disc of a galaxy

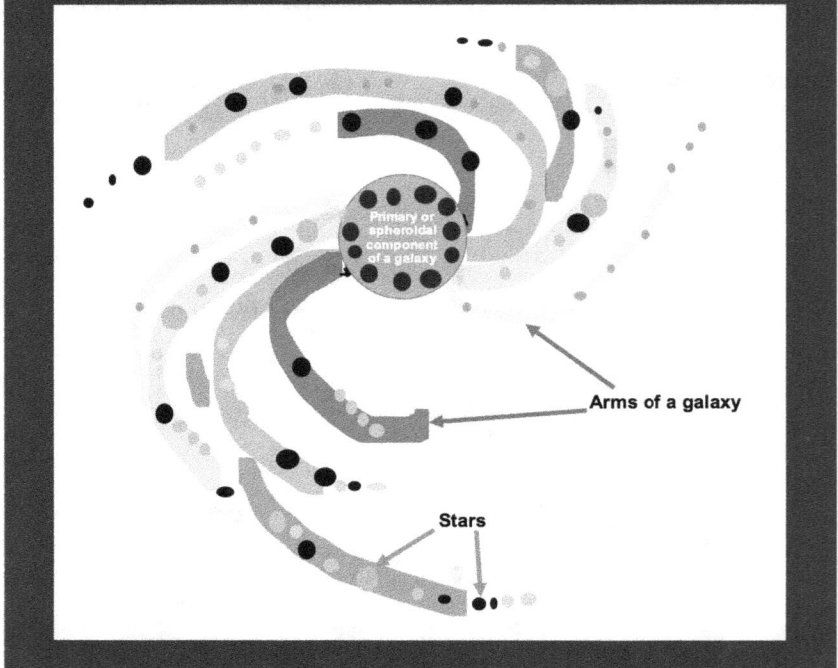

Primary or spheroidal component of a galaxy

Arms of a galaxy

Stars

Another Book by Nathanael-Israel Israel:
RECONCILING SCIENCE AND CREATION ACCURATELY

THERE IS ONLY ONE SIMPLE, COMPELLING, SOLUTION-DIRECTED SCIENTIFIC FORMULA ACCURATE ENOUGH TO RATIONALLY EXPLAIN HOW GOD CREATED THE UNIVERSE

"Reconciling Science and Creation Accurately" is a landmark book in universe-origin writing from a rare perspective by one of the most respected minds of our time. It scientifically explores the most challenging questions of all times that believers, nonbelievers, and all freethinkers are interested in: How can we rationally demonstrate, without checking our brain at the door in the name of faith, that God created the universe? How did the universe begin and what processes did God use to create it? Are these processes still operating in the universe or not?

Can believers abandon wrong theories if they think it is impossible for science to literally prove the Genesis story, or if they think that science is evil and diametrically opposed to faith, or if they compromisingly embrace scientific theories that contradict the Biblical account of creation written before the scientific era? What can believers do to help the skeptics believe in the Biblical narrative of creation?

Lucky you, Dr. Nathanael-Israel Israel successfully navigated all those questions with an accuracy that both scientists and nonscientists have been applauding across the globe. After reading *"Reconciling Science and Creation Accurately"*, you will confidently:

- Scientifically prove the Biblical account of the creation of the universe and the existence of God in a way that makes the head of those who deny God to spin faster than a DJ's turntable

- Know how to rationally talk to anti-creationists, evolutionists, Big Bang proponents, atheists, skeptics, and other freethinkers about the universe-formation and they will beg you to know more about God, the Creator, that they mistakenly rejected

- Discover very accurate, rare, and factual truths about the universe-origin that will save you time and money, and get you much closer to the better and joyful life you want to live today and forever

- Improve your health and faith by knowing that the existence of God can be scientifically justified using Science180 Cosmology and particularly Science180 Creationism

- Enter a new area of freedom and power by crushing the head of and breaking free from the suffocating expectations of all wrong theories that have highjacked secular and religious education, and that have held the Biblical account of creation captive for almost 3500 years

- Break free from the suffocating expectations of some forms of creationism that have sequestered the mind of some believers for a long time

- Uncompromisingly, intelligently, and scientifically explode the myth of those who, instead of literally taking the Biblical days of creation as 24-hours consecutive days, think that they were millions of years, or were representative of long ages, or that millions of years existed before them or were positioned between them

- Understand the accurate standard to interpret the Biblical account of creation thanks to Science180's breakthrough that transformed science and laid a foundational bedrock for the inerrancy of Scripture

Now that Genesis (the oldest manuscript in the world, written before science and most religions were born) is scientifically proven to be correct (Science180.com/biblical), what unstoppable, jaw-dropping paradigm shift will the discovery of the perfect alignment between science and the Bible bring into the religious, rational, and secular world today? Get this thoughtful book now to figure out what happened at the beginning, what is coming up, and why it is time to urgently rethink everything you have been told about the universe-origin so you don't eventually regret! Don't say nobody told you!

Founder of Science180 Academy, **Dr. Nathanael-Israel Israel** is acknowledged worldwide as the discoverer of the all-in-one, proven, and simple scientific formula that accurately cracked the origin of the universe, of life, and of chemicals, and that scientifically unearthed the holy grail at the intersection of science and the Biblical account of creation. Learn more at Israel120.com.

CHAPTER 29

CAN THE FUNDAMENTAL FORCES IN NATURE AND THE MATHEMATICAL ERRORS APPLIED TO CHEMISTRY TALK THEMSELVES OUT OF DOUBTING THE TRUTH ABOUT CHEMICALS-ORIGIN (IF WE DON'T DENOUNCE THEM NOW)?

29.1. Fundamental forces in nature

Before my discoveries about the origin of the universe, existing theories postulated that 4 fundamental forces exist in nature:

- the strong force also called the strong nuclear interaction,
- the electromagnetic force,
- the weak force also called the weak nuclear interaction, and
- gravity.

These co-called forces made me wonder:

- Are these interactions the only (fundamental) forces in the universe?
- Are they even worth calling forces?
- Do they really explain reality as seen in nature?
- What is their real explanation in light of my turbulence discoveries?

Because I was unsatisfied with the existing theories, I crafted a new theory on the fundamental forces in nature, using my discoveries about the mother of all turbulences that I detailed in my books on the origin of the universe. In those books, I extensively tackled other aspects of the real meaning of the so-called fundamental forces in nature, beginning with gravity. But, because gravity is much more concerned with celestial bodies, I did not detail it in this book. Therefore, if you want to learn more about the origin and real meaning of gravity, you may have to refer to that book.

In my book on the formation of the universe, I demonstrated that:

- Gravity is not responsible for the orbit of the celestial bodies and does not control the trajectory of the bodies;
- Gravity is not the process that caused the secondary bodies to orbit their primary bodies and it is not a force of attraction acting between all matters;
- Gravity is not responsible for the mass, the structures, and the development of the universe;
- Gravity is not responsible for the configuration and constitution of atoms and subatomic particles.

Although I will not detail the gravity of celestial bodies here, I will address why it could not have been responsible for the configuration and constitution of chemical particles. Indeed, like I proved in *"Turbulent Origin of the Universe"*, celestial bodies with the highest gravity are not dominated by the heaviest atoms. In fact, the greater the gravity, the lighter or smaller the dominant atoms. For instance, the Sun has the greatest gravity in the Solar System, yet, it is not dominated by heavy chemical elements. Similarly, although Jupiter is the planet with the highest gravity in the Solar System, it is not dominated by heavy elements. In contrast, most terrestrial planets that abound with heavy elements do not have the greatest gravity in the Solar System. The diversity and variation of the characteristics of atoms and subatomic particles and even of celestial bodies were defined by the impact of the process their precursor went through during and after their split gathering from their mother precursors and from adjacent daughter bodies of their mothers. Gravity is not the force which confers a certain configuration to microscopic particles.

Considering the scales of the systems of bodies in the universe, I think that gravity may end up not being the weakest known force in nature. According to the scale of interest (e.g. planets, stars, galaxies, clusters of galaxies, clusters of clusters of galaxies, etc.), other forces may exist and contribute to maintaining the force of the central body, or primary systems of bodies, or the core of systems of bodies in the universe with an intensity smaller than what can be attributed to gravity. The misunderstanding of gravity has caused some people to think that it played the most significant role in the internal properties of matter.

In my book on the origin of celestial bodies, I also explained how the mass of particles and celestial bodies cannot be properly explained without an intimate understanding of the process that split-gathered the precursors of the bodies in the universe. For instance, as particles were formed, some were clustered together to an extent that their clusters were significant enough to generate mass. The mass of a particle (small or big) was defined by the amount of the turbulent prima materia which was amassed and squeezed to yield it. The extent to which the original particles were gathered together into matter defined the density of the matter and system of matter. In other words, if the turbulent prima materia was not split-gathered into bodies under the influence of turbulence, nothing could have gained a mass and if the precursors of the matter and system of matter were not squeezed or compressed differently, the density of everything in the universe could have been the same. But because the bulk of the turbulent prima materia could not have stayed stable without being scattered, the series of split-gathering occurred and mass was differently gained by the clusters of particles on many scales.

In *"Turbulent Origin of the Universe"*, I demonstrated that gravity relates to the vorticial structures in the fluid flow during the formation of the celestial bodies. In fact, gravity could be the consequence of the vortices of the precursors of the celestial bodies during the turbulence that formed the universe. Then, after the formation of the universe, the impact was left as a gravity field around each celestial body. If the processes that formed the universe could be reversed and the celestial bodies unwound, the universe can be brought back to a state near that of its origin. For the bodies in the universe to be formed, a process was needed to split and pull together the precursors of bodies into different "compartments".

The way the fluid layers and the vorticial bodies formed in them were moved, sheared, revolved, rotated, tilted, and elongated had aggregated some precursors into bigger particles and others into smaller ones. Before some particles got their size, fluids in their surroundings were accumulated as they were moving until they reached a position where their volume could no longer allow them to be sheared or moved away any farther by the processes (e.g. flow) which displaced them. Therefore, these fluids were kind of "stationed" at specific distances which agreed with the shear rate, their speed, energy, etc. In other words, the force that was displacing the fluids was not able to move their bodies farther away beyond certain limits. Therefore, each particle was positioned at a location matching its size, movement, constitution, and the history of its precursor. Just as wind can easily move a leaf but not a heavy house made of concrete, so also when a big particle was formed in the fluid flow by the processes that displaced the precursors of the particles, the ability of the flow to position it depended on many factors including the energy or stress in the flow, the energy of the body itself. But if the particle contains more energy, it may not be moved by the flow beyond a certain distance. Hence, the largest particles are not usually very far from their mother precursors. Similarly, the energy that could have caused the shear flow could have pulled or pushed things in its way or into its surroundings.

Without a mechanism, process, or force to combine together or assemble the precursors of particles, most daughter particles could not have been able to form a body after their split from their mothers. The gathering together of the matter in the precursors of particles as they started splitting from their mother was not initially controlled by gravity. It was later that gravity and the other fundamental forces in nature were gradually established. This means that, before the so-called fundamental forces in nature were established, forces and interactions were already acting on the precursors of bodies and particles to help gather together their matters into unified bodies and systems of bodies. The fundamental forces known in nature today are like the matured or last state of expression of some of the forces that collected and dispersed matter in the precursors of bodies and particles.

Under the influence of turbulence, vorticial structures present in the fluid flow and/or fluid layers were progressively squeezed, amassed, and wound up into bigger particles as fluid layers were pulled and pushed while movements were being born. Some precursors of particles started rotating more than others. Fundamental forces were progressively established as the fluid layers of the precursors were amassed under the influence of their rotation, rotational-like or spiraling motion. The forces involved in the shear flow of the fluid and in their squeezing or compression acted on the precursors according to the intensity of their rotation and their size and consequently, affected the magnitude of the fundamental forces of these bodies. Some fundamental forces (e.g. gravity and nuclear force) and their precursors allowed the precursors of some bodies and particles not to be dispersed but to form unique or unified entities. Another way of explaining this complex process is that, as the precursors of the particles were being split and moved, a squeezing force started compressing and shaping them at the same time that other internal processes were also gathering together their constitutive clusters of particles inside of them. Because the intensity of the fundamental forces (e.g. gravity) depends on the position and movement of the bodies, their value as of today could not have been established before the bodies got their current shape. In other words, as the precursors of the particles were getting their shape, size, and speed (including rotational speed), the process of their gathering established the fundamental forces. Just as all bodies and particles today had their own precursors, so also all forces and interactions in nature had their forerunners. Similarly, gravity and the other fundamental forces had their precursors, which were embedded into how turbulence allowed the bodies to be split-gathered. By the time that the size and movement were set and the celestial bodies and their particles were put on their "final trajectory" or orbit, all fundamental forces including gravity were fully established.

Without a mechanism to match the fundamental forces with the size, movement, and composition of the bodies they are applied to, the systems of gathering together on all scales of the precursors could have highly compressed some bodies or particles but scattered or dispersed others. As they were moving inside the fluid layers of the precursors of the bodies and particles, vorticial structures could have pulled some bodies just as, even as of today, a rotating vortex can pull some matter. By the time the bodies or particles were formed, a zone of a seemingly "attractive" field was created around them and most things in that zone were forced to move toward that body. On a small scale also, all bodies have a field that was formed to exert a force upon matter to define and maintain the form of these bodies. Hence, even at the atomic and subatomic level, there is a field around particles. It is this field that allowed these particles to exist, meaning that without it, matter would have never existed.

Some fundamental forces are like the left-over of the force that defined the existence and organization of the celestial bodies and their particles. In *"Turbulent Origin of the Universe"*, I drew lessons from living things to explain this concept. As far as nonliving things are concerned, matter is equipped with a law allowing it to defend its environment. Atoms know how to maintain their integrity and not voluntary engage into reactions or interactions that will destroy them. And when a stranger enters into their environment, "frustrations" are expressed in the form of reactions, including radioactive reactions, which can hurt some nearby living organisms. Likewise, larger scales (e.g. satellite, planets, and stars), celestial bodies know how to defend their environment or propriety. When a stranger enters their environment, celestial bodies can attract or pull them toward their surface using gravity. In other words, gravity is a mechanism that celestial bodies use to maintain and protect their environment or property. The impartation of energy, motion, and other characteristics of anything in nature was not meant to be lost easily, nor the force to reverse these laws is cheap!

On the scale of subatomic particles and atoms, the process that formed and defined their limits also locked them into compartments in such a way that some theorists mistakenly think that those particles are attracting one another. In other words, just as some theories on gravity make people believe in an "action at a distance" (e.g. hypothetical attraction between a primary body and its secondary bodies), some people mistakenly believe that nucleons are attracting the electrons, and vice versa.

If the fundamental forces could be reversed for all bodies in the universe, matter may not hold together freely and the celestial bodies cannot maintain their state, and in the end, clusters of matter will be spread out in the universe into smaller particles similar to the first particle in the early universe. I also showed in *"Turbulent Origin of the Universe"* that, anything that can reverse the direction or sense of the rotation and revolution of the celestial bodies, meaning causing celestial bodies to rotate and orbit in a direction and sense contrary to what they are following today, could also reverse gravity and the other so-called fundamental forces. For the systems or processes that progressively established gravity were also connected to the ones that forged the rotation, revolution, and other properties of the celestial bodies and their particles.

At this point, I would like to summarize the generic establishment of the fundamental forces in nature: nuclear forces, electromagnetism, and gravity. Indeed, as a general rule, I demonstrated in *"Turbulent Origin of the Universe"* that, as the turbulent prima materia was pushed into movement, destabilized, and entered turbulence, different fields were birthed and formed across the compartments of matter that were changing. Some of these fields could have been the precursors of some interactions that were twisted by theorists to invent the fundamental forces. For instance, the processes that compressed or squeezed the subatomic particles and even the atoms may be perceived as forces between them. But in reality, no subatomic particles carry any force just as the presumed graviton is not responsible for gravity. In other words, although it can be ok to talk about forces inside atoms, I am not convinced that they are mediated by particles, but by the processes that determined how matter was formed. This way of thinking agrees with the explanation of gravity, and does not imply that celestial bodies are attracting one another. For instance, applying my hands to compress or squeeze something does not mean that the constituents of that thing in my hands are attracting one another. The interactions between human beings through love and sometimes hate are strong and may be perceived as a force of love or force of hatred although they are not mediated by chemical particles even if the settings of the human heart may play a role.

The so-called fundamental forces were not forged in one step or instantly. However, as the structures and characteristics of the precursors of the bodies (on different scales) in the early universe were changing, so also did the intensity and nature of their interactions between them, some of which would later be called fundamental forces. The characteristics of these forces and the things that mediate them evolved until a kind of equilibrium was reached and most of the constituents of matter (small and big) were locked into systems which dynamism is much smaller than what prevailed at the beginning of the world. In other words, the changes that occurred inside bodies or systems of bodies in the universe today is much smaller than those which occurred during the formation of the universe. Some of the modifications currently occurring inside the bodies and systems of bodies are not very significant and could not even be perceived of course, but many things in the universe are changing. The magnitude of those changes may depend on the size and localization of the bodies bearing them.

Concerning the electromagnetic force, for more than a century, it has been proven that the movement of electrons in a medium can generate electricity. Electric fields can induce a magnetic field perpendicular to it. This suggests that the movement of electrons around the atomic nucleus can generate an electric field, which can induce a magnetic field. Consequently, an electromagnetic force can be formed with a tendency of attaching the electrons to the atomic nucleus they belong to. In other words, the electromagnetic force found in atoms can be the manifestation of physical forces arising from the movement of electrons and which tend to "bind" the electrons to or around the nucleus. Unlike what people think, it is not the protons that are mediating the electromagnetic force in atoms.

As I carefully thought about the interactions between the subatomic particles, I realized that, bosons may not even be responsible for a fundamental force.

Current estimations of the so-called fundamental forces in nature suggest that electromagnetism is stronger than the weak force, which at its turn is stronger than gravity. Although some scientists think that bosons are responsible for or are the carriers of the nuclear force, I do not believe in that. As for me, the fundamental forces were caused by the ways the precursors of particles (of various sizes) were compressed by the turbulence that took place during the genesis of all matters. Without the impact of turbulence on the precursor of matters, bosons and other particles could not have been compressed to force dense matter. Bosons found in nuclei are just particles sandwiched between others by the processes that split-gathered particles. For instance, if I can squeeze the Solar System tightly, some innermost planets and asteroids may be like a trace of matter. Similarly, some subatomic particles found in the nucleus are just sandwiched there by the process that compressed the nucleus. In *"Turbulent Origin of the Universe"*, more details about the fundamental forces are provided as well as how large scales of vorticial structures were aggregated to form astronomical systems of bodies.

Another Book by Nathanael-Israel Israel:
ORIGIN OF THE SPIRITUAL WORLD

ONLY ONE ANCIENT BLUEPRINT HAS THE RELIABLE POWER TO HELP YOU TO ACCURATELY DECRYPT THE SPIRITUAL ORIGIN AND HISTORY OF EVERYTHING IN THE UNIVERSE

Countless books talk about the origin of the universe and of life, but this amazing book is the first and the only one that has undeniably explained how the formation of the universe and everything in it was truly revealed in the rejected and hidden scriptures such as the Books of Enoch and others. In *"Origin of the Spiritual World"*, you will:

- Discover deep rejected secrets that have prevented humankind from unearthing the beginning of the universe
- Plainly see the scientific proof (hidden in scriptures) of the formation of the Earth, the Moon, and the Sun in a matter of days, a historic revelation that bizarrely and shockingly matches the scientific data as scientifically proved in *"From Science to Bible's Conclusions"*, a popular book written by Dr. Nathanael-Israel Israel
- Properly use the lost and rejected scriptures to articulate the process by which the universe was formed, and use that insight to improve your understanding of the Bible, innovate in your domain of interest, and improve your life perpetually
- Empower and align yourself with the historic breakthrough that has done what no other discovery has ever done: accurately unlock and decode mysteries concerning the origin of the cosmos and its content using scientific keys revealed in ancient scriptures that some elites have concealed (Science180.com/pseudepigraphic)
- Discover and apprehend the complex formation of the universe and life without leaving out the challenging questions that people of all ages have been struggling to answer for thousands of years, while the answers were hidden
- Find more joy in life through a clear interpretation of old and fresh revelations about the creation of the universe astonishingly backed by modern science, which some people wrongly think opposes the Bible
- Make a difference and blaze new trails for those who depend on your leadership

If you believe in God, have some origin-related questions which answers you cannot find anywhere, not even in the Bible, and if you want to tap into historically neglected revelations to answer fundamental universe and life questions, then be sure to get a copy of *"Origin of the Spiritual World"* today.

Dr. Nathanael-Israel Israel happens to be the discoverer of the historic mathematical equations that scientifically demonstrated that the Earth was formed 2.82 days, the Moon 3.32 days, and the Sun 3.69 days after the beginning of the universe, therefore confirming the Biblical account of creation that revealed about 3500 years ago that the formation of the Earth was completed on the 3rd day, while that of the Moon and the Sun was completed on the 4th day of creation. Nathanael-Israel Israel is referred to as the "Undisputable Specialist of all Questions at the Intersection of Science and Biblical Creation". Learn more about this rare scientist at Israel120.com.

29.2. Fundamental mathematical errors applied to chemistry

During my investigation, I understood that scientists have made many mistakes in their efforts to understand the universe. Here, I will use statistical mechanics and quantum mechanics to explain how some scientists have erred by deducing the characteristics of a body from the study of its constituents. Indeed, although the aim of statistical mechanics is noble, its approach of *"predicting and explaining the measurable properties of macroscopic systems on the basis of the properties and behavior of the microscopic constituents of those systems"* (Encyclopaedia Britannica, 2018; Gordon, 2018) is highly misleading. For bodies (small and big) in the universe do not always behave just according to the characteristics of their constituents. Bodies in the universe interact with their neighbors differently and those interactions do not just depend on the individual properties of their constituents. For instance, the properties of a nation do not always depend on the properties of its population. How nations interact with one another for instance does not always reflect the viewpoint of their population and vice versa. Although nations can elect their president to represent them, it is frequent that elected officials do not always work for the best interest of their constituents. Similarly, depending on the conditions, the properties of a substance can be significantly different from those of its constituents.

The way the constituents (cells, tissues, organs, and systems) of an organism interact with one another is not the same way that the organism itself interacts with other organisms in its neighborhood. The proprieties of an organism are different from that of its constituent cells. That is partially why, to avoid the harsh stress from their environment, cells, tissues, organs, systems, and organisms protect themselves with membranes. A membrane can allow internal reactions to occur without too much affect from external factors, which are not always 100% avoidable. For instance, the skin that protects the external parts of a human being does not have the same characteristics as the membranes that surround cells. A living cell can die immediately in an environment where the organism from which that cell was extracted can live for days. In other words, in circumstances where an organism can survive very harsh stimuli, a cell can die immediately. For example, a living cell exposed to heat or cold can die immediately, whereas a human being from which that cell is extracted can live in that condition for days without dying. A human being can live in a desert like that of the Sahara and Arizona for days, even without eating or drinking anything. However, a living cell extracted from that human being and exposed to that desert environment can die immediately, suggesting that the properties of microorganisms cannot always measure those of its constituents very well. From a nonliving viewpoint, the way satellites interact with their primary planets is not the same for all planetary systems. The way the constituents of each chemical element interact with one another and with other chemical elements is not always similar.

To try to explain the same concept differently, most countries have borders that define their territories. Cities and villages inside a country also have limits. However, the characteristics of a national border are not always the same as those of the internal borders (e.g. city limits, village limits). National borders can have border patrols, military presence, and strong walls to prevent outsiders from entering into that country "illegally". Similarly, how each city or village within a country defines and defends itself is different from how a country does. In other words, the understanding of the constituents of a body (living and nonliving) is not enough to properly understand the properties of the body itself. Similarly, the understanding of the properties and behavior of the microscopic constituents of a system cannot always be enough to understand the macroscopic properties of that system itself. As particles interact with others, they can behave differently in response to the stimuli of their environment. This is a kind of "adaptation" that matter has and which can explain why it is almost impossible to replicate some experiments in nature and get the same results. For environmental conditions are changing all the time and because, unlike what people think, the laws that govern nature are not always the same everywhere, and the results of reactions should not be expected to remain the same. The preciseness, sensitivity, calibration, and details of the laws that were used to form and sustain nature can never allow getting the same results from significantly different reactants. Otherwise, the universe and its constituents would not have been diverse. Consequently, many experiments, for instance in flow instability and turbulence, usually yield different results despite scientific efforts to make them repeatable and reproducible. I cannot understand why some scientists always want their experiments to be repeatable and reproducible, and why some are very frustrated when they cannot achieve that goal. Why should we keep expecting the behavior of the world to conform to what we want or to what we are familiar with or to what we "found" in the past? The business of scientific research is not supposed to be about collecting data and describing the microscopic and macroscopic world according to how scientists imagine it or would like it to be. Science may be better off focusing on facts instead of on the belief and hope (scientific hypotheses) of some scientists who, sometimes, claim that they are not believers or religious believers. Yet, some scientific hypotheses and theories require more faith than that expected from some religions. What will I say if I can elaborate on the refutability and palpability criteria that some scientists evoke while addressing their methodology of research?

It is the wrong assumption that natural laws are the same everywhere that causes some people to reject or to be shocked by some data in quantum mechanics, which, as Encyclopedia Britannica said it, *"frequently conflict with common-sense notions derived from empirical observations of the everyday world"* (Encyclopaedia Britannica, 2018; Gordon, 2018). Why should we expect the behavior of the atomic world to always conform to that of the familiar, large-scale world? At the same time, some scientists do not want to realize that, just as it is impossible to make omelets without breaking eggs, it can also be impossible to use physical means to try to understand details on the supra microscopic world without breaking certain things. Therefore, some of the scientific results collected in nuclear physics for instance can be the remnant of a demolished house instead of the house itself. For some scientific methodologies destroy the subject they want to investigate. Similarly, some data collected on the macroscopic large scale allude to a half blind person trying to get a deep insight into a night sky. Therefore, scientists should be very tolerant, modest, humble, nondeterministic, and careful at the way they interpret their data. For most of what we think we know are partial and insignificant compared to the unlimited data hidden from a mere human comprehension.

Therefore, not fully understanding the underlying laws that can explain the variation in their results, some scientists just think that some natural processes are random or connected to hazard or chance. In reality, randomness can be a mysterious code that testifies of the inability of human beings to ever fully comprehend natural laws and use them against themselves and nature. For in general, what human beings don't comprehend, they call it by various names, but what they comprehend, they usually destroy or replicate it for their selfish desires, while usually neglecting others, nature, and God the Creator of all things, who many people refuse to acknowledge. If some mysteries revealed in my discoveries were known long ago, let's suppose 2000-3000 years ago, when Greek thinkers like Pythagoras, Plato and his teacher Socrates, and Aristotle (the most famous student of Plato), were laying the foundations of philosophy and science (or let's say the forerunner of science), this world could have already been destroyed by some crazy actions or reactions (e.g. nuclear bombs that some countries are trying to prevent other countries from developing, while themselves have tons of them). Can you imagine how this world would have been if ancient people who live thousands of years ago knew details I revealed in my books on the origin of the universe? That is why, knowing what I am pushing into the world via my books, I know the end of this world is very near. For, using some codes I revealed in my books, a lot of things which used to be considered very complex will be broken down and resolved very fast.

One of the huge mistakes scientists have made and which put science in a big mess is the belief that natural laws are constant. Therefore, they created many things called "fundamental constants" of the universe that they tried to use most of the time to justify their theories. Although their data does not add up, some scientists always try to force them to say what they want to hear, not knowing that by doing so, they are simply throwing away what would expand (instead of shrinking) their understanding of the world. Based on my findings, the laws of physics are not the same everywhere and throughout history. Things are changing and will continue to change until the end of the world that none can comprehend unless by referring to some religion.

In other books, I explained how the conventional mathematical operators have also created big problems, which have been preventing the understanding of reality. This is one of the reasons I have been working on the new mathematics that can better help to grasp reality.

'Science180 Academy' Success Strategy:
SCIENCE180 INTERVIEW REPORT (AKA SCIENCE180 INTERNET-TV-RADIO INTERVIEW REPORT)

Science180 Interview Report is the newsletter to read for guests and unconventional show ideas at the intersection of science and faith. Indeed, many hot questions are still unanswered on the road leading to the correct understanding of the origin of the universe, of life, and of chemicals. But most people don't know where to find the accurate answers to those challenging questions. What if with one simple call you can accurately answer all of those questions. You need to get in touch with or interview Dr. Nathanael-Israel Israel on your show, radio, tv, podcast, and even website, or invite him for a live presentation at your organization if your audience can benefit from any of the following show, talk, speaking, or interview ideas:

- Can a single variable play a crucial role in cracking the universe?
- Can the distance separating celestial bodies give a clue to the universe's birthdate?
- What is the master key to crack the universe-origin and chemicals-origin?
- How to deal with the fear of not knowing the origin of the universe, chemicals, and life?
- Can you be really free from doubt about the universe-origin and chemicals-origin?

- Can Mathematics and Science collide to accurately explain the creation of chemical particles?
- Can mathematics help Science to crack the code of the universe formation and chemicals-origin?
- Can mathematics help science to crack the code of the origin of chemical particles?
- Can we scientifically demonstrate without a doubt that the Moon and the Sun were really formed on the 4th day of creation like the Bible says?
- Can mathematics and science rescue Christians in their efforts to rationally prove the existence of God, the Creator?
- Did the Quran and any other religious book make any gigantic error about the universe creation that any scientific formula proves the Bible got right?
- How to scientifically crack the code of turbulence
- How can scientists know which theory to trust to understand turbulence?
- How can people benefit from understanding the mother of all turbulences?
- Can anyone really be scientifically 100% sure and prove that God created the universe using physics and chemistry?
- Can we explain the formation of the universe and its chemicals through natural processes without evoking evolution and Big Bang?
- Is it a waste of time to attempt to prove the Biblical creation using chemistry or historical investigation?
- Is there any need to prove the Biblical creation to be true?

I know you may be tempted to answer these questions by yourselves, but avoid landing yourself on wrong paths that caused some people to lose contact with reality, it is better to get the accurate answer from the know-how expert, Dr. Nathanael-Israel Israel, the author of many books on the origin of the universe, of life, and of chemicals, and the standout expert who accurately decoded the scientific formula that forces science to bow to the truth. If you would like to register to Science180 Interview Report to periodically receive show ideas and opportunities related to the origin of the universe, of life, and of chemicals particles, please visit Science180Interviews.com for more details. If you want to invite Nathanael-Israel Israel to your organization to hear his answers to any of those questions, please contact him at Israel120.com.

CHAPTER 30

WHY SMART SCIENTISTS ARE ABANDONING ALL EXISTING THEORIES ON CHEMICALS FORMATION TO EMBRACE "SCIENCE180 MODEL OF THE ORIGIN OF CHEMICAL PARTICLES"

For the first time in history, the findings presented in this book cracked the code of the formation of chemical particles. I showed that the process that birthed the chemical particles is similar to the one that birthed the celestial bodies in the universe. Hence, a lot of similarities exist between the proprieties of the chemical particles and celestial bodies in the universe. I showed that, in the beginning, an original particle that I called turbulent prima materia mysteriously appeared in the universe out of nothing, and then through very complex, turbulent processes, it was molded into all types of matter or bodies in the universe today. The bulk of the initial matter was huge and occurred a wide and deep portion of space. That original matter was like an "undetermined" particle, which had the potential of becoming anything, and was programmed to yield different types of matter depending on the conditions that subdued it. The turbulent prima materia went through fragmentations simultaneously associated with modifications of the characteristics of its daughter bodies. The state of the turbulent prima materia was none of the major states of matter known today (e.g. solid, liquid, gas, and plasma). It was progressively that the original matter went through complex processes that transformed it into the current states of matter.

Soon at its appearance, the turbulent prima materia was broken open by a violent explosion that I termed the original mysterious scattering. In other words, the original mysterious scattering split the turbulent prima materia into pieces or blocks of matter, exposing its content and propelling the broken pieces into various directions. As they started moving, the clusters or pieces of the turbulent prima materia went through turbulent changes and processes of differentiation including breakup or fragmentation, squeezing or compression, and others, which constitute the basis of the story in *"Turbulent Origin of the Universe"*: fluid flow and instability, split of fluid bodies, formation of the precursors of bodies; birth and transfer of energy; initiation of movement (e.g. revolution and rotation), formation of fluid layers, transfer of momentum, initiation and development of turbulence, fluid breakup and mixing, separation of fluids, fluid gathering, squeezing of precursors, birth and strengthening of fundamental "forces", etc. Although I detailed these processes in *"Turbulent Origin of the Universe"*, here, I focused on aspects relating to chemical elements.

During its movement, the initial matter reached a state of fluid. These fluids were organized into layers, which interact with one another. As the fluid layers were moving under the influence of the original mysterious scattering, they were destabilized and their instability quickly developed into a major turbulence. Vorticial structures of various sizes were born in the precursors. The turbulence occurred at different intensities and on various scales: subatomic scale, composite particle scale, atomic scale, molecular scale, mineral scale, rock scale, and celestial body scales. In my books on the origin of the universe, I addressed the turbulence on the scales of celestial bodies. In this book. I focused on the scales of chemical particles only. I showed that the precursors of chemical particles were shaped by turbulence, which caused them to spit-gather into precursors of primary particles and precursors of secondary particles. The same process also occurred in the precursors of the celestial bodies by split-gathering them into precursors of primary bodies and secondary bodies. The precursors of rocks, minerals, mineraloids were split-gathered into the precursors of chemical compounds, molecules, and atoms, which split-gathered into the precursors of subatomic particles, which split-gathered into the precursors of the particles smaller than subatomic particles that may never be scientifically discovered. In other words, as the precursors of minerals, mineraloids, and rocks were being split-gathered, precursors of composite particles such as atoms, molecules, and chemical compounds were formed. The precursors of some atoms and molecules were split-gathered differently to form specific atoms some of which are not ordinary or conventional atoms, but exotic ones, having characteristics not seen in ordinary atoms on Earth.

The precursor of the system of secondary bodies in each atomic system was split-gathered into electrons placed around the nucleus. I showed that the precursors of the electrons escaped the precursors of the nucleus of the atoms they belong to. Just as some asteroids and planets have no satellite, so also some chemical particles have no electron "orbiting" their nucleus. For instance, electrons and other types of leptons (e.g. muon) can also be found in nature not around a nucleus, but in free and isolated forms. The precursors of some particles could not have been initially organized into layers but were just formed as the tube or filament or ligament of their precursors were split-gathering. The fluid layers did not have the same size, thickness, or depth, or speed. Some could have been tiny and others deep. Some layers moved faster than others. The top layers could have had a speed higher than that of the bottom layers. The speed of the fluid layers in the precursors of the chemical particles was later translated into orbital and rotational speed of the particles. The direction of the fluid flow of the layers could explain the direction of the movement of their daughter chemical particles.

The particles were molded according to the "impartation" of proprieties they received from their mother precursors. As I demonstrated in *"Turbulent Origin of the Universe"*, the fluid layers of the precursors of bodies in the most developed turbulences can be divided into regions that I called turbulence zones. Similar turbulence zones existed for chemical particles. Considering what I did for the celestial bodies, the fastest electrons may be found in Zone 1 of atoms. Likewise, the largest electrons may be found in Zone 3. In other words, I do not think that the electrons size or cloud will be the same for all atoms. The difference in the turbulence of the precursors of the electrons could explain why all electrons are not and do not behave the same. In the same manner, the slowest electrons could be in Zone 5 which are the outermost electrons for the atoms which precursors had a fully developed turbulence.

The turbulence that each precursor of bodies went through after splitting from their mother precursors defined the "final" chemical composition of their daughter bodies. Therefore, the similarities and differences between the chemical composition of the bodies in the universe is due to the processes that molded their precursors. One of the last steps of the formation of chemical particles was the gathering together of the fluids in the layers of their precursors into a unique body or system of bodies after their separation. The organization of the fluids of the precursors of the particles as they were moving under the influence of the turbulence caused them to collect into specific bodies. The chemical particles were not collected the same. Some were wrapped up tightly, while others were gathered together loosely. The precursors of electrons around the nucleus could have also been organized into tiny layers according to the scale of the turbulence that occurred at their level.

Among all the variables I studied on electrons, ionization energy was the one that gave me a better insight into potential behaviors of fluid layers of electrons around the nuclei according to their potential position. Unlike celestial bodies where layers of precursors of secondary bodies were mostly converted into secondary bodies orbiting their primary bodies, it is the case for electrons, the fluid layers of their precursors could not have always been fully wrapped around or gathered together to form unified electrons. The inability of some layers of the precursors of electrons to collect together into a single body could explain why electrons are postulated not as orbiting the nucleus but spread over an electronic shell. I also showed that, it is possible that even in the electronic shell, some electrons may be surrounded by some smaller particles, which belonged to their mother precursors and which were not compressed or collected together with the electron itself. In other words, even in an electronic shell, the size and particles of the constitutive bodies may differ according to their position with respect to the nucleus of their atoms. Considering the properties of electrons and nucleons found in atoms, it may not be fully appropriate to view the precursors of all atoms as fluids like liquid or gas, but as a kind of matter between fluid and waves. It is to simplify the presentation of my writing that I adopted the term fluid layers when I talked about atoms and subatomic particles. In other words, when I talked about fluid layers of the precursor of electrons, nucleons, or atoms, understand that I am not necessarily talking about conventional liquid and gas.

As some fluid layers originally stacked one on top of the other were separating, a space appeared between them. Putting it another way, after their separation, 2 originally consecutive fluid layers became surrounded by a space. Above the space between them could have been the layers which were on top and below the space between them was the fluid layer that was below. Some fluid bodies of the precursors could have leaked into these layers, therefore causing some spaces to be filled with particles while others can be "empty". The number of particles between bodies in space could have been defined by the amount of fluid that was leaked into them or which landed into them. By the time the bodies of fluids were gathered together, the space that appeared between them became the space where the secondary bodies of the fluid layers were born. As the fluid layers of the precursors of the secondary particles were going through their own split-gathering into their daughter secondary particles, additional space was created between them. These new spaces became the precursors of the space around their secondary particles.

Unlike some celestial bodies that have gigantic atmospheres around them, atoms and subatomic particles do not have much space around them. Yet, when they were being shaped, "empty voids" were formed between their constitutive particles. In the case of atoms surrounded by electrons, the space around the nucleus where the electrons are located is called "electron cloud". For each particle, the size of that space defined the size of what could be called "atmosphere" (if they were celestial bodies), but which in the case of microscopic particles is the space between the electrons and the nucleons. Just as the atmosphere around a celestial body is under the influence of the later yet all the space between 2 celestial bodies is not always under the influence of neither of them, so also between 2 atoms or subatomic particles, a space near them can be under their respective influence but another space existing between them is not controlled by either of them. The length of the space between atoms and subatomic particles can affect their "bonding" to one another and their reactivity with others.

As the precursors of particles were being assembled and their movement set, they were dynamically led into their orbit or orbital, where they will spend the rest of their "lifespan". This implies that, when the particles were being formed, the position of their precursors was changing in space and at the end of their formation, the position of their "orbital" movement inside their system was "locked down". The time it took to position the particles at their position in orbit or orbitals could have been affected by the longitudinal stretch and detachment of the ligaments of the precursors from their mother.

The precursors of the subatomic particles were diverse according to their size, movement, speed, and location. Just as the precursors of the celestial bodies, the precursors of the subatomic particles could have been at one point organized in tiny layers or tiny pockets of matter, which, upon tightly folding, wrapping up, winding up, packing, or spiraling, yielded the plethora of subatomic particles according to their characteristics. The characteristics of the subatomic particles depended on where and how their precursors were shaped. The size, mass, or radius of the subatomic particles depend on the amount of matter in their precursors as they were compressed or gathered into their daughter bodies. Some particles found in the nucleus of atoms (e.g. protons, neutrons) could be supercoiled structures of the turbulent prima materia separated by other subatomic particles. Hence the density of some subatomic particles is very high. The difference in the size, winding up, spiraling, coiling, supercoiling, and twisting of the precursors of the subatomic particles may explain the difference in density. Because the packing of the subatomic particles was differently done from one type of subatomic particles to another, some subatomic particles were tightly packed, while others were loosely packed according to the environmental conditions of where the packaging occurred. From one celestial body to the others, the subatomic particles could have been tightly or loosely packed differently. Subsequently, the atoms and the compounds made from them could have also been organized differently. Some bosons which are known to be very heavy (e.g. Higgs bosons) could have been tightly packed hence their higher density and mass. In contrast, electrons and neutrinos could have been loosely packed hence their smaller mass.

The diversity of the turbulence that the precursors of the atoms and subatomic particles went through outlined the proprieties of their daughter bodies. For instance, the distribution of electrons around the nucleus of atoms would depend on the characteristics of the precursors of the atoms. The configuration or the organization of particles is not static, but dynamic and can be affected by the environment. When different subatomic particles are placed in the vicinity of other particles, they can change the configuration or organization of one another. Some of these changes are responsible for some chemical reactions. Similarly, on the nuclear scale and much smaller scales, when different particles are forced into the environment of other particles, the nature and characteristics can change depending on the amplitude of the induced changes. Electrons around a nucleus were not placed by chance, but according to the turbulence which split-gathered their precursors into electrons differently positioned. I showed that a relationship exists between the size of a primary body and the maximum size of a secondary body which can orbit it. When the relationship is broken, the structure of the primary body and the secondary body forced to orbit it can change. It is a similar problem that protons, muons, and electrons are facing in the proton radius puzzle. And because the physics of the formation of the universe and its constituents has not been understood before my discoveries, the error has never been corrected before. When an electron is replaced by a muon (which is 100 times heavier than an electron), the characteristics of the atom must change to account for the changes of the law. The revolution of a muon around a proton must change the way the proton is wound up for the electron and proton constituting a protium for instance were being gathered together as the protium itself was being formed. Meaning that the action of the movement of the electron around a proton is equilibrated with the characteristics of the proton. When the electron is replaced by a muon, the proton can be squeezed more, hence the decrease in the radius of the proton orbited by a muon.

In the precursors of the celestial bodies, larger clusters of atoms were split-gathered into aggregates of one or more compounds held together by bonds. Some of these clusters of compounds are called minerals. At the same time, the precursors of rocks embedding the precursors of minerals also went through changes which shaped them. The variation of the changes that the precursor of rocks, minerals went through defined the diversity of these particles today. I showed that the chemical particles could have been formed within a few days. For, if despite their gigantic size, the celestial bodies were formed within days (as I extensively demonstrated in my books on the origin of the universe), it is possible for the chemical particles to have been formed within days as well.

Using the formula and the abundance of the chemical compounds present in the atmosphere of the planets, and the characteristics of the planets, I was able to understand how the environmental conditions during the formation of chemical particles affected the split of their precursors into entities and how the resulting products could have been combined or bonded to others to yield different forms of matter (e.g. particles, atoms, molecules, and compounds) according to the scale of their precursors. For instance, in Earth's atmosphere, some oxygen atoms combined with atoms of other chemical elements to form different compounds. Indeed, as atoms were mixing after their formation, some oxygen atoms associated with carbon to form CO_2. Other oxygen atoms associated with hydrogen atoms to form ordinary water (H_2O). At a lower altitude, oxygen atoms associated with themselves to form oxygen gas (O_2), while in the upper altitude they associated with one another to form the ozone (O_3). I demonstrated that the Earth and the Moon have a similar bulk composition and also share the same isotopic fingerprint because they both descended from a common precursor: the precursor of the Earth-Moon system, which split gathered to birth the precursor of the Earth and the precursor of the Moon. The difference in the composition of the Moon and the Earth can be related to the additional process that the precursor of the Moon and Earth went through after splitting from one another.

I would like to thank you for the time and consideration you put into this book, and for reading it until this point. By reaching this stage, you have shown that you really want to understand the underlying factors of the formation of the chemical particles in the universe. If you really read this book in its entirety to this point, I am confident you got a new insight about the origin of chemical particles.

CHAPTER 30: CONCLUSION–SCIENCE180 MODEL OF THE ORIGIN OF CHEMICAL PARTICLES

If there is any question you have that I did not address, I would like to hear from you. If you like the book or have any comments, please let me know. If you are interested in knowing more about the books I wrote on the origin of the universe, you can visit www.Science180.com, the website associated with my books where I will also be posting updates and more resources. If you want to be in touch with me and get regular updates, you can visit my personal website (www.Israel120.com).

NEXT STEPS OF THE JOURNEY

Get free resources on Science180.com

If you have finished reading this book and would like to learn more about my discoveries and how they can help you, you are at the right place. Indeed, I am really committed to helping you address any questions that you may still have concerning the origin, functioning, and fate of the universe, and how you can partner or collaborate with me for greater results.

To get free resources that will help you understand other aspects of the universe formation not covered in this book, visit Science180.com and my personal website Israel120.com. On those sites, I will be sharing guides and strategies to get the most out of my initiatives. I will also be sharing my favorite references, tips, next-steps readings and other important things in the pipeline that will help you regardless of your field of expertise, interest, and needs.

Subscribe to "Science180 Newsletter": The only accurate universe-origin, life-origin, and chemicals-origin newsletter in the whole world!

Be a part of decoding the universe-origin, life-origin, and chemicals-origin! Get origin-related news, information, discoveries, updates, announcements, news, reviews, articles, educational materials, and opportunities, from a holistic perspective not available anywhere else so you can participate in and enjoy decoding the origin, current state, and fate of the universe and its content. You will also receive priceless tips about how Nathanael-Israel thinks, what are his secrets and initiatives, what he has accomplished, and what he recommends. Without any delay, sign up for Science180 Newsletter today at Science180.com/newsletter. It is free!

Speaking engagement

In addition to writing groundbreaking books and engaging in other business endeavors, Nathanael-Israel Israel is a renowned speaker, who you can invite to speak at your organization.

Values that Dr. Nathanael-Israel Israel can add to your life include:

- Unquestionably rare expertise and tips that will increase your abilities
- Usefulness that will advance your impact regardless of your field of expertise
- Understanding of the world that will sharpen your perspective

- Critical information that will positively change your life, will save you time and launch you into a zone of unlimited opportunities
- Experiences turned into insight that will motivate and guide you
- Enlightenment that will help people to start using their brain instead of just praying and expecting God to do everything for them

For speaking inquiries, including how you can get Dr. Nathanael-Israel Israel to speak to your organization or at an event, visit Science180.com/speaking for more details.

As the standout scientific authority who accurately decoded the universe, Nathanael-Israel Israel has been helping countless people across the globe to discover and understand the complex origin of the universe without leaving out the challenging questions that people of all ages have been struggling to answer for thousands of years! As the true go-to expert when it comes to the formation of the universe and of life, Nathanael-Israel believes that, regardless of age, background, culture, religion, profession, everyone deserves to understand how the universe and life were formed and how they can leverage on that knowledge to improve lives nonstop. Therefore, his groundbreaking discoveries of the formation of the universe, life, and chemicals have been broken down into books tailored to scientists (including physicists, chemists, biologists, mathematicians), laypeople or general public, believers and freethinkers, philosophers, children, etc., therefore maximizing the benefits to humanity. These historic, internationally-acclaimed origin books include:

- "Turbulent Origin of the Universe"
- "Reconciling Science and Creation Accurately"
- "Turbulent Origin of Chemical Particles"
- "From Science to Bible's Conclusions"
- "Turbulent Origin of Life"
- "Origin of the Spiritual World"
- "How Baby Universe Was Born"
- "How God Created Baby Universe"
- "Science180 Accurate Scientific Proof of God"

When you hire Nathanael-Israel Israel to speak at your organization, you will:

- get specific in-depth knowledge, up-to-the-minute information, ideas, and insights about the universe-origin, life-origin, and chemicals-origin so that you expand your market, cut useless costs, stop wasting time on inadequate projects, and start focusing on the profitable solutions
- get relevant universe-origin stories that are specific to your field of expertise

- learn from a cooperative, flexible, and an easy to work with expert who will respond to your universe formation needs and position you to stay on top of your competitors
- interact with a renowned expert that will not just lecture you, but that will help you sort out your origin-related questions using strategies to tap into deep secrets you ignore
- listen to an experienced expert who discovered outstanding secrets about the origin of all there is
- learn authentic information not from someone who just reads you a PowerPoint, but from the true go-to expert (when it comes to critical cosmological problems) who will share with you both his mistakes and successes that will help you get much closer to the better life you want to live
- revolutionize every origin-related domain with your accurate understanding of the universe-origin
- hear Dr. Nathanael-Israel Israel's personal selection and teaching of key topics that will help you break the code of the universe formation and functioning, and strategically enlighten you, guide you to navigate and filter the massive data collected on the universe and its content so you know how to answer the world's most challenging origin questions, remove any scientific and philosophical cataracts that may be blocking you, and help bring you many steps closer to your best life today and forever
- hear the greatest scientific and philosophic lessons of some top scientists, philosophers, thinkers, and public figures who have realized historic mistakes they made in life (concerning the origin of the universe, life, and chemicals), and that they corrected thanks to the discoveries of Nathanael-Israel Israel, who founded Science180, and who is acknowledged as the scientist that truly decrypted the universe-origin for the first time
- Get world key lessons successful people have learned in life and how people can learn from their experiences to improve lives instead of repeating their mistakes that many people still ignore at their own perils

To book Dr. Nathanael-Israel Israel for a speaking engagement purpose, visit Science180.com/speaking.

How you can make money by joining the affiliate program to sell Nathanael-Israel Israel's books

Greetings,

Do you want to make easy money by selling the #1 universe-origin, life-origin, and chemicals-origin books on your website, newsletter, and by mail? You can start making big money as you help sell Science180 Books including this one on your website and network. Indeed, by now, you know that I operate a website called Science180.com, specialized in helping people across the globe to scientifically decode and understand the formation of the universe, life, and chemicals.

Your contacts, site, blog, forum, podcast, and newsletter may be admired among my target audience. Some of my products and services may be of interest to your audience. My books are the first in history to scientifically demonstrate the match between science and Biblical creation in a way that satisfies both believers and nonbelievers, a historic achievement and discovery that is revolutionizing our view of the origin of the universe, life, and chemicals for the benefit of humankind.

Imagine you have a website where you can talk to people about my books and services and get a great percentage of every purchase they do on my site? Imagine you send a certain link about my books to your friends or network and, when any of your contacts buy a copy of my books, you get a percentage or a certain amount of what they pay on my sites. Imagine you can email your friends and spread the good news about my books and when anyone uses that link to buy my books, I give you something. Well! This is what the affiliate program is about. Apply today or learn more about it at Science180.com/affiliate. Likewise, if you own a website, you can apply for Science180's affiliate program, and I will send you a specific affiliate link that you will place on your website and newsletter, and if people click on it to buy my books, they will be led to my page and after they buy, I will pay you a certain amount, sharing the profit with you instead of just verbally saying thank you.

Would you be interested in reviewing some of my products and services with the aim of becoming an affiliate? We have a wonderful affiliate program and commissions are paid quickly and accurately.

If you are satisfied by the quality of our products and services, I am convinced you will also be impressed by our affiliate program.

I look forward to hearing from you

Nathanael-Israel Israel, PhD

Collaborate or partner with Nathanael-Israel Israel

If you have any lawful idea, initiative, or suggestion for a genuine partnership with Dr. Nathanael-Israel Israel or Science180, please visit Science180.com/partner to inform us.

How to be trained or mentored by or have a one-on-one consulting with Dr. Nathanael-Israel Israel

Hire Nathanael-Israel Israel to train you or your organization in the best ways to conduct yourself and your organization to align your initiatives with the real understanding of the origin of the universe, of life, and of chemical particles in a way that you will not hear anywhere else. Nathanael-Israel Israel offers training through the program called "Science180 Academy". For training purposes, please visit Science180Academy.com.

Visit Nathanael-Israel Israel's personal website to get for free great resources you won't find anywhere else

To stay in touch with, Dr. Nathanael-Israel Israel, and to get updates directly from him, please visit his website, Israel120.com, and sign up for his popular newsletter at Israel120.com/newsletter for free.

Ask for review

If you are a book reviewer or a professional wanting to review this book or others written by Nathanael-Israel Israel, please contact us at Science180.com/AskForReview

Donate and support Nathanael-Israel Israel's efforts and initiatives

To help humankind accurately understand the real origin of the universe and its content, like I have done in the groundbreaking books I published after 12 years of sacrifice, I need your financial support. Please consider donating to me or to Science180 by visiting Israel120.com/donate or Science180.com/donate.

Your donation will be used to help me continue doing what I did to birth these books that you enjoyed and that you know will help many people across globe. No amount of money is too small or too big. Whatever you can give, please give.

Quantity discounts: Purchase Science180 books including this one in bulk at a special discount

To purchase Science180 books including this one in bulk at a special discount for sales promotion, corporate gifts, fund-raising, or educational purposes or to create special editions to specifications, contact specialsales@science180.com or visit Science180.com/discount.

Buy a copy of Nathanael-Israel Israel's books for your friends, family, or someone

If this book has been a blessing to you, and we know it has, please consider getting another copy and giving it to a friend, a family member, or someone you think it may help or challenge. If you want to get many copies, we can even give you a discount; just contact us as we previously explained.

Recommend Nathanael-Israel Israel's books to your organization

Because I know this book has been a blessing to you, I ask that you recommend it and others that I wrote to your organization, class, workplace, church, school, network, or clubs. Recommending this book will help others to tap into the blessing and opportunities that my books will open for them.

Share Nathanael-Israel Israel's groundbreaking discovery with others

To improve more lives, please share the findings of Nathanael-Israel Israel's books with others, for many people out there still do not understand how the universe was formed and sharing your experience of reading this book will help them. If you enjoy Nathanael-Israel Israel's books (www.Israel120.com/books), please help other people find them by writing a book review on your blog or on online bookstores, or write it and share it with us. Likewise, share and mention this book on your social media platforms (e.g. Facebook, Twitter, YouTube, etc.).

Follow Nathanael-Israel Israel on social media

In our modern world, social medias have become a huge part of how messages spread across the globe today. To ensure more people hear about the good news revealed in my books, I need you to follow me and share my contents on your social medias and in your network. To know the full list of my social media accounts and follow me please visit Science180.com/socialmedia.

Share your feedback, critics, testimony, experience, adventures, story, or comment about this book with me

How has Nathanael-Israel Israel's books and services at Science180 improved your life? I would love to hear from you.

To help me know how I can better help you next and encourage others, I need to know and capture your testimony or critics. Please visit the feedback page, science180.com/feedback, to tell me:

- how this book impacted you or will impact you
- what you like or dislike or disagree with

Science180: Understand the Origin of Chemicals. Increase Your Glory and Peace of Mind

- what you think, wish, or dream that I need to work on next
- what you wish to see in this book but that was absent
- what shocked you the most
- what got your heart pumping as you were reading this book
- what you found more insightful or thought-provoking
- what you want to do to be a part of my journey
- how my work changed your life or someone else's life

Message from the publisher of this book

Just like Nathanael-Israel Israel, you can publish your book(s) with us too. To get started and see how we may help you, please visit Science180Publishing.com today.

To contact Nathanael-Israel Israel or Science180

For any suggestions or questions, please visit Science180.com/contact and Nathanael-Israel Israel's personal website: Israel120.com. Feel free to ask me any questions you have about the universe formation, life formation, and chemicals formation.

REFERENCES

About.com Education (2017). "Quark-gluon plasma is the most primordial state of matter". About.com Education. Archived from the original on 2017-01-18. Retrieved on 2017-01-16, from http://physics.about.com/od/physicsqrot/fl/Quark-Gluon-Plasma.htm.

Allen, Robert (2006). "A guide to rebound hardness and scleroscope test". Archived from the original on 2012-07-18 Retrieved 2008-09-08. URL: http://www.articlestree.com/science/a-guide-to-rebound-hardness-and-scleroscope-test-tx301428.html.

AMS (2015). "Particle". AMS Glossary. American Meteorological Society. Retrieved 2015-04-12.

Amsler C. et al. (Particle Data Group) (2008). "Review of Particle Physics" (PDF). Physics Letters B. 667 (1): 1–1340. Bibcode:2008PhLB..667....1A. doi:10.1016/j.physletb.2008.07.018.

Amsler, C; Doser, M; Antonelli, M; Asner, D; Babu, K; Baer, H; Band, H; Barnett, R; et al. (2008). "Review of Particle Physics" (PDF). Physics Letters B. 667 (1–5): 1. Bibcode:2008PhLB..667....1A. doi:10.1016/j.physletb.2008.07.018. Summary Table.

Anicin, Ivan V.; Aprili, P.; Baiboussinov, B.; Baldo Ceolin, M.; Benetti, P.; Calligarich, E.; Canci, N.; Centro, S.; Cesana, A.; Cieślik, K.; Cline, D.B.; Cocco, A.G.; Dabrowska, A.; Dequal, D.; Dermenev, A.; Dolfini, R.; Farnese, C.; Fava, A.; Ferrari, A.; Fiorillo, G.; Gibin, D.; Gigli Berzolari, A.; Gninenko, S.; Guglielmi, A.; Haranczyk, M.; Holeczek, J.; Ivashkin, A.; Kisiel, J.; Kochanek, I.; et al. (2012). "Measurement of the neutrino velocity with the ICARUS detector at the CNGS beam". Physics Letters B. 713 (1): 17–22. arXiv:1203.3433. Bibcode:2012PhLB..713...17A. doi:10.1016/j.physletb.2012.05.033.

Anchordoqui L.; T. Paul; S. Reucroft; J. Swain (2003). "Ultrahigh Energy Cosmic Rays: The state of the art before the Auger Observatory". International Journal of Modern Physics A. 18 (13): 2229–2366. arXiv:hep-ph/0206072. Bibcode:2003IJMPA..18.2229A. doi:10.1142/S0217751X03013879.

Antognini et al. (2013). Science 339, 417.

Antonello, M.; Aprili, P.; Baiboussinov, B.; Baldo Ceolin, M.; Benetti, P.; Calligarich, E.; et al. (2012). "Measurement of the neutrino velocity with the ICARUS detector at the CNGS beam". Physics Letters B. 713 (1): 17–22. arXiv:1203.3433. Bibcode:2012PhLB..713...17A. doi:10.1016/j.physletb.2012.05.033.

Arndt, Markus; Nairz, Olaf; Vos-Andreae, Julian; Keller, Claudia; Van Der Zouw, Gerbrand; Zeilinger, Anton (2000). "Wave-particle duality of C60 molecules". Nature. 401 (6754): 680–682. Bibcode:1999 Natur.401..680A. PMID 18494170. doi:10.1038/44348.

Aspinall, Helen C. (2001). Chemistry of the f-block elements. CRC Press. p. 8. ISBN 978-90-5699-333-7.

ATLAS collaboration (2018). "Observation of H→bb decays and VH production with the ATLAS detector". Physics Letters B. 786: 59–86. arXiv:1808.08238. doi:10.1016/j.physletb.2018.09.013.

Bahcall, N. A. (1988). "Large-scale structure in the Universe indicated by galaxy clusters". Annual Review of Astronomy and Astrophysics. 26 (1): 631–686. Bibcode:1988ARA&A..26..631B. doi:10.1146/annurev,aa.26.090188.003215.

Beringer J. et al. (Particle Data Group) (2012). "Review of Particle Physics". Physical Review D. 86 (1): 010001. Bibcode:2012PhRvD..86a0001B. doi:10.1103/PhysRevD.86.010001.

Bettelheim FA, Brown WH, Campbell MK, Farrell SO & Torres OJ (2016). Introduction to general, organic, and biochemistry, 11th ed., Cengage Learning, Boston, ISBN 978-1-285-86975-9.

Bondi (1964). "van der Waals Volumes and Radii". J. Phys. Chem. 68: 441. doi:10.1021/j100785a001.

Brucat, Philip J. (2008). "The Quantum Atom". University of Florida. Archived from the original on 7 December 2006. Retrieved 4 January 2007.

Burns, J.A.; Hamilton, D.P.; Showalter, M.R. (2001). "Dusty Rings and Circumplanetary Dust: Observations and Simple Physics" (PDF). In Grun, E.; Gustafson, B. A. S.; Dermott, S. T.; Fechtig H. (eds.). Interplanetary Dust. Berlin: Springer. pp. 641–725. Bibcode:2001indu.book..641B. ISBN 3-540-42067-3.

Carlson, C.E. (2015). The proton radius puzzle. Progress in Particle and Nuclear Physics, 82, pp.59-77.

CERN (2012). OPERA experiment reports anomaly in flight time of neutrinos from CERN to Gran Sasso (UPDATE 8 June 2012)". CERN press office. 8 June 2012. Retrieved 19 April 2013.

Chemistry Libre Texts (2019). "Bond Energies". Chemistry Libre Texts. Retrieved 2019-02-25.
https://chem.libretexts.org/Bookshelves/Physical_and_Theoretical_Chemistry_Textbook_Maps/Supplemental_Modules_(Physical_and_Theoretical_Chemistry)/Chemical_Bonding/Fundamentals_of_Chemical_Bonding/Bond_Energies.

Nathanael-Israel Israel: Acclaimed at the Standout Scientific Authority Who Accurately Decoded the Universe-Origin

REFERENCES

Chesterman, C.W.; Lowe, K.E. (2008). Field guide to North American rocks and minerals. Toronto: Random House of Canada. ISBN 0-394-50269-8.

Chibisov, G V (1976). "Astrophysical upper limits on the photon rest mass". Soviet Physics Uspekhi. 19 (7): 624. Bibcode:1976SvPhU..19..624C. doi:10.1070/PU1976v019n07ABEH005277.

Clark, Jim (2005). "Atomic and Physical Properties of the Group 1 Elements". chemguide. Retrieved January 30, 2012.

Clementi E.; D.L.Raimondi; W.P. Reinhardt (1967). "Atomic Screening Constants from SCF Functions. II. Atoms with 37 to 86 Electrons". J. Chem. Phys. 47: 1300. Bibcode:1967JChPh..47.1300C. doi:10.1063/1.1712084.

CMS collaboration (2018). "Observation of Higgs Boson Decay to Bottom Quarks". Physical Review Letters. 121 (12): 121801. arXiv:1808.08242. Bibcode:2018PhRvL.121l1801S. doi:10.1103/PhysRevLett.121.121801. PMID 30296133.

Cottingham W. N. and Greenwood D. A. (2007). An introduction to the standard model of particle physics. Cambridge University Press. 2nd edition, p. 1. ISBN 978-0-521-85249-4.

Cotton, Simon (2006). Lanthanide and Actinide Chemistry. John Wiley & Sons Ltd.

Danby G.; J.-M. Gaillard; K. Goulianos; L. M. Lederman; N. B. Mistry; M. Schwartz; J. Steinberger (1962). "Observation of high-energy neutrino reactions and the existence of two kinds of neutrinos". Physical Review Letters. 9 (1): 36. Bibcode:1962PhRvL...9...36D. doi:10.1103/PhysRevLett.9.36.

Dayah, M. (1997). Dynamic Periodic Table. 1997, October 1. Retrieved December 4, 2014, from Ptable: http://www.ptable.com.

Demtröder, Wolfgang (2006). Experimentalphysik. 1 (4 ed.). Springer. p. 101. ISBN 978-3-540-26034-9.

Dubinski, J. (1998). "The Origin of the Brightest Cluster Galaxies". The Astrophysical Journal. 502 (2): 141–149. arXiv:astro-ph/9709102. Bibcode:1998ApJ...502..141D. doi:10.1086/305901. Archived from the original on May 14, 2011. Retrieved January 16, 2007.

Dyar, M.D.; Gunter, M.E. (2008). Mineralogy and Optical Mineralogy. Chantilly, Virginia: Mineralogical Society of America. ISBN 978-0-939950-81-2.

Elders A. Wilfred (1989). Exploring the Deep Continental Crust by Drilling. Eos, Vol. 70, No. 21, May 23, 1989. PAGES 609, 616-617.

Encyclopædia Britannica (2008). "Noble Gas". Encyclopædia Britannica.

http://www.britannica.com/eb/article-9110613/noble-gas.

Encyclopædia Britannica (2010). Lepton (physics)". Encyclopædia Britannica. Retrieved 2010-09-29.

Encyclopaedia Britannica (2018). Statistical mechanics. Retrieved on 2018/5/29 from https://www.britannica.com/science/statistical-mechanics.

ESA Staff (2019). "How Many Stars Are There In The Universe?". European Space Agency. Retrieved September 21, 2019.

Esposito, L. W. (2002). "Planetary rings". Reports on Progress in Physics. 65 (12): 1741–1783. Bibcode:2002RPPh...65.1741E. doi:10.1088/0034-4885/65/12/201.

Figgis, B.N.; Lewis, J. (1960). Lewis, J.; Wilkins, R.G., eds. The Magnetochemistry of Complex Compounds. Modern Coordination Chemistry. New York: Wiley Interscience. pp. 400–454.

Firestone, Richard B. (2000). "Radioactive Decay Modes". Berkeley Laboratory. 22 May 2000. Archived from the original on 29 September 2006. Retrieved 7 January 2007.

Fricke, Burkhard (1975). "Superheavy elements: a prediction of their chemical and physical properties". Recent Impact of Physics on Inorganic Chemistry. 21: 89–144. doi:10.1007/BFb0116498. Retrieved 4 October 2013.

Goobar, Ariel; Hannestad, Steen; Mörtsell, Edvard; Tu, Huitzu (2006). "The neutrino mass bound from WMAP 3-year data, the baryon acoustic peak, the SNLS supernovae and the Lyman-α forest". Journal of Cosmology and Astroparticle Physics. 2006 (6): 019. arXiv:astro-ph/0602155. Bibcode:2006JCAP...06..019G. doi:10.1088/1475-7516/2006/06/019.

Gordon L. Squires (2018). Quantum mechanics. Encyclopedia Britannica. Retrieved on 2018/5/29, from https://www.britannica.com/science/quantum-mechanics-physics.

Gott III, J. R.; et al. (2005). "A Map of the Universe". The Astrophysical Journal. 624 (2): 463–484. arXiv:astro-ph/0310571. Bibcode:2005ApJ...624..463G. doi:10.1086/428890.

Greenwood, Norman N.; Earnshaw, Alan (1997). Chemistry of the Elements (2nd ed.). Butterworth-Heinemann. ISBN 978-0-08-037941-8).

Greiner, Walter (2001). "Structure of vacuum and elementary matter: from superheavies via hypermatter to antimatter." In Arias, J.M.; Lozano, M. (eds.). An Advanced Course in Modern Nuclear Physics. Lecture Notes in Physics. 581. pp. 316–342. doi:10.1007/3-540-44620-6_11. ISBN 978-3-540-42409-3.

REFERENCES

Gray, Theodore (2007). "The Elements", Black–Dog & Leventhal, Chicago 2007: "Lanthanum" and "Cerium" entries Ch 57 & 58, pp 134-7.

Guijón, R.; Henríquez, F.; Naranjo, J.A. (2011). "Geological, Geographical and Legal Considerations for the Conservation of Unique Iron Oxide and Sulphur Flows at El Laco and Lastarria Volcanic Complexes, Central Andes, Northern Chile". Geoheritage. 3 (4): 99–315. doi:10.1007/s12371-011-0045-x.

Harlov, D.E. et al. (2002). "Apatite–monazite relations in the Kiirunavaara magnetite–apatite ore, northern Sweden". Chemical Geology. 191 (1–3): 47–72. Bibcode:2002ChGeo.191...47H. doi:10.1016/s0009-2541(02)00148-1.

Hill, Christopher T.; Lederman, Leon M. (2013). Beyond the God Particle. Prometheus Books. ISBN 978-1-6161-4801-0).

Hoffman, Darleane C.; Lee, Diana M.; Pershina, Valeria (2006). "Transactinides and the future elements". In Morss; Edelstein, Norman M.; Fuger, Jean (eds.). The Chemistry of the Actinide and Transactinide Elements (3rd ed.). Dordrecht, The Netherlands: Springer Science+Business Media. ISBN 1-4020-3555-1.

Holman, S. W.; Lawrence, R. R.; Barr, L. (1895). "Melting Points of Aluminum, Silver, Gold, Copper, and Platinum". 1 January 1895. Proceedings of the American Academy of Arts and Sciences. 31: 218–233. doi:10.2307/20020628. JSTOR 20020628.

Huheey J.E.; E.A. Keiter & R.L. Keiter (1993). Inorganic Chemistry: Principles of Structure and Reactivity (4th ed.). New York, USA: HarperCollins. ISBN 0-06-042995-X.

Hut, P.; Olive, K.A. (1979). "A cosmological upper limit on the mass of heavy neutrinos". Physics Letters B. 87(1–2): 144–146. Bibcode:1979PhLB...87..144H. doi:10.1016/0370-2693(79)90039-X.

INIST (2008). Rheological properties of basaltic lavas at sub-liquidus temperatures: laboratory and field measurements on lavas from Mount Etna". cat.inist.fr. Retrieved 19 June 2008.

Israel Nathanael-Israel (2025a). Turbulent Origin of the Universe. Science180, Augusta, USA 683 pages.

Israel Nathanael-Israel (2025b). From Science to Bible's Conclusions. Science180, Augusta, USA 170 pages.

Israel Nathanael-Israel (2025c). Reconciling Science and Creation Accurately. Science180, Augusta, USA 299 pages.

Israel Nathanael-Israel (2025d). Turbulent Origin of Chemical Particles. Science180, Augusta, USA 397 pages.

Israel Nathanael-Israel (2025e). Turbulent Origin of Life. Science180, Augusta, USA 370 pages.

Israel Nathanael-Israel (2025f). Origin of the Spiritual World. Science180, Augusta, USA 151 pages.

Israel Nathanael-Israel (2025g). How Baby Universe Was Born. Science180, Augusta, USA 130 pages.

Israel Nathanael-Israel (2025h). How God Created Baby Universe. Science180, Augusta, USA 224 pages.

Israel Nathanael-Israel (2025i). Science180 Accurate Scientific Proof of God. Science180, Augusta, USA 214 pages.

IUPAC (1997). Compendium of Chemical Terminology, 2nd ed. (the "Gold Book") (1997). Online corrected version: (2006) "Electron affinity". doi:10.1351/goldbook.E01977.

IUPAC (2006). Compendium of Chemical Terminology, 2nd ed. (the "Gold Book") (1997). Online corrected version (2006) "Electronegativity". doi:10.1351/goldbook.E01990. Url: https://goldbook.iupac.org/E01990.html.

James A.M. & M.P. Lord (1992). Macmillan's Chemical and Physical Data. MacMillan. ISBN 0-333-51167-0.

Kolena (2021). Forces. Duke University. http://webhome.phy.duke.edu/~kolena/modern/forces.html#005.

Krebs, Robert E. (2006). The History and Use of Our Earth's Chemical Elements: A Reference Guide. Westport, Conn.: Greenwood Press. ISBN 978-0-313-33438-2.

L'Annunziata, Michael F. (2003). Handbook of Radioactivity Analysis. Academic Press. pp. 3–56. ISBN 978-0-12-436603-9. OCLC 16212955.

Legoas Miguel (2019). The Augusta Chronicle, Article "Cold War Patriots Instruct SRS Workers on Possible Benefits" written by Miguel Legoas, posted on December 11, 2019.

Liang et al. (2018). Observation of three-photon bound states in a quantum nonlinear medium. Science 16 Feb 2018, Vol. 359, Issue 6377, pp. 783-786. DOI: 10.1126/science.aao7293.

Lide, D. R., ed. (2003). CRC Handbook of Chemistry and Physics (84th ed.). Boca Raton, FL: CRC Press.

Lindsay, Don (2000). "Radioactives Missing From The Earth". 30 July 2000. Don Lindsay

REFERENCES

Archive. Archived from the original on 28 April 2007. Retrieved 23 May 2007.

Los Alamos Science (1997). The Reines-Cowan Experiments: Detecting the Poltergeist" (PDF). Los Alamos Science. 25: 3. 1997. Retrieved 2010-02-10.

MacDonald, Matthew R.; Bates, Jefferson E.; Ziller, Joseph W.; Furche, Filipp; Evans, William J. (2013). "Completing the Series of +2 Ions for the Lanthanide Elements: Synthesis of Molecular Complexes of Pr, Gd, Tb, and Lu". Journal of the American Chemical Society. 135 (26): 9857–9868. doi:10.1021/ja403753j. PMID 23697603.

Mackie, Glen (2002). "To see the Universe in a Grain of Taranaki Sand". Centre for Astrophysics and Supercomputing. February 1, 2002. Retrieved January 28, 2017.

Makhijani, Arjun; Saleska, Scott (2001). "Basics of Nuclear Physics and Fission". Institute for Energy and Environmental Research. 2 March 2001. Archived from the original on 16 January 2007. Retrieved 3 January 2007.

Manthey, David (2001). "Atomic Orbitals". Orbital Central. Archived from the original on 10 January 2008. Retrieved 21 January 2008.

Mantina M.; A.C. Chamberlin; R. Valero; C.J. Cramer; D.G. Truhlar (2009). "Consistent van der Waals Radii for the Whole Main Group". J. Phys. Chem. A. 113 (19): 5806–12. Bibcode:2009JPCA..113.5806M. doi:10.1021/jp8111556. PMC 3658832. PMID 19382751.

Matthew Philips (2014). "A botched plan to turn nuclear warheads into fuel. Businessweek. Bloomberg. 24 April 2014. Retrieved 26 April 2014.

Miessler, G. L. and Tarr, D. A. (2010). Inorganic Chemistry 3rd ed., Pearson/Prentice Hall publisher, ISBN 0-13-035471-6.

Mortimer, Charles E. (1975). Chemistry: A Conceptual Approach (3rd ed.). New York: D. Van Nostrad Company.

Mulliken, Robert S. (1967). "Spectroscopy, Molecular Orbitals, and Chemical Bonding". Science. 157 (3784): 13–24. Bibcode:1967Sci...157...13M. doi:10.1126/science.157.3784.13. PMID 5338306.

NASA (2018). Planetary fact sheets. Fact sheets of the Sun, planets, satellites, rings and selected asteroids in the Solar System. Author/Curator: Dr. David R. Williams, NASA Goddard Space Flight Center, Greenbelt, MD, USA. Retrieved November 19, 2018, from http://nssdc.gsfc.nasa.gov/planetary/factsheet/.

Ockert, M. E.; Cuzzi, J. N.; Porco, C. C.; Johnson, T. V. (1987). "Uranian ring photometry: Results from Voyager 2". Journal of Geophysical Research. 92(A13): 14, 969–78. Bibcode:1987JGR....9214969O. doi:10.1029/JA092iA13p14969.

Olive K.A.; et al. (Particle Data Group) (2014). "Review of Particle Physics". Chinese Physics C. 38 (9): 090001. arXiv:1412.1408. doi:10.1088/1674-1137/38/9/090001.

Oxford Dictionary (2005). "Particle". Oxford English Dictionary (3rd ed.). Oxford University Press. September 2005.

Pauling, Linus (1960). Nature of the Chemical Bond. Cornell University Press. pp. 88–107. ISBN 978-0-8014-0333-0.

Periodictable.com (2014a). Abundance in the Universe of the elements. Retrieved on December 4, 2014, from https://periodictable.com/Properties/A/UniverseAbundance.html.

Periodictable.com (2014b). Abundance in the Sun of the elements. Retrieved on December 4, 2014, from https://periodictable.com/Properties/A/SolarAbundance.html.

Periodictable.com (2014c). Abundance in Humans of the elements. Retrieved on December 4, 2014, from https://periodictable.com/Properties/A/HumanAbundance.html.

Perl, M. L.; et al. (1975). "Evidence for Anomalous Lepton Production in e+e− Annihilation". Physical Review Letters. 35 (22): 1489. Bibcode:1975PhRvL..35.1489P. doi:10.1103/PhysRevLett.35.1489.

Pfeffer, Jeremy I.; Nir, Shlomo (2000). Modern Physics: An Introductory Text. Imperial College Press. pp. 330–336. ISBN 978-1-86094-250-1. OCLC 45900880.

Physics Letters (2001). Observation of tau neutrino interactions". Physics Letters B. 504 (3): 218–224. 2001. arXiv:hep-ex/0012035. Bibcode:2001PhLB..504..218D. doi:10.1016/S0370-2693(01)00307-0.

Pinkerton, H.; Bagdassarov, N. (2004). "Transient phenomena in vesicular lava flows based on laboratory experiments with analogue materials". Journal of Volcanology and Geothermal Research. 132 (2–3): 115–136. Bibcode:2004JVGR..132..115B. doi:10.1016/s0377-0273(03)00341-x.

Pohl et al. (2010). Nature 466, 213.

Pohl R. et al. (2016). Laser spectroscopy of muonic deuterium. Science 12 Aug 2016, Vol. 353, Issue 6300, pp. 669-673. DOI: 10.1126/science.aaf2468. Visited on 8/12/2016 at http://science.sciencemag.org/content/353/6300/669.

Porco, Carolyn (2012). "Questions about Saturn's rings". CICLOPS web site. Archived from the original on 2012-10-03. Retrieved 2012-10-05.

Porterfield W.W. (1984). Inorganic chemistry, a unified approach. Reading Massachusetts, USA: Addison Wesley Publishing Co. ISBN 0-201-05660-7.

Nathanael-Israel Israel: Acclaimed at the Standout Scientific Authority Who Accurately Decoded the Universe-Origin

REFERENCES

Raymond, David (2006). "Nuclear Binding Energies". New Mexico Tech. Archived from the original on 1 December 2002. Retrieved 3 January 2007.

Riedel S.; P.Pyykkö, M. Patzschke; Patzschke, M (2005). "Triple-Bond Covalent Radii". Chem. Eur. J. 11 (12): 3511–3520. doi:10.1002/chem.200401299. PMID 15832398. Mean-square deviation 3pm.

Royal Society of Chemistry (2012a). "Visual Elements: Group 1 – The Alkali Metals". Visual Elements. Royal Society of Chemistry. Archived from the original on 5 August 2012. Retrieved 13 January 2012.

Royal Society of Chemistry (2012b). "Visual Elements: Group 2–The Alkaline Earth Metals". Visual Elements. Royal Society of Chemistry. Archived from the original on 5 October 2011. Retrieved 13 January 2012.

Sanderson R.T. (1962). Chemical Periodicity. New York, USA: Reinhold.

Showalter, Mark R.; Burns, Joseph A.; Cuzzi, Jeffrey N.; Pollack, James B. (1987). "Jupiter's ring system: New results on structure and particle properties". Icarus. 69 (3): 458–498. Bibcode:1987Icar...69..458S. doi:10.1016/0019-1035(87)90018-2.

Shultis, J. Kenneth; Faw, Richard E. (2002). Fundamentals of Nuclear Science and Engineering. CRC Press. pp. 10–17. ISBN 978-0-8247-0834-4. OCLC 123346507.

Schulze-Makuch D and Irwin LN (2008). Life in the Universe: Expectations and constraints, 2nd ed., Springer-Verlag, Berlin, ISBN 9783540768166.

Sills, Alan D. (2003). Earth Science the Easy Way. Barron's Educational Series. pp. 131–134. ISBN 978-0-7641-2146-3. OCLC 51543743.

Slater J.C. (1964). "Atomic Radii in Crystals". J. Chem. Phys. 41: 3199. Bibcode:1964JChPh..41.3199S. doi:10.1063/1.1725697.

Smirnov, Boris M. (2003). Physics of Atoms and Ions. Springer. pp. 249–272. ISBN 978-0-387-95550-6.

Smith, Bradford A.; Soderblom, Laurence A.; Johnson, Torrence V.; Ingersoll, Andrew P.; Collins, Stewart A.; Shoemaker, Eugene M.; Hunt, G. E.; Masursky, Harold; Carr, Michael H. (1979). "The Jupiter System Through the Eyes of Voyager 1". Science. 204 (4396): 951–972. Bibcode:1979Sci...204..951S. doi:10.1126/science.204.4396.951. ISSN 0036-8075. PMID 17800430.

Smith, B. A.; Soderblom, L. A.; Banfield, D.; Barnet, C; Basilevsky, A. T.; Beebe, R. F.; Bollinger, K.; Boyce, J. M.; Brahic, A. (1989). "Voyager 2 at Neptune: Imaging Science Results". Science. 246 (4936): 1422–1449. Bibcode:1989Sci...246.1422S. doi:10.1126/science.246.4936.1422. ISSN 0036-8075. PMID 17755997.

Staff (2007). "ABC's of Nuclear Science". Lawrence Berkeley National Laboratory. Archived from the original on 5 December 2006. Retrieved 3 January 2007.

Steurer W (2007). "Crystal structures of the elements" in JW Marin (ed.), Concise encyclopedia of the structure of materials, Elsevier, Oxford, pp. 127–45, ISBN 0080451276.

Strassler Matt (2013). The Strengths of the Known Forces. Retrieved from https://profmattstrassler.com/articles-and-posts/particle-physics-basics/the-known-forces-of-nature/the-strength-of-the-known-forces/. Posted on May 31, 2013 on the blog of Prof Matt Strassler.

Sukys P (1999). Lifting the scientific veil: Science appreciation for the nonscientist, Rowman & Littlefield, Oxford, ISBN 0847696006.

Sutton L.E., ed. (1965). "Supplement 1956–1959, Special publication No. 18". Table of interatomic distances and configuration in molecules and ions. London, UK: Chemical Society.

Tanabashi, M.; Hagiwara, K.; Hikasa, K.; Nakamura, K.; Sumino, Y.; et al. (Particle Data Group) (2018). "Review of Particle Physics". Physical Review D. American Physical Society (APS). 98 (3): 030001. Bibcode:2018PhRvD..98c0001T. doi:10.1103/physrevd.98.030001. ISSN 2470-0010.

Taylor Larry (2019). The Augusta Chronicle, Article Plutonium Shipped From Savannah River Site to Nevada, Court Filing States, written by Larry Taylor, posted and updated on January 30, 2019.

Taylor Lucas (2014). "Observation of a new particle with a mass of 125 GeV". 4 Jul 2014. CMS. Retrieved 2012-07-06.

The T2K Collaboration (2019). Neutrinos in the Standard Model". Retrieved 15 October 2019.

Throop, H. B.; Porco, C. C.; West, R. A.; et al. (2004). "The Jovian Rings: New Results Derived from Cassini, Galileo, Voyager, and Earth-based Observations" (PDF). Icarus. 172 (1): 59–77. Bibcode:2004Icar..172...59T. doi:10.1016/j.icarus.2003.12.020.

Tiscareno, Matthew S. (2013). "Planetary Rings". In Oswalt, Terry D.; French, Linda M.; Kalas, Paul (eds.). Planets, Stars and Stellar Systems. Springer Netherlands. pp. 309–375. arXiv:1112.3305. doi:10.1007/978-94-007-5606-9_7. ISBN 9789400756052.

University of Alberta (2008). "Solid Helium". University of Alberta. Archived from the original on February 12, 2008. Retrieved 2008-06-22. http://www.phys.ualberta.ca/~therman/lowtemp/projects1.htm.

Nathanael-Israel Israel: Acclaimed at the Standout Scientific Authority Who Accurately Decoded the Universe-Origin

REFERENCES

Watson E. B., M. F. Hochella, and I. Parsons (editors) (2006). Glasses and Melts: Linking Geochemistry and Materials Science, Elements, volume 2, number 5, (October 2006) pages 259–297.

Wheeler, John C. (1997). "Electron Affinities of the Alkaline Earth Metals and the Sign Convention for Electron Affinity". Journal of Chemical Education. 74 (1): 123–127. Bibcode:1997JChEd..74..123W. doi:10.1021/ed074p123.; Kalcher, Josef; Sax, Alexander F. (1994). "Gas Phase Stabilities of Small Anions: Theory and Experiment in Cooperation". Chemical Reviews. 94 (8): 2291–2318. doi:10.1021/cr00032a004.

Wikipedia (2016a). Alkali metal. Retrieved on 2016-11-25, from https://en.wikipedia.org/wiki/Alkali_metal.

Wikipedia (2016b). Noble gas. Retrieved on 2016-11-25, from https://en.wikipedia.org/wiki/Noble_gas.

Wikipedia (2020a) Nuclear fission. Retrieved on 2020-4-22, from http://en.wikipedia.org/wiki/Nuclear_fission.

Wikipedia (2020b). List of quasiparticles. Retrieved on 2020-4-22, from https://en.wikipedia.org/wiki/List_of_quasiparticles..

Wikipedia (2020c). Cosmic ray. Retrieved on May 26, 2020, from https://en.wikipedia.org/wiki/Cosmic_ray.

Wikipedia (2020d). Metal. Retrieved on 2020-4-22, from https://en.wikipedia.org/wiki/Metal.

Wikipedia (2020e). Specific heat capacity. Retrieved on 2020-4-22, from https://en.wikipedia.org/wiki/Specific_heat_capacity.

Wikipedia (2020f). Hardness. Retrieved on 2020-4-22, from https://en.wikipedia.org/wiki/Hardness.

Wikipedia (2020g). Bulk modulus. Retrieved on 2020-4-22, from https://en.wikipedia.org/wiki/Bulk_modulus.

Wikipedia (2020h). Atomic radii of the elements (data page). Retrieved on 2020-4-22, from https://en.wikipedia.org/wiki/Atomic_radii_of_the_elements_(data_page)

Wikipedia (2020i). State of matter. Retrieved on 2020-4-22, from https://en.wikipedia.org/wiki/State_of_matter.

Wikipedia (2020j). Deuterium. Retrieved on June 5, 2020, from https://en.wikipedia.org/wiki/Deuterium.

Wikipedia (2020k). Chemical compound. Retrieved on 2020-4-22, from https://en.wikipedia.org/wiki/Chemical_compound.

Wikipedia (2021a). Photoelectric effect. Retrieved on 2021-10-17, from https://en.wikipedia.org/wiki/Photoelectric_effect.

Wikipedia (2021b). Molar ionization energies of the elements. Retrieved on 2020-4-22, from http://en.wikipedia.org/wiki/Molar_ionization_energies_of_the_elements and
https://en.wikipedia.org/wiki/Ionization_energies_of_the_elements_(data_page).

Wikipedia (2022a). Photon. Retrieved on May 6, 2022, from https://en.wikipedia.org/wiki/Photon.

Wikipedia (2022b). Chemical bond. Retrieved on March 29, 2022, from http://en.wikipedia.org/wiki/Chemical_bond.

Wilczek, Frank (2010). The Lightness of Being: Mass, Ether, and the Unification of Forces. Physics Today. 62. Basic Books. p. 212. Bibcode:2009PhT....62d..61W. doi:10.1063/1.3120899. ISBN 978-0-465-01895-6.

Williams, E.; Faller, J.; Hill, H. (1971). "New Experimental Test of Coulomb's Law: A Laboratory Upper Limit on the Photon Rest Mass". Physical Review Letters. 26 (12): 721. Bibcode:1971PhRvL..26..721W. doi:10.1103/PhysRevLett.26.721.

Winter, Mark (2008). "Helium: the essentials". University of Sheffield. Retrieved 2008-07-14.

World Nuclear News (2019). NRC terminates US MOX plant authorization. World Nuclear News, 13 February 2019. Retrieved 14 February 2019.

Wredenberg, Fredrik; PL Larsson (2009). "Scratch testing of metals and polymers: Experiments and numerics". Wear. 266 (1–2): 76. doi:10.1016/j.wear.2008.05.014.

Yao W.-M.; et al. (Particle Data Group) (2006). "Review of Particle Physics" (PDF). Journal of Physics G. 33 (1): 1. arXiv:astro-ph/0601168. Bibcode:2006JPhG...33....1Y. doi:10.1088/0954-3899/33/1/001.

Yoder CH, Suydam FH & Snavely FA (1975). Chemistry, 2nd ed, Harcourt Brace Jovanovich, New York, ISBN 978-0-15-506470-6.

Yonezawa, F (2017). Physics of Metal-Nonmetal Transitions. Amsterdam: IOS Press. p. 257. ISBN 978-1-61499-786-3.

Yndurain F. (1995). "Limits on the mass of the gluon". Physics Letters B. 345 (4): 524. Bibcode:1995PhLB..345..524Y. doi:10.1016/0370-2693(94)01677-5.

INDEX

A

Abundance ...7, 1, 6, 32, 37, 62, 74, 89, 94, 111, 188, 189, 190, 191, 192, 193, 194, 195, 196, 197, 200, 201, 202, 204, 205, 206, 207, 208, 209, 211, 212, 213, 214, 216, 218, 219, 220, 222, 225, 226, 227, 228, 232, 233, 235, 236, 237, 244, 245, 246, 249, 255, 256, 257, 258, 259, 260, 261, 262, 263, 264, 265, 266, 270, 274, 278, 294, 295, 296, 305, 306, 360, 376

Actinides 104, 105, 110

Actinium 104, 186

Actinoid 6, 145, 192

Actinoides 110

Actinoids 104, 115, 124, 126, 139, 159, 160, 174, 176

Aerosol .. 217, 228, 235, 257, 259, 260, 261, 265, 267

Aiken 66, 67

Albert Einstein 7, 43, 44, 63, 70, 73, 74, 98

Alkali metal 6, 105, 138, 139, 142, 148, 172, 176, 189, 196

Alkali metals 105, 106, 110, 160

Alkaline earth metal ... 6, 145, 189, 196

Alkaline earth metals..... 107, 110, 173, 192, 245

Allotropes 299, 300

Alpha decay 91

Aluminum .. 89, 94, 109, 111, 124, 140, 152, 189, 190, 192, 195, 196, 197, 199, 232, 245, 246, 248, 249, 252, 256, 373

Amalgamation 34, 54, 55

American 369, 371, 373, 375, 378, 396

Americium 104, 105, 126, 175, 185, 186

Ammonia 217, 258, 259, 265, 295

Ampere 60

Ampere-hour 60

Amphiboles......................... 94, 95, 298

Angular momentum 42, 46, 47, 49, 59, 60

Animals . 38, 44, 58, 224, 225, 234, 293

Anions..................................... 92, 112

Antimony 110, 127, 178

Antineutrino 65, 92

Argon 111, 112, 124, 152, 174, 176, 184, 189, 190, 195, 196, 199, 236, 268, 270

Arsenic........................... 110, 192, 212

Astatine . 107, 133, 157, 176, 177, 184, 185, 186, 192

Asteroids.....1, 5, 18, 19, 21, 25, 26, 27, 35, 40, 47, 51, 55, 58, 61, 62, 91, 95, 104, 151, 211, 214, 243, 286, 312, 313, 315, 316, 326, 336, 346, 356, 375, 396

Atlantic Ocean 67

Atmosphere... 6, 32, 34, 37, 38, 44, 65, 87, 94, 99, 112, 114, 187, 188, 193, 194, 196, 197, 198, 199, 200, 201, 203, 204, 205, 206, 207, 209, 211, 212, 214, 217, 218, 219, 220, 221, 222, 223, 224, 225, 226, 228, 230, 231, 232, 233, 234, 235, 236, 237, 238, 239, 243, 244, 245, 246, 249, 252, 255, 256, 257, 258, 259, 260, 261, 262, 263, 264, 265, 266, 267, 268, 270, 271, 274, 278, 283, 287, 295, 303, 304, 305, 306, 313, 314, 336, 358, 360

Atomic mass 6, 126

Atomic nuclei. 5, 42, 56, 71, 74, 82, 85, 87, 89, 90, 98, 99, 101, 278, 286

Atomic nucleus... 22, 25, 26, 41, 48, 71, 151, 283, 284, 286, 287, 346

Atomic number 5, 119

Atoms1, 5, 9, 12, 13, 14, 17, 18, 19, 21, 22, 23, 25, 26, 29, 30, 31, 33, 34, 35, 36, 37, 38, 39, 40, 42, 43, 47, 48, 49, 50, 51, 55, 57, 59, 61, 62, 65, 71, 73, 74, 82, 89, 90, 91, 92, 94, 95, 97, 98, 100, 101, 102, 104, 105, 107, 110, 111, 112, 113, 115, 116, 119, 124, 142, 148, 150, 151, 152, 159, 169, 170, 172, 174, 176, 177, 178, 180, 182, 183, 184, 189, 198, 200, 204, 205, 208, 209, 219, 220, 221, 225, 226, 228, 229, 230, 235, 237, 238, 243, 244, 245, 249, 252, 253, 256, 257, 258, 259, 260, 261, 262, 263, 264, 265, 266, 267, 271, 274, 275, 276, 277, 278, 280, 281, 282, 283, 284, 285, 286, 287, 288, 290, 291, 293, 294, 295, 296, 297, 298, 299, 300, 301, 304, 305, 306, 314, 326, 336, 342, 345, 346, 356, 357, 358, 359, 360

Augusta 3, 4, 66, 68, 374, 378

Axial tilt 52, 53, 301, 332

B

Baby Universe 2, 363, 374, 397

Barium 107, 160, 174, 176

Baryon 5, 82, 83, 85, 89, 372

Basalt 195, 302

Bending 34

Beninese 396

Berkelium 104, 105, 158, 185, 186

Beryllium 107, 141, 144, 145, 146, 161, 162, 174, 212

Beta decay 92

Bible 2, 363, 373, 397

Big Bang 2, 3, 4, 5, 6, 8, 9, 52

Biochemical 10

Biological 44, 114

Biological systems 10

Biology 13

Biosphere 38

Bismuth 109, 128, 178

Bohrium 115, 165, 185

Boiling point 7, 6, 62, 106, 112, 113, 114, 116, 182, 187

Bombs 66, 125, 285, 351

Boron 110, 124, 138, 139, 140, 141, 143, 162, 174, 176, 178, 192, 196, 223

Bose 63, 70

Bose-Einstein statistics 63, 70

Bosons 5, 26, 61, 63, 69, 70, 71, 72, 75, 78, 87, 88, 97, 275, 276, 278, 280, 281, 283, 286, 346, 359

Bottom quark 5, 69, 70, 78, 80, 276

Braiding 34, 54

Brinell hardness .6, 137, 138, 139, 172, 249

Bromine .107, 131, 133, 142, 143, 184, 192, 196, 240

Bulk modulus 6, 142, 143, 172

Burst asunder 9

C

Cadmium 115, 155

Caesium 106

Calcium 94, 107, 124, 174, 189, 190, 192, 193, 195, 196, 197, 237, 245, 249, 252

Calculated radius 6, 170, 171, 172, 173, 249, 256, 263

Californium 104, 105, 185, 186

Carbohydrates 93, 224

Carbon.. 61, 92, 94, 111, 113, 114, 124, 125, 128, 133, 138, 140, 143, 152, 163, 164, 173, 174, 175, 176, 184, 189, 190, 191, 193, 195, 196, 204, 205, 206, 214, 219, 224, 225, 226, 228, 236, 252, 255, 256, 257, 258, 259, 260, 261, 262, 263, 264, 265, 266, 290, 294, 295, 298, 300, 360

Carbon dioxide .92, 204, 206, 214, 219, 226, 228, 260, 261, 264, 265, 295

Carbon monoxide ...219, 225, 260, 261,

265

Carbonates 299

Cascade of breakup 12

Cascade of explosions 10

Cascade of fragmentation 10

Cations.....................92, 106, 107, 109

Celestial bodies.7, 1, 4, 5, 7, 10, 12, 14, 17, 19, 21, 26, 29, 30, 31, 32, 34, 35, 36, 37, 38, 40, 41, 43, 45, 46, 47, 48, 49, 52, 54, 55, 59, 60, 61, 62, 91, 94, 95, 100, 103, 150, 151, 179, 188, 189, 190, 193, 194, 205, 206, 226, 233, 244, 245, 274, 275, 278, 280, 281,282, 283, 286, 291, 292, 294, 295, 296, 297, 298, 300, 301, 302, 303, 304, 305, 306, 307, 312, 313, 314, 315, 316, 325, 326, 331, 332, 334, 335, 341, 342, 343, 344, 345, 346, 355, 356, 357, 358, 359, 360

Cells 8, 9, 10, 114, 349, 350

Cellular division 10

Cerium 108, 373

CERN 68, 370

Cesium... 105, 138, 139, 142, 145, 148, 156, 159, 172, 174, 176, 178

Charm quark 5, 69, 70, 78, 80, 275

Chemical bonds 42, 43, 90, 92, 94, 112, 136, 290, 291, 294

Chemical elements .6, 7, 4, 5, 7, 10, 12, 17, 18, 21, 32, 36, 37, 38, 40, 42, 43, 44, 48, 49, 61, 62, 68, 89, 90, 94, 103, 104, 106, 107, 108, 109, 110, 111, 112, 113, 114, 116, 119, 124, 125, 126, 127, 128, 129, 131, 132, 133, 136, 137, 138, 139, 140, 142, 143, 144, 145, 148, 149, 151, 152, 159, 160, 162, 169, 170, 172, 174, 175, 182, 183, 184, 185, 186, 188, 189, 190, 191, 193, 194, 195, 196, 197, 199, 201, 203, 204, 206, 207, 208, 209, 211, 212, 213, 214, 217, 222, 225, 226, 227, 228, 233, 234, 237, 243, 244, 256, 257, 258, 264,

274, 278, 282, 283, 284, 285, 292, 294, 298, 299, 300, 304, 305, 306, 307, 315, 342, 350, 356, 360

Chemical reactions 42, 74, 111, 292, 359

Chemistry 5, 7, 2, 3, 4, 7, 13, 17, 59, 106, 109, 111, 188, 274, 312, 322, 341, 349, 376

Chemists 48, 285

Chlorine . 107, 142, 143, 174, 184, 191, 192, 193, 196, 235

Chromium114, 144, 146, 153, 190, 212

Clumping 34

Clusters of galaxies 27

Clyde Lorrain Cowan Jr 67

Cobalt 115, 116, 212

Color 6

Color charge 63, 64, 69, 75

Composite particles... 6, 7, 1, 5, 14, 19, 26, 55, 59, 61, 64, 65, 69, 71, 75, 81, 82, 87, 89, 93, 99, 101, 189, 274, 276, 277, 278, 281, 282, 356

Compression 10, 50, 55, 137, 142, 280, 307, 344, 356

Conductivity 6, 62, 106, 111, 113, 129, 283

Copernicium 115, 124, 158, 185

Copper 115, 127, 148, 178, 212, 373

Corpuscle 58

Cosmic radiation...............................65

Cosmic rays..................... 65, 70, 87, 99

Coulomb 60

Covalent bonds... 92, 93, 290, 291, 294

Covalent radius.. 6, 170, 173, 174, 249, 264, 271

Creation 2, 45, 363, 373, 397

Creationism 74

Crystal structure 6, 34, 299, 300

Curium 104, 105, 157, 185, 186

Cysts 68

D

Dalton 60, 73, 126

Darmstadtium 115, 185

Daughter bodies...9, 10, 12, 14, 17, 18, 19, 21, 23, 26, 32, 35, 37, 41, 42, 43, 47, 50, 51, 53, 56, 57, 276, 277, 278, 281, 284, 300, 301, 305, 316, 323, 325, 332, 334, 342, 355, 357, 359

Daughter precursors 31, 47

Daughters of the turbulent prima materia 9, 10

Deformation..........53, 54, 56, 137, 138

Delta baryons 83, 277

Density 6, 33, 52, 55, 57, 58, 62, 94, 106, 112, 113, 114, 116, 131, 132, 133, 136, 151, 205, 220, 225, 226, 233, 236, 244, 249, 252, 256, 257, 263, 270, 271, 280, 281, 283, 286, 287, 291, 305, 306, 307, 316, 317, 323, 325, 332, 342, 343, 359

Density at 293 K 6, 132, 133, 249

Determined cells 10

Deuterium 102, 124, 125, 217, 219, 220, 221, 228, 255, 258, 259, 261, 266, 267, 285, 295, 376, 379

Diamond..........137, 291, 294, 299, 300

Discovery year............................ 6, 127

Diseases 67, 68, 180

DNA 2, 9, 93, 125, 234, 279, 280

Down quark....5, 65, 69, 70, 78, 80, 86, 275, 277

Downstream 53, 54, 56, 324

Dubnium 115, 124, 163, 185

Dusts ...36, 59, 322, 323, 324, 325, 326

Dysprosium 108, 156

E

Early universe..................................... 9

Earth... 6, 19, 31, 38, 43, 51, 58, 65, 68, 70, 85, 89, 94, 95, 99, 104, 108, 111, 112, 114, 125, 160, 183, 188, 193, 194, 195, 196, 197, 198, 199, 200, 201, 204, 206, 211, 212, 214, 217, 218, 219, 222, 223, 224, 225, 226, 228, 229, 230, 232, 233, 234, 237, 238, 239, 240, 241, 259, 260, 261,

262, 263, 265, 266, 270, 277, 278, 282, 283, 285, 287, 292, 297, 301, 302, 303, 304, 305, 306, 312, 314, 315, 316, 326, 331, 334, 336, 356, 360, 374, 377, 378, 379, 396

Earth's crust .6, 95, 108, 111, 114, 188, 193, 194, 195, 196, 197, 211, 212, 214, 217, 218, 223, 224, 226, 227, 228, 229, 230, 232, 233, 234, 237, 238, 239, 240, 241, 297, 301, 304, 306

Eccentricity.........................52, 53, 301

Eddies.. 33, 55

Einsteinium 104, 105, 185

Electric charge 5, 59, 60, 61, 64, 65, 66, 68, 69, 72, 73, 75, 78, 86

Electric conductivity.................. 6, 129

Electric field 60, 346

Electricity 27, 43, 109, 111, 114, 346

Electro negativity 6, 54, 62

Electromagnetic force 7, 60, 73, 91, 97, 98, 287, 341, 346

Electromagnetism 7, 42, 43, 44, 60, 64, 69, 70, 71, 73, 74, 87, 91, 92, 97, 98, 100, 281, 286, 287, 294, 341, 345, 346

Electron 6, 5, 25, 26, 35, 37, 44, 47, 48, 54, 61, 64, 65, 66, 67, 68, 78, 79, 80, 86, 89, 90, 91, 92, 94, 97, 98, 101, 102, 104, 106, 108, 109, 112, 113, 114, 136, 148, 150, 151, 170, 220, 255, 258, 266, 275, 284, 285, 291, 293, 303, 304, 356, 357, 358, 360

Electron affinity......6, 54, 62, 112, 113, 114, 136, 374

Electron neutrino 68

Electronic shell 6

Electrons ..6, 20, 22, 23, 25, 26, 27, 30, 31, 34, 35, 37, 40, 41, 42, 43, 44, 45, 47, 48, 49, 50, 53, 56, 57, 59, 60, 62, 65, 66, 74, 89, 90, 91, 92, 97, 98, 99, 101, 102, 104, 106, 107, 110, 111, 113, 115, 116, 119, 136, 148, 149, 150, 151, 152, 159, 170, 182,

Nathanael-Israel Israel: Member of the and Soil Science Society of America

183, 275, 276, 278, 281, 282, 283, 284, 287, 290, 291, 292, 293, 304, 326, 335, 345, 346, 356, 357, 358, 359

Electrostatic force 71

Electroweak force.............................. 7

Elementary charge 60, 69

Elementary particles 6, 1, 5, 14, 20, 55, 59, 61, 63, 66, 70, 73, 74, 76, 78, 79, 81, 82, 87, 93, 96, 97, 101, 275, 276, 277, 278

Elliptical 331, 333, 335

EMC debate 7, 96

Empirical radius .. 6, 170, 171, 175, 176

Energy...6, 9, 10, 11, 12, 14, 33, 40, 41, 42, 43, 44, 45, 46, 47, 48, 49, 50, 57, 60, 70, 73, 74, 87, 90, 91, 92, 96, 97, 98, 99, 101, 104, 107, 111, 112, 114, 136, 138, 148, 150, 151, 152, 159, 180, 189, 219, 225, 234, 249, 252, 257, 258, 262, 263, 264, 271, 281, 286, 287, 288, 292, 305, 313, 314, 333, 334, 335, 343, 345, 356, 371

Energy levels............. 6, 44, 74, 91, 151

Energy of ionization...................... 6, 62

Enrico Fermi 63, 66

Epigenetics 10

Equilibrium 30, 56, 184, 187, 243, 244, 280, 335, 346

Erbium 108

Ethane 217, 224, 259, 261, 295

Europium 108, 140

Exotic particles5, 19, 65, 82, 88, 89, 90, 98, 102, 277, 278, 356

Explosion 9, 10, 105, 355

F

Faraday constant 60

Feldspars 94, 95, 298

Feldspathoids 95, 298

Fermi-Dirac statistics 63

Fermions.5, 61, 63, 70, 82, 87, 88, 277, 278

Fermium 104, 105, 124, 185

Filaments 54, 332

Fission....................................... 42, 375

Flerovium...................... 116, 125, 185

Fluid breakup. 6, 10, 17, 18, 19, 20, 21, 47, 50, 51, 52, 55, 56, 57, 245, 300, 336, 356

Fluid dynamics................... 10, 50, 107

Fluid flow 10, 54, 57, 304, 336, 343, 344, 356, 357

Fluid layers5, 10, 13, 29, 30, 31, 35, 36, 37, 38, 41, 53, 54, 282, 298, 303, 316, 317, 323, 324, 335, 336, 343, 344, 356, 357, 358

Fluid mechanics. 10, 98, 101, 349, 351, 372

Fluid mixing 10

Fluids 10, 13, 14, 29, 30, 31, 32, 35, 37, 39, 41, 42, 43, 50, 52, 53, 54, 55, 56, 57, 282, 293, 301, 304, 312, 313, 315, 316, 323, 325, 336, 343, 356, 357, 358

Fluorine . 107, 108, 124, 166, 167, 173, 174, 175, 178, 184, 192, 193, 195, 196, 230

Foundational matter 8

Fragmentation 10, 17, 49, 55, 245, 356

Fragmentations 9, 17, 355

Francium. 105, 124, 136, 176, 186, 192

Frederick Reines 68

Fundamental forces..... 7, 1, 7, 42, 285, 286, 341, 343, 344, 345, 346, 347

Fusion 42, 43, 91, 222

G

Gadolinium 108

Galactic systems 27

Galaxies 7, 1, 5, 6, 10, 12, 13, 18, 19, 21, 27, 40, 55, 58, 59, 61, 62, 91, 95, 104, 114, 151, 183, 188, 189, 190, 195, 211, 212, 214, 217, 238, 243, 274, 278, 287, 314, 331, 332, 333, 334, 335, 336, 342, 370, 396

Gallium............................ 109, 154, 212

Gamma decay 92

Gamma rays 42, 91

Gas ... 9, 55, 90, 92, 103, 108, 111, 112, 113, 114, 133, 138, 145, 159, 160, 172, 173, 174, 176, 177, 182, 183, 184, 189, 191, 194, 196, 198, 199, 200, 205, 206, 217, 220, 222, 224, 228, 230, 236, 252, 260, 262, 263, 264, 265, 267, 268, 270, 271, 283, 299, 305, 306, 313, 314, 326, 355, 358, 360, 372, 379

Gases... 182

Gauge bosons..................5, 69, 70, 276

Gene expression........................ 10, 280

Generation of leptons 64

Generation of quark......................... 69

Generations of breakup 18, 19

Generations of gathering together .. 18

Generations of turbulent split-gathering 18, 19, 20

Georgia.................................3, 4, 66, 68

German 74, 105

Germanium 110, 212

Ghost fields 63

Giant planets.................................... 206

Gluons 5, 22, 26, 69, 70, 71, 72, 75, 78, 79, 86, 88, 89, 97, 276, 283, 286

God.. 2, 3, 4, 75, 99, 351, 363, 373, 397

God particle................................ 75, 99

Gold.... 12, 89, 108, 115, 127, 131, 143, 157, 298, 299, 373, 374

Granite 94, 195, 302

Graphene 300

Graphite 294, 299, 300

Gravity.. 1, 7, 97, 98, 99, 100, 286, 292, 306, 313, 341, 342, 343, 344, 345, 346

Groups 6, 13, 26, 59, 61, 62, 63, 64, 70, 72, 82, 83, 90, 92, 94, 95, 103, 104, 106, 109, 110, 126, 132, 138, 176, 252, 276, 277, 298, 299, 300, 305, 331

H

Hadrons.... 5, 65, 69, 75, 82, 87, 88, 89, 98, 277, 278

Hafnium 115, 141, 156

Halides 299

Halogen 6, 107, 131, 138, 139, 141, 142, 145, 173, 176, 184, 196, 299

Halogens 107

Hardness 6, 62, 94, 137, 138, 139, 140, 143, 144, 172, 249, 316, 369, 379

Harvard University 39

Hassium... 115, 124, 132, 158, 166, 185

Heavy 39, 40, 48, 65, 66, 67, 80, 86, 89, 91, 99, 102, 107, 109, 126, 159, 161, 169, 183, 189, 199, 200, 201, 205, 206, 219, 229, 230, 232, 233, 236, 237, 238, 244, 249, 252, 255, 256, 257, 258, 259, 260, 262, 267, 270, 277, 285, 293, 305, 360

Helium.. 43, 91, 99, 107, 109, 111, 112, 114, 126, 159, 160, 172, 174, 178, 184, 189, 190, 198, 205, 206, 209, 214, 222, 226, 236, 252, 256, 257, 259, 265, 268, 270, 271, 283, 299, 305, 307, 312, 314, 315, 378, 380

Higgs boson.........71, 72, 75, 78, 80, 97

High-energy physics 59

Holmium 108, 156

Humans 2, 6, 10, 11, 17, 20, 31, 38, 43, 58, 62, 68, 90, 92, 94, 113, 180, 181, 188, 191, 192, 193, 211, 212, 214, 217, 224, 225, 226, 227, 228, 234, 235, 240, 244, 274, 275, 283, 291, 293, 296, 334, 346, 350, 351, 396

Hydrocarbons.217, 224, 225, 265, 266, 267

Hydrogen ... 26, 43, 66, 67, 90, 99, 102, 105, 107, 109, 113, 114, 124, 125, 126, 132, 139, 152, 159, 173, 174, 175, 176, 178, 184, 189, 190, 191, 193, 195, 196, 199, 204, 205, 206, 209, 214, 217, 218, 219, 220, 221, 222, 225, 226, 228, 252, 255, 256, 257, 258, 259, 260, 261, 262, 263,

264, 265, 266, 267, 282, 284, 285,
290, 294, 295, 305, 307, 312, 314,
315, 360

Hydrogen bond 294

Hydrogen deuteride 220, 228, 295

Hydrogen Deuteride 217, 258, 259,
266, 267

Hydrogen Deuterium Oxygen. 217, 258

Hyperons 85

Hypothetical particles.. 6, 7, 20, 96, 99,
100

I

Ice 34, 217, 224, 228, 257, 258, 260,
265, 267, 274, 316, 322

Igneous rocks 195, 302

Indian 70

Indium 109, 155

Innermost 30, 31, 35, 48, 150, 151,
200, 204, 230, 244, 245, 255, 256,
260, 261, 262, 270, 306, 314, 316,
322, 323, 324, 326, 333, 346

Instability. 10, 12, 13, 30, 31, 184, 281,
345, 350, 356

Intermittence 5, 17, 21, 22, 30, 282,
333

Invisible 26, 62

Iodine..... 107, 124, 133, 143, 184, 185,
192

Ionic radius 6, 106, 109, 170

Ionization energy 6, 35, 40, 48, 106,
113, 136, 148, 149, 150, 151, 152,
159, 160, 161, 162, 163, 164, 165,
166, 167, 168, 169, 357

Iridium ... 115, 131, 132, 141, 143, 144,
145

Iron .. 94, 115, 116, 127, 140, 143, 144,
145, 189, 190, 192, 193, 195, 196,
197, 239, 245, 249, 252, 256, 298,
299, 303, 373

Irregular 333, 334

Isaac Newton 7, 71

Isotopes 6, 19, 36, 62, 90, 101, 105,

112, 114, 119, 124, 125, 266, 281,
284, 285

Israel 2, 4, 2, 4, 362, 363, 364, 365,
366, 367, 368, 373, 374, 396

Italian.. 63, 66

J

Jupiter...... 35, 194, 200, 205, 206, 214,
220, 257, 259, 260, 265, 267, 307,
312, 314, 342, 377

K

Karman line 201

Kinking 34

Knobbing 34

Krypton.. 111, 112, 154, 184, 212, 268,
270

L

Lambda baryons 83, 84, 277

Lanthanides 108, 109, 110

Lanthanoid.......... 6, 108, 109, 125, 191

Lanthanoids ... 108, 109, 110, 115, 124,
126, 132, 133, 140, 159, 160, 173,
174, 176

Lanthanum 108, 109, 174, 373

Lava 188, 195, 196, 301, 302, 313, 315,
316, 376

Lawrencium 104, 105, 158, 185

Lead 90, 109, 128, 192, 193, 285

Leptons 5, 25, 26, 60, 64, 65, 68, 69,
78, 80, 96, 97, 178, 183, 275, 283,
356

Life 5, 1, 2, 3, 4, 7, 68, 87, 93, 105, 111,
114, 125, 176, 180, 193, 224, 225,
227, 228, 234, 264, 362, 363, 364,
365, 366, 367, 368, 396, 397

Ligament 29, 55, 56, 57, 356

Ligaments 39, 52, 55, 56, 57, 359

Liquid 6, 9, 13, 34, 55, 90, 110, 112,
131, 132, 182, 183, 184, 185, 187,
193, 220, 291, 355, 358

Liquid density 6, 131, 132

Liquids 182
Lithium 89, 105, 131, 160, 161, 212
Livermorium 116, 125, 185
Location 6
Lumping 34
Lutetium 108
Lymphoma 68

M

Macromolecules 93, 125
Magma 94, 95, 195, 196, 301, 302, 313, 315, 316
Magnesium 94, 107, 152, 189, 190, 192, 193, 195, 196, 197, 199, 232, 245, 246, 248, 249, 252
Manganese 115, 153, 191, 193, 195, 239
Mars ... 35, 94, 193, 200, 204, 206, 214, 219, 220, 221, 224, 225, 226, 230, 255, 258, 259, 260, 262, 264, 265, 266, 270, 312, 315, 316
Martin Heinrich Klaproth 105
Mass 6, 5, 42, 43, 46, 47, 48, 49, 56, 58, 59, 60, 61, 62, 64, 65, 66, 69, 70, 71, 72, 73, 74, 75, 76, 79, 80, 84, 85, 86, 90, 99, 100, 102, 112, 119, 126, 150, 151, 152, 159, 160, 174, 180, 185, 189, 190, 191, 192, 195, 196, 200, 201, 206, 214, 220, 233, 243,244, 245, 246, 249, 252, 255, 256, 259, 260, 263, 270, 271, 276, 281, 284, 286, 292, 305, 313, 314, 342, 343, 359, 371, 372, 373, 378, 380
Massachusetts Institute of Technology
.... 39
Mathematics 275
Meitnerium 115, 124, 158, 185
Melting point ... 7, 6, 62, 106, 113, 114, 116, 182, 187, 225
Mendeleev 90
Mendelevium 104, 105, 124, 185
Mercury .. 115, 127, 184, 185, 200, 204, 206, 209, 214, 218, 222, 226, 228, 230, 231, 232, 237, 239, 244, 245,

249, 252, 259, 260, 261, 262, 265, 266, 270, 312, 315, 316, 317, 326
Meson 5, 82, 87, 88, 90
Mesons 71, 72, 87, 88, 89, 277
Metal .. 6, 103, 105, 109, 110, 125, 131, 132, 139, 140, 142, 148, 184, 185, 189, 214, 237, 245, 246, 249, 290, 313, 379
Metallic radius 6, 170
Metalloid 6, 111, 125, 138, 140, 143, 189, 196
Methane 205, 217, 219, 224, 256, 257, 258, 259, 261, 263, 264, 265, 266, 267, 295
Methane ice 217
Micas 94, 95
Microscope 12, 26, 58, 62
Microscopic .. 19, 21, 26, 30, 34, 36, 37, 40, 59, 98, 101, 106, 204, 300, 304, 316, 335, 336, 342, 349, 350, 351, 358
Mineral 14
Mineraloids 18, 19, 21, 94, 301, 356
Minerals . 1, 5, 9, 18, 19, 21, 22, 34, 36, 55, 58, 61, 94, 95, 108, 109, 189, 195, 274, 282, 296, 297, 298, 299, 300, 301, 302, 303, 356, 360, 371
Modulus 6, 62, 137, 142, 144, 145, 146, 172, 316, 379
Mohs hardness ... 6, 138, 139, 140, 172, 249
Mole 60, 148
Molecules 1, 5, 9, 12, 13, 18, 19, 21, 22, 26, 29, 33, 34, 36, 39, 55, 59, 74, 82, 89, 90, 91, 92, 93, 183, 184, 189, 196, 205, 220, 258, 259, 261, 264, 266, 274, 278, 291, 293, 294, 295, 300, 301, 356, 360, 370, 378
Molybdenum 115, 141, 146, 155
Mono-disperse 57
Moon 6, 58, 94, 188, 193, 197, 198, 199, 201, 239, 282, 302, 303, 306, 360, 396
Moscovium 116, 125, 185

Nathanael-Israel Israel: Member of the and Soil Science Society of America

Mother precursors .. 17, 18, 23, 26, 30, 31, 32, 35, 50, 51, 323, 334, 342, 343, 357

Muon ..5, 25, 64, 65, 68, 69, 78, 80, 89, 90, 97, 101, 102, 275, 278, 293, 356, 360

N

Nathanael-Israel Israel... 2, 4, 2, 4, 362, 363, 364, 365, 366, 367, 368, 373, 374, 396

Neck.............................. 55, 56, 57, 257

Neodymium 108, 156

Neon 104, 111, 112, 124, 159, 167, 168, 172, 174, 183, 184, 189, 190, 198, 230, 268, 270

Neptune. 205, 206, 214, 220, 222, 257, 259, 265, 266, 267, 313, 377

Neptunium 104, 105, 186

Netting....................................... 34, 301

Neutrino ...5, 64, 65, 66, 67, 68, 69, 78, 79, 80, 92, 97, 100, 275, 369, 371, 372, 376

Neutrinos. 64, 65, 66, 68, 69, 275, 281, 359, 370, 371, 373, 378

Neutrons...6, 22, 26, 56, 59, 62, 67, 71, 72, 85, 86, 87, 89, 90, 91, 92, 97, 98, 100, 101, 104, 105, 119, 124, 125, 220, 277, 278, 280, 281, 283, 284, 285, 304, 335, 359

Nevada 67, 378

Nickel.............. 115, 116, 189, 190, 240

Nihonium........................ 116, 125, 185

Niobium..115

Nitrogen 113, 125, 140, 152, 164, 165, 173, 174, 175, 178, 184, 189, 190, 191, 193, 196, 206, 226, 255, 256, 259, 261, 262, 263, 264, 295

Nitrogen oxide........................ 228, 295

NMR..59

Nobel Prize 44, 67, 68, 75

Nobelium 104, 105, 124, 158, 185

Noble gas... 6, 108, 111, 112, 125, 126, 131, 138, 139, 141, 143, 145, 159, 184, 189, 191, 195, 196, 236, 252, 256, 270, 271

Noble gases 113

Nodules 68, 303

Nonliving things...... 12, 31, 44, 58, 345

Non-metal 6

Nonmetals 109, 110, 111, 113, 114, 131, 133, 138, 142, 143, 145, 159, 172, 173, 174, 177, 184, 189, 191, 195, 197, 200, 244, 245, 252, 253, 255, 256, 257, 258, 259, 262, 263, 305

Non-silicate................................... 298

Nuclear energy 42

Nuclear fission 42, 91

Nuclear magnetic resonance............ 59

Nuclear reaction 42

Nuclei..35, 85, 86, 89, 91, 99, 182, 183, 277, 346, 357

Nucleons.26, 37, 40, 48, 49, 50, 56, 71, 72, 83, 84, 85, 87, 90, 97, 98, 178, 183, 277, 284, 285, 293, 304, 345, 358

Nucleus...22, 23, 24, 25, 26, 27, 30, 34, 35, 36, 37, 40, 42, 44, 47, 48, 49, 50, 53, 56, 60, 67, 71, 74, 80, 85, 89, 90, 91, 92, 97, 98, 100, 101, 102, 104, 105, 119, 150, 151, 152, 159, 170, 183, 220, 258, 277, 278, 280, 281, 283, 284, 285, 286, 287, 292, 293, 303, 326, 346, 347, 356, 357, 358, 359

Nuclides...90

Number of neutrons.................. 6, 124

Number of protons........................... 6

O

Occurrence ... 6, 62, 182, 185, 186, 188

Oceans. 6, 58, 114, 188, 195, 196, 197, 211, 212, 214, 217, 220, 223, 224, 228, 236, 238, 240

Oganesson 111, 116, 125, 126, 185

Olivines 94, 95

Omega baryons 84, 277

Orbital angular momentum .46, 49, 59, 60, 83

Orbital inclination .52, 53, 91, 301, 332

Orbital moment of inertia 46, 49

Orbital speed..40, 41, 45, 53, 150, 151, 204, 245, 256, 260, 262, 284, 303, 316, 317

Organic.. 299

Organisms 9, 10, 43, 44, 58, 92, 93, 114, 139, 176, 193, 228, 279, 283, 293, 302, 345, 349

Origin of Chemical Particles 373

Origin of Life 374

Origin of name 62

Original mysterious scattering 9, 13, 355, 356

Osmium...........115, 131, 132, 139, 144

Outermost....30, 31, 48, 106, 107, 150, 151, 152, 159, 170, 205, 206, 255, 262, 282, 284, 314, 357

Oxygen92, 94, 104, 108, 110, 111, 113, 114, 125, 140, 165, 166, 173, 174, 175, 176, 178, 184, 189, 190, 191, 193, 195, 196, 197, 198, 199, 200, 204, 206, 214, 219, 220, 221, 225, 228, 229, 233, 234, 235, 236, 252, 255, 256, 258, 259, 260, 261, 262, 263, 264, 265, 266, 267, 283, 294, 295, 297, 305, 360

P

Palladium 115

Particle physics............................... 59

Paul Dirac 63, 70

Pauli exclusion principle.................. 63

Pauling electro negativity 6

Pentaquark................................. 88, 89

Periodic table ...90, 105, 106, 108, 112, 115, 126, 170

Philosophical 1, 2, 45, 73, 364

Phosphates.................................... 299

Phosphorus 113, 125, 128, 140, 143,

176, 185, 191, 192, 193, 195, 234

Photoelectric............................. 44, 45

Photons 5, 39, 40, 42, 44, 45, 70, 73, 74, 75, 78, 79, 91, 94, 97, 98, 101, 276, 281, 283, 286, 287, 288, 314, 371, 374

Physicist ... 5, 1, 63, 66, 70, 74, 90, 109, 177

Physics.. 6, 7, 13, 20, 35, 48, 52, 59, 60, 63, 66, 73, 75, 96, 97, 98, 102, 106, 151, 245, 274, 288, 292, 312, 351, 352, 359, 369, 371, 372, 378

Pinching off 55, 56

Planck constant 59, 74

Planetary systems 13, 18, 21, 27, 40, 47, 48, 50, 51, 55, 61, 89, 91, 95, 99, 104, 114, 151, 204, 211, 243, 276, 277, 278, 281, 282, 284, 296, 305, 307, 313, 322, 323, 325, 326, 331, 336, 350

Planets ... 1, 5, 6, 18, 19, 22, 25, 26, 27, 34, 40, 47, 49, 50, 51, 55, 58, 62, 91, 95, 104, 106, 114, 150, 151, 179, 183, 188, 193, 194, 195, 196, 198, 200, 201, 202, 203, 204, 205, 206, 207, 209, 211, 212, 214, 215, 217, 218, 219, 220, 222, 224, 225, 226, 228, 230, 232, 233, 234, 235, 236, 237, 238, 239, 241, 243, 244, 245, 252, 255, 256, 257, 258, 259, 260, 261, 262, 264, 265, 266, 267, 268, 270, 271, 281, 282, 286, 292, 301, 304, 305, 306, 312, 313, 314, 315, 316, 317, 322, 323, 325, 326, 333, 336, 342, 345, 346, 350, 356, 360, 375, 396

Plant Vogle 66

Plants 38, 44, 58, 224, 234, 293

Plasma.. 9, 13, 34, 55, 90, 94, 182, 183, 315, 355, 369

Plasmas 182, 183

Platinum....89, 115, 131, 132, 157, 373

Plato 351

Pluto.......179, 204, 206, 214, 219, 225,

226, 255, 259, 260, 261, 262, 265, 266, 267, 270, 271, 283, 284, 312

Plutonium ... 43, 67, 104, 105, 126, 378

Polonium 109, 157, 186

Polymorphs 298, 299, 300

Potassium 94, 105, 106, 124, 144, 153, 174, 176, 178, 191, 192, 193, 195, 196, 237, 245, 246, 248, 249, 252, 265, 298

Praseodymium............................. 108

Precursors of bodies 10, 12, 13, 22, 29, 31, 39, 41, 57, 190, 284, 305, 325, 343, 356, 357

Pressure... 9, 33, 53, 57, 109, 112, 113, 137, 142, 182, 184, 187, 228, 280, 292, 302

Primary bodies... 22, 23, 25, 26, 27, 30, 35, 37, 40, 47, 48, 50, 53, 56, 179, 180, 278, 283, 284, 285, 292, 293, 297, 331, 332, 333, 342, 345, 356, 357, 359

Primary particles 56

Promethium 108, 156, 186

Protactinium........... 104, 157, 161, 186

Proteins 93, 193

Protium...... 26, 90, 125, 220, 221, 258, 285, 293, 360

Proton radius puzzle 6, 7, 96, 101, 102, 178, 180, 280, 286, 291, 292, 293, 359, 370

Protons 22, 86, 96, 101

Pulling.................................34, 53, 336

Pushing 34, 53, 334, 336, 352

Pyroxenes 94, 95, 298

Pythagoras.................................... 351

Q

Quantum mechanics 59

Quantum physics.............................. 35

Quantum-statistics 5, 59, 61

Quarks 5, 69, 70, 71, 72, 75, 78, 82, 83, 84, 85, 86, 87, 88, 89, 97, 98, 99, 275, 276, 277

Quartz................................ 94, 95, 298

Quasiparticles..... 5, 59, 61, 93, 94, 379

R

Radioactivity 42, 66, 67, 68, 91, 100, 101, 104, 105, 107, 124, 125, 192, 276, 285, 345

Radium 107, 176, 186, 191

Radius ...6, 7, 21, 35, 37, 52, 54, 55, 62, 71, 90, 96, 101, 102, 105, 106, 112, 170, 171, 172, 173, 174, 175, 176, 177, 178, 179, 180, 194, 206, 236, 249, 252, 256, 263, 264, 271, 276, 280, 284, 286, 287, 291, 292, 293, 303, 322, 335, 359, 360, 370

Radon 111, 124, 177, 184, 186, 270

Random 10, 49, 315, 351

Reactive nonmetals 113

Reactivity ... 37, 92, 106, 108, 111, 112, 113, 283, 358

Recoil 56

Rest mass.. 88

Revolution . 10, 39, 180, 293, 345, 356, 360

Rhenium 115, 131, 132, 141, 142, 144, 145

Rhodium 115, 142, 144, 146

Rings 7, 5, 18, 35, 91, 298, 306, 312, 322, 323, 324, 325, 326, 336, 372, 375, 376

RNA.. 93, 125

Rocks 1, 5, 9, 18, 19, 21, 34, 36, 58, 61, 94, 95, 112, 189, 195, 196, 197, 274, 296, 301, 302, 303, 312, 313, 322, 323, 324, 325, 326, 356, 360, 371

Roentgenium 115, 125, 185

Rotation..10, 25, 30, 33, 39, 40, 41, 42, 45, 46, 47, 49, 52, 53, 54, 59, 60, 198, 283, 292, 301, 316, 317, 332, 336, 344, 345, 356, 357

Rotational angular speed 40, 41, 42, 53, 54, 60, 283, 292

Roughness 57

Science180: The Place Where the Accurate Interpretation of Chemicals-Origin Data Matters

Rubidium..89, 105, 155, 174, 176, 178, 212
Ruthenium 115, 144, 146
Rutherfordium 115, 124, 185

S

Samarium 108
Satellite systems 27
Satellites..... 1, 5, 18, 19, 21, 22, 25, 26, 27, 31, 35, 40, 41, 47, 48, 49, 50, 51, 55, 61, 62, 91, 95, 104, 106, 150, 151, 179, 195, 215, 281, 282, 286, 292, 312, 314, 315, 316, 322, 323, 324, 325, 326, 336, 350, 375, 396
Saturn.....200, 205, 206, 214, 220, 222, 257, 259, 265, 267, 270, 312, 314, 324, 376
Satyendra Nath Bose....................... 70
Savannah River....................66, 67, 378
Savannah River Site.......66, 67, 68, 378
Scalar bosons 5, 70, 78, 276
Scandium............................. 114, 212
Schrödinger 101
Science ... 6, 1, 2, 20, 22, 43, 45, 47, 61, 101, 106, 170, 351, 352, 365, 369, 372, 374, 375, 376, 377, 396
Science180 ..... 2, 3, 4, 7, 4, 11, 45, 129, 136, 187, 355, 361, 362, 363, 364, 365, 366, 367, 368, 396, 397
Seaborgium 115, 124, 164, 185
Secondary bodies.....22, 23, 24, 25, 26, 27, 34, 35, 37, 47, 48, 50, 56, 151, 179, 180, 183, 284, 292, 293, 297, 323, 324, 325, 326, 332, 333, 342, 345, 356, 357, 358
Secondary particles..22, 31, 37, 41, 56, 356, 358
Sedimentary rocks 303
Selenium 113, 126, 131, 133, 138, 139, 143, 145, 185, 212
Shear modulus 6, 144, 145
Shearing 10, 34, 41, 53, 336
Shell......25, 35, 48, 104, 109, 110, 111, 170, 312, 357

Sigma baryons...............83, 84, 85, 277
Silica 94, 233, 297, 298, 302, 303
Silicate.............195, 196, 297, 298, 303
Silicon.......94, 110, 111, 113, 124, 178, 189, 190, 192, 195, 196, 197, 199, 214, 226, 233, 234, 297, 298, 304, 305
Silver 115, 127, 373
Smoothness 57
Sodium89, 94, 105, 106, 107, 112, 168, 178, 189, 190, 192, 193, 195, 196, 199, 204, 206, 214, 231, 245, 249, 252, 262, 263, 265, 290, 298
Solar System... 6, 10, 18, 19, 31, 32, 44, 46, 51, 60, 85, 89, 99, 114, 179, 188, 190, 193, 194, 200, 201, 204, 205, 215, 225, 228, 230, 235, 245, 252, 277, 278, 281, 282, 284, 285, 288, 303, 304, 307, 313, 314, 315, 316, 326, 332, 333, 342, 346, 375
Solid ... 9, 29, 74, 90, 94, 110, 112, 137, 183, 184, 185, 195, 226, 312, 313, 314, 315, 355
Solids... 182
South Carolina........................... 66, 67
Space.. 9, 29, 35, 36, 37, 38, 39, 41, 58, 90, 99, 106, 112, 136, 182, 183, 189, 201, 211, 228, 233, 277, 283, 287, 288, 304, 313, 323, 324, 334, 355, 358
Special heat...........................6, 62, 129
Spectroscopy........................... 94, 375
Speed of light 40, 66, 73, 74, 99, 287
Spin 5, 54, 59, 63, 64, 65, 70, 75, 83, 86, 87, 336
Spiral 35, 189, 325, 332, 333, 334, 335, 336
Spiraling34, 41, 42, 275, 279, 280, 323, 325, 333, 334, 335, 344, 359
Spiritual 2, 17, 20, 36, 38, 363, 397
Spiritual World 374
Split-gathering....18, 19, 20, 21, 34, 37, 39, 42, 46, 47, 54, 56, 211, 284, 286, 297, 301, 315, 325, 326, 333, 334,

343, 356, 358

Squashing 53

Squeezing 10, 22, 33, 41, 55, 182, 183, 200, 224, 233, 249, 263, 279, 280, 286, 293, 294, 300, 307, 316, 333, 342, 344, 345, 356, 360

Stability.......................... 5, 57, 59, 125

Stacking 41, 53

Standard Model..... 6, 7, 63, 96, 97, 98, 274, 378

Standard potential............................. 6

Standard temperature and pressure. 6, 62, 106, 107, 184, 185

Stars.. 7, 1, 5, 10, 18, 19, 21, 26, 27, 40, 51, 58, 59, 62, 95, 104, 150, 183, 190, 211, 214, 215, 224, 236, 243, 278, 287, 312, 314, 315, 331, 332, 333, 334, 336, 342, 345, 396

State at standard temperature and pressure 6

State at STP.............................. 6, 182

Stellar systems... 12, 13, 18, 19, 21, 27, 31, 51, 55, 61, 85, 89, 95, 99, 104, 114, 151, 183, 211, 228, 243, 276, 277, 278, 281, 285, 287, 296, 335, 336

Strange quark 5, 69, 70, 78, 80, 85, 275

Stratified fluid.................................. 53

Stretching13, 39, 43, 52, 53, 54, 55, 57, 301, 336

Strong force ... 7, 70, 71, 72, 75, 82, 86, 97, 286, 341

Strong interaction........... 64, 69, 75, 87

Strong nuclear interaction.......... 7, 341

Strontium 107, 168, 174, 176, 196, 241

Subatomic particles5, 9, 12, 13, 17, 18, 19, 20, 21, 22, 26, 27, 29, 33, 34, 35, 36, 37, 47, 48, 51, 55, 57, 59, 60, 61, 62, 69, 82, 89, 90, 91, 95, 98, 100, 102, 113, 183, 189, 211, 243, 245, 274, 275, 276, 278, 279, 280, 281, 284, 285, 286, 288, 290, 291, 300, 301, 304, 305, 306, 326, 335, 336,

342, 345, 346, 347, 356, 358, 359

Sulfates 299

Sulfides 299

Sulfur 113, 127, 131, 138, 143, 185, 189, 190, 192, 193, 195, 196, 235, 255, 256, 258, 259, 260, 262, 263, 295

Sulfur dioxide.......................... 228, 295

Sun....6, 34, 44, 46, 50, 58, 91, 99, 179, 183, 188, 190, 191, 194, 195, 209, 211, 212, 214, 217, 218, 222, 223, 224, 225, 226, 228, 230, 237, 238, 282, 287, 292, 302, 303, 305, 307, 313, 314, 315, 316, 323, 336, 342, 375, 376, 396

Superclusters 27, 278, 331, 332

Superconductors 73

System-additive variables 5, 46, 47, 48, 49, 51, 80

T

Tantalum 115, 157

Tanzania 195

Tau....5, 64, 65, 66, 68, 78, 80, 97, 275, 278, 376

Technetium 115, 155, 186

Telescope.............. 12, 58, 62, 334, 336

Tellurium 110, 124, 125

Temperature.... 6, 34, 57, 62, 106, 107, 109, 112, 113, 114, 142, 182, 184, 185, 187, 194, 225, 301, 302, 313, 314

Tennessine.............. 107, 116, 125, 185

Terbium 108

Terrestrial planets 19

Thallium.. 109

Thermal conductivity.......... 6, 113, 129

Thorium 104, 157, 162

Thulium.. 108

Thyroid 67, 68, 193

Tin................... 109, 125, 127, 178, 192

Titanium 114, 195, 197, 238

Top quark .5, 69, 70, 75, 78, 79, 80, 85,

276

Topology 30, 278

Transcriptomic 10

Transition elements 115

Transition metal................................. 6

Transition metals ...114, 115, 116, 126, 132, 138, 139, 140, 143, 145, 174, 176, 193

Translation 10, 40, 42, 45, 46, 47, 49

Tritium.................66, 67, 124, 125, 285

Tungsten 115, 131, 139, 140, 143, 144, 145, 146

Turbulence ... 5, 7, 2, 10, 12, 13, 14, 30, 31, 32, 35, 36, 40, 41, 42, 43, 46, 49, 51, 52, 53, 54, 56, 57, 60, 91, 111, 112, 151, 183, 189, 190, 196, 200, 206, 211, 219, 224, 228, 236, 243, 260, 281, 282, 283, 284, 285, 286, 292, 293, 294, 304, 305, 307, 313, 314, 317,322, 323, 325, 333, 334, 335, 336, 341, 342, 343, 344, 345, 346, 350, 356, 357, 359, 396

Turbulence zones 31

Turbulent intermittence 21, 22

Turbulent multipliers ..5, 46, 49, 50, 51

Turbulent original substance 8

Turbulent prima materia....8, 9, 10, 12, 13, 18, 20, 21, 22, 34, 42, 55, 274, 275, 279, 280, 342, 343, 345, 355, 359

Turbulent split-gathering 5, 17, 18

Twisting.......34, 39, 279, 280, 345, 359

U

Undetermined...................8, 9, 10, 355

Universe 5, 7, 1, 2, 3, 4, 5, 6, 7, 8, 9, 10, 12, 13, 14, 17, 18, 19, 20, 21, 22, 26, 27, 30, 31, 32, 34, 36, 37, 39, 40, 42, 43, 44, 45, 48, 49, 50, 51, 52, 53, 55, 61, 62, 66, 72, 73, 75, 82, 85, 90, 91, 96, 98, 99, 100, 114, 180, 184, 188, 189, 190, 195, 211, 212, 214, 217, 222, 223, 224, 225, 228, 230, 236, 237, 238, 240, 243, 244, 252, 262,

274, 275, 278, 280, 281, 284, 285, 286, 287, 288, 292, 293, 294, 302, 303, 304, 305, 312, 331, 333, 334, 335, 336, 341, 342, 343, 345, 346, 349, 350, 352, 355, 356, 357, 359, 360, 361, 362, 363, 364, 365, 366, 367, 368, 396, 397

Universe-origin. 2, 3, 4, 7, 96, 362, 363, 364

Up quark 5, 65, 69, 70, 78, 80, 86, 275, 277

Upstream 53, 56, 106, 324, 326

Uranium 43, 59, 104, 105, 131, 139, 141, 144, 145, 192, 283

Uranus....105, 205, 206, 214, 220, 257, 259, 265, 266, 267, 313

USA... 373, 374

V

Valence 6, 62, 82, 87, 89, 107, 111, 148

van der Waals bonds...................... 294

van der Waals radius..6, 170, 171, 177, 178

Vanadium.............................. 114, 212

Venus 200, 201, 204, 206, 214, 222, 224, 225, 226, 230, 235, 255, 258, 259, 260, 262, 265, 266, 312, 315, 316, 317, 326

Vickers hardness .6, 137, 138, 140, 141

Viscosity . 30, 52, 55, 57, 137, 195, 301, 304

Volcano 94, 95, 111, 195, 196, 301, 302, 315, 316

Volume. 46, 48, 49, 50, 57, 58, 66, 129, 142, 182, 183, 200, 205, 219, 255, 258, 260, 261, 263, 343, 379

Vortex 52, 53, 54, 55, 281, 335, 336, 344

Vortices .. 33, 39, 52, 53, 54, 55, 56, 57, 335, 336, 343

Vorticity..................................... 53, 54

W

W boson 70, 72, 80

Nathanael-Israel Israel: Member of the and Soil Science Society of America

INDEX

W bosons 5, 72

Water .. 6, 13, 34, 44, 58, 106, 107, 176, 194, 195, 196, 199, 217, 219, 220, 221, 224, 228, 257, 258, 259, 260, 261, 263, 264, 266, 267, 274, 295, 303, 305, 322, 360

Water ice 261

Weak force 7, 70, 97, 286, 341, 346

Weak interaction 63, 64, 69, 70, 71, 75, 87

Weak nuclear interaction 7, 341

Winding 34, 41, 275, 279, 287, 333, 359

Wolfgang Pauli 68

Wrapping ... 22, 33, 35, 41, 43, 54, 105, 233, 280, 284, 286, 287, 306, 316, 332, 335, 357

X

Xenon 111, 112, 125, 155, 178, 184, 268, 270, 283

Xi baryons 84, 277

Y

Young's modulus 6, 142, 145, 146, 172

Ytterbium 108

Yttrium 109, 115, 167

Z

Z boson 5, 72, 80

Zinc . 115, 128, 153, 178, 192, 193, 240

Zirconium 115

Science180: The Place Where the Accurate Interpretation of Chemicals-Origin Data Matters

ABOUT THE AUTHOR

Dr. Nathanael-Israel Israel is the founder of Science180, the American company which mission is to improve the current and future state of human beings by accurately decoding and teaching them the real origin and formation of the universe, of life, and of chemicals, and meaningfully engaging business, nonprofit, political, academic, civil society leaders and followers to properly shape local and global agendas that authentically value the truth. Known as the creator of the Universe Turbulent Origin Formula™. Dr. Nathanael-Israel Israel has revolutionized the way billions of people around the world think about the origin of the universe, of life, and of chemicals. Nobody understands and teaches the formation of everything in the universe (e.g. the Milky Way Galaxy, the Sun, the Earth, the Moon, and all other galaxies, stars, planets, satellites, and asteroids) better than Nathanael-Israel Israel. Individuals and organizations across the globe have been calling him so he helps them scientifically unlock the code of the universe-formation, helping veterans and newcomers to have the real keys to decrypt the universe and turbulence (one of the top biggest unsolved mysteries in science) from the historic, unique, accurate, simple, easy-to-understand, nonconformist, trailblazing perspective that anybody can quickly learn at Science180 Academy (Science180Academy.com). Science180 Academy delivers outstanding value, insight, and lessons to assist people to accurately understand the true origin of the universe, chemicals, and life, so they can tap into that knowledge to improve lives perpetually. Nathanael-Israel's goal is to give you practicable and undeniable proofs of the formation of the universe so you can be fired up to become the best version of you, and to cause positive changes to your initiatives that will profit you today and forever. For Nathanael-Israel, accurately decoding and teaching the origin of the universe and everything in it is not a job, but his life mission, and helping others to fully understand that brings him closer to his assignment.

Dr. Israel earned his PhD in Plant, Insect, and Microbial Sciences in the USA, where he graduated first of his class of hundreds of PhD candidates. This Beninese-American is a member of the American Chemical Society, American Association for the Advancement of Science, American Society of Agricultural and Biological Engineers, American Society for Microbiology, American Society of Biochemistry and Molecular Biology, Ecological Society of America, American Society of Agronomy, Crop Science Society of America, and Soil Science Society of America. A scientist, a mathematician, a consultant, and the owner of Global Diaspora News (www.GlobalDiasporaNews.com), a news company in the USA. Dr. Israel is the author of the popular books:

ABOUT THE AUTHOR

- Turbulent Origin of Chemical Particles
- Turbulent Origin of Life
- From Science to Bible's Conclusions
- How Baby Universe Was Born
- How God Created Baby Universe
- Science180 Accurate Scientific Proof of God
- Turbulent Origin of the Universe
- Reconciling Science and Creation Accurately
- Origin of the Spiritual World

If you want to accurately understand the origin of anything, then be sure to get a copy of these amazing books. You cannot afford to ignore the greater, better, faster, simpler, cheaper, easier, and accurate formulas unlocked in these important books that successfully cracked the origin of the universe, of life, and of chemicals in a language that scientists, laypeople, adults, children, believers, skeptics, and anybody else can properly understand and enjoy.

Visit Israel120.com today to connect with this historic discoverer of the all-in-one proven and uncomplicated formula that accurately decoded the origin of the universe, of life, and of chemicals.

www.ingramcontent.com/pod-product-compliance
Lightning Source LLC
Chambersburg PA
CBHW071316210326
41597CB00015B/1251